PROGRESSIVE RELAXATION

A PHYSIOLOGICAL AND CLINICAL IN-
VESTIGATION OF MUSCULAR STATES
AND THEIR SIGNIFICANCE IN PSY-
CHOLOGY AND MEDICAL PRACTICE

By
EDMUND JACOBSON

THE UNIVERSITY OF CHICAGO PRESS
CHICAGO & LONDON

THE UNIVERSITY OF CHICAGO PRESS, CHICAGO & LONDON
The University of Toronto Press, Toronto 5, Canada

THE UNIVERSITY OF CHICAGO
MONOGRAPHS IN MEDICINE

PROGRESSIVE RELAXATION

TO MY MOTHER

PREFACE TO THE FIRST EDITION

THERE has been a long-felt need on the part of the medical and surgical profession for a method of study and management of the nervous element that appears in a large variety of diseases. Less clearly realized has been the want of an approach to problems of fatigue, debility and lowered resistance occurring in patients who are not properly called neurotic but whose energy output in muscular terms might properly be economized in the interests of their general state of health.

It is the hope of this volume to draw attention to these problems and to present a method that will interest the general practitioner, the internist and the surgeon, no less than the neurologist. Time, thoroughness and patience will be required of the physician who employs this method. Needless to say, he will in every instance take a complete medical history of the patient, make a general physical as well as an adequate clinical laboratory examination and fully apply such proper measures of medicine and surgery as are indicated in the light of current knowledge.

In what direction is it more natural to look in order to meet the foregoing problems than that of rest? Rest has been found useful in treatment throughout the practice of medicine. Nature often enforces it where the physician does not order it. It is commonly prescribed in some form in various acute and chronic infectious diseases, in the more severe metabolic and nervous disorders, in gastrointestinal and general systemic affections, in asthenia due to new-growths and many other causes, and in a large variety of surgical conditions.

Oddly enough, in spite of the apparently vast importance

of rest, and although several books have been written on the subject, the field has remained practically unexplored from a scientific standpoint. While devoting huge efforts to the development of other investigative and therapeutic measures, Medicine has used this, perhaps her oldest remedy, naïvely and with little attempt at systematic study.

However, if straws show which way the wind blows, there are signs of awakening curiosity. This is well illustrated in a recent letter from a practicing physician:

> For some time, I have been trying to make my therapeutics more rational: to separate the chaff from the grain. But rest is nearly always a part of any therapeutic procedure, and I find myself unable to evaluate my results because the repose factor is not known. I should like to know what rest alone can accomplish in the treatment of disease.

Our eyes are invited to turn toward the goal indicated by this writer. A scientific study of rest must seek, so far as possible, to isolate the effects of rest alone. It must therefore differentiate between rest, which is a physiological state, and suggestion or other psychotherapeutic measures. This will be attempted in the following pages.

The task is no easy one. We know from experience the difficulties of securing proof in clinical therapeutics: how readily after a so-called "remedy" has been applied, a change which occurs *post hoc* is without warrant assumed to be *propter hoc*. Yet in Medicine we must often sail between Scylla and Charybdis. If on the one hand lie dangers of overenthusiasm, gullibility and the glamor of unproved remedies, such as "pluriglandular" products, yet on the other hand lie the dangers of overconservatism and opposition to change and progress. Clinicians as a body are perhaps less prone to adopt novelties in their field than are laboratory workers, possibly because the latter have the courage which comes from the possession and ready application of reliable testing methods.

PREFACE TO THE FIRST EDITION

Our choice of methods for the present studies must therefore favor laboratory investigations, bringing as they do a greater measure of certainty than do clinical therapeutic tests. Yet we should not pass to the other extreme and underemphasize clinical consequences, since in the words of Billings, Coleman and Hibbs,

> It is generally recognized that the end-results of treatment of patients who suffer from a recognized disease entity are a fair measurement of the success of the management and the treatment, and also the correctness of the ascribed cause and of the character of the morbid process.

In any event, our aim must be to avoid that speculation which has marked the course of functional neurology during past decades. This implies that interest in the neuromuscular states and the forms of rest described in the present volume should not be merely practical; rather, the effort should be to help to open the way to the physiological understanding of neurosis, psychoneurosis and other conditions so far as they involve neuromuscular disorder. For clearly the causes of even a nervous malady are not completely comprehended by a series of mental events unearthed by the psychologist or analyst. Explanations of the occurrence, character and influence of mental events must admittedly be sought in the light of the total physiology of the organism. In this direction the following studies suggest that the conditions of neuromuscular tonus and contraction are exceedingly important. If we would build firmly in such fields, it must evidently be step by step upon the basis of carefully controlled investigations.

During neurosis, as examined by the methods of the present investigation, there is failure to relax. Recovery, by whatever route attained, generally is characterized by a return to a fairly normal relaxed state. The various methods to this end heretofore have been indirect in the measures employed, perhaps because the importance of physiological relaxation has not been fully realized. Therefore I have sought to test directly the effects of cultivating relaxation during neurosis.

With the cautions above-mentioned in view, a method will presently be described to quiet the nervous system, including the mind. The history of this method will be given and the principles discussed. Because of reflex connections, the nervous system cannot be quieted except in conjunction with the muscular system. In fact it becomes evident that the whole organism rests as neuromuscular activity diminishes. Therefore the possible range of usefulness of the present method of relaxation should not be narrowly restricted to neurology, since it may conceivably be applied wherever rest is useful in the practice of medicine.

The present studies of neuromuscular tension and relaxation, although begun twenty years ago, are still in the early stages of development. Further investigations are under way, and very many more must follow on the various systems before the range of the physiological effects of relaxation and of the clinical applications can be fully stated. It is hoped that these shortcomings will not greatly detract from the interest that adheres to the various problems, nor obscure the possible importance of diminishing voluntary and reflex activities in disease and in prophylaxis.

I am indebted to Professor A. J. Carlson for his interest and encouragement, for frequent advice and constructive criticism, and for a careful reading of the text with suggestions leading to an extensive revision; and to Professor Nathaniel Kleitman for a critical reading of chapters xiii, xiv and xv.

EDMUND JACOBSON

CHICAGO
May 10, 1928

PREFACE TO THE SECOND EDITION

SINCE 1929, when the first edition of this volume was published, sprouts of interest in relaxation have appeared in various quarters. Physiologists investigating reflexes, muscular contraction, sleep and divers other functions have given it a certain attention, which has increased with the development of technics of electrical measurement. Psychologists concerned with emotion, attention, imagination and other general functions have found in the present field the possibility of newer explanations as well as the opportunity for further investigations by new but definite procedures. Many physicians have felt that the objective study of relaxation is a continuation of methods which they have been accustomed to employ. Their interest in rest, as prescribed in manifold states and conditions, has been shown to have a scientific basis permitting a possible development of greater service than heretofore. General practitioners have informed the author that a very considerable proportion of the patients they see each day present symptoms chiefly due to states which would doubtless not appear if they were relaxed. Accordingly, there has seemed to be a growing tendency among some physicians to add the methods of progressive relaxation to their practice. We are encouraged to believe that in time the methods of investigation and of practice herein outlined will find their place in the hands of the practitioner who sees daily the "tense" patient in general practice; the internist concerned with "irritable colon" and with other spastic visceral states, as well as with arterial hypertension; the neurologist interested in "functional nervous disorders"; and the specialist in each field where muscular states play a significant rôle.

PREFACE TO THE SECOND EDITION

Various procedures in physical diagnosis may be hindered by tense muscles. In consequence, the physician has always meant something very definite when, during an examination, he has requested his patient to "relax." This instruction is the germ of the present methods. In principle, also, the physician's customary prescription of rest leads directly toward the technical field of study and application of muscular relaxation. Notwithstanding these facts, there is, from another approach, greater novelty in this new field than would at first appear.

The family doctor commonly treated the whole patient and advised him with a regard to his entire life. In contrast, as physicians have indicated, scientific medicine has tended at times to treat the patient more or less as a mere assembly of organs and tissues (cf. Herrick, 1907; Lapham, 1937). It has studied these parts in point of their responses to environment, including reactions to disease, and particularly the results on these parts of chemical and surgical intervention. In short, the patient has been studied and treated mostly in terms of the systems which compose his organism. There was missing a scientific procedure based on the completed recognition that he is all this and something more, namely, an active individual, autonomously regulating his own actions and in consequence *subject to disorders in his functions as active agent.* In a sense, certain newer schools of psychiatry set out to meet this need on the part of physicians for a viewpoint and methodology from which to regard disorders of autonomous action; but the assumptions of such schools failed, as a rule, to become connected with those principles and practices which have commonly been inculcated in modern medical schools.

On the other hand, physiological studies of the nervous and muscular systems and their relations with organic

processes form the basis of the present studies and lead directly to methods of relaxation. In this way we come to be able,
through current scientific procedures, to arrive finally at a
point of regard for the patient as an integrated whole, but as
an active agent.

On rational grounds no less than on the basis of current
physiological knowledge, it seems permissible to say that *to
be excited and to be fully relaxed are physiological opposites.*
Both states cannot exist in the same locality at the same
time. The rule or law thus stated would seem to apply to
many conditions familiar in medical practice and would suggest that the rôle of applied relaxation might be directly
efficient.

In the treatment of nervous states, but particularly in the
approach to spastic visceral conditions and to arterial hypertension, there is increasing evidence that the subsidence of
symptoms becomes most marked upon the disappearance of
residual tensions. This is remarkable because of the small
proportion of caloric energy represented by residual tensions.
However, it becomes comprehensible by analogy: a flame
spreading to the vicinity will leave an incendiary residue
unless it be extinguished approximately completely.

In the present edition it has not been possible to give a
complete account of the extensive developments in related
fields since 1929, but some of the more important references
have been included, while others are mentioned in the new
bibliography. Three new chapters have been added concerning the electrical measurement of nervous and muscular
states. These methods afford a direct and effective approach
to many fundamental problems not otherwise accessible.
They have rendered it possible to measure directly states of
tension, or relaxation; in consequence, they constitute an
accurate test and guide to therapy.

[xv]

PREFACE TO THE SECOND EDITION

According to determinations outlined in chapter xvii, the participation of muscular components during imagination, reflection, emotion and other "mental activities" can no longer be regarded as mere theory. The failure of specific contractions to occur results in the absence of the mental activities. If so, there is definite light on the old problem of "mind and body," significant to the psychologist. A certain practical familiarity with this problem results from dealing with mental difficulties through drill in the relaxation of ocular, speech and other tensions. Evidence that these methods afford direct routes to the treatment of fears, anxieties, over-emotion and certain other psycho-pathological states will be outlined. The more complete discussion of these subjects, including psychological interpretations, must be postponed to a later volume.

It seems well to explain briefly, however, that cultivation of the muscle-sense does not encourage morbid self-consciousness. Indeed, the highly distracted or nervous person often finds it difficult to detect the sensations from muscular contractions. There is little likelihood of such persons becoming overattentive to their tensions. It is only as they become more relaxed that they succeed in recognizing tensions better. Thus, hypochondriasis is reversed.

The hypochondriac is characterized by his almost incessant seeking of reassurance. In present methods, this reassurance is not given. Instead, he is trained to look for tensions at moments when he finds himself discussing his ills and to relax these tensions. In my experience, he can in this way be re-educated.

EDMUND JACOBSON

CHICAGO
January 3, 1938

CONTENTS

[xvii]

CHAPTER I
REST AND RELAXATION IN THE PRACTICE OF MEDICINE

THERE is perhaps no more general remedy than rest. Examples abound in the current literature. It is used, along with symptomatic treatment, in various acute infectious diseases where no special remedy has been found; but even in others, like diphtheria, it remains an important part of treatment. It has a place in peptic ulcer and other serious gastrointestinal disorders, and is commonly considered essential in states of exhaustion, in divers cardiac maladies, in nephritis with edema or high blood-pressure and in various systemic diseases where general recuperation is needed. Rest has been systematically applied to hyperthyroidism, and of late to chronic pulmonary tuberculosis, where it is now believed by many to be the most important element of treatment. It has been said to be an indispensable element in the treatment of chronic arthritis (Pemberton, 1926). In surgery there has been a growing recognition of its importance after serious operations and frequently in preparation for them.

Particular instances are not far to seek. Every clinician prescribes rest along with digitalis for cardiac incompetency. This is because of repeated experience that the alkaloid plus rest proves more effective than the alkaloid alone. Osler, who often dwelt to his students on the importance of rest in various diseases, wrote in 1910 that "the ordinary high-pressure business or professional man suffering from angina pectoris may find relief, or even cure, in the simple process of slowing the engines." In arterial hypertension, every clinician has

[1]

seen striking examples of the fall of blood-pressure with protracted rest. Equally familiar are the ill effects which so
often follow the return to duties too soon after exhausting
diseases such as influenza.

Experience has shown that for the convalescent who is not
confined to bed, there is no conflict in prescribing physical
exercise to alternate with rest. The one prepares for the
other and the degree and extent of relaxation are likely to be
increased after moderate exercise.

Analysis of the conditions for which it is generally prescribed indicates that rest (1) repairs fatigue or exhaustion,
thereby increasing the general resistance of the organism to
infection and other noxious agents; (2) decreases the strain
on the heart and blood-vessels; (3) diminishes the energy output and thus also the required caloric intake; (4) quiets the
nervous system, thus tending to relieve excitement, heightened reflexes and often spastic states; and (5) diminishes the
motion of the affected part or parts, thereby averting possible
strain and injury.

If useful for so many and varied conditions, it would seem
but reasonable to seek by scientific means the most effective
form of rest. This brings to attention the fact that the
patient advised to remain in bed often fails to get the desired
restful effects. He may not know how to relax and the restlessness may be increased by distress; therefore he may shift
and fidget in bed; lie stiffly or uncomfortably, owing to tense
muscles; and may be worried, impatient or otherwise over-
active in mind. In effect the physician's purpose in prescribing rest in bed may be nullified. This doubtless explains the
many failures which are commonly reported concerning the
so-called "rest-cure."

It is generally admitted that there is a lack of adequate
treatment for what is commonly called "nervousness." Often

the patient is told that there is nothing the matter with him—
that he should go home and "forget it!" If there is an apparent pathological condition, as in postoperative cases, as a rule
relatively little is done to quiet the nervous system, although
the suffering here may outlast the organic cause. We forget
that relief from functional ailments may aid the organism in
its struggle against other disorders. In consequence of repeated but futile efforts to get relief through the doctor, the
public often turns to pseudo-religious cults, medical quacks
or charlatans.

Physicians have generally considered the Wier-Mitchell
rest-cure a step in the right direction. Yet there seems to be
a strange neglect on the part of that writer and his followers
of the underlying physiology; his method apparently was
used without penetration to the cause of its success; for, as
we shall see, the goal should be to achieve habitual relaxation, but this was not fully realized. The title of his principal
book, *Fat and Blood and How To Make Them* (1879), indicates that he stressed nutrition in the treatment of nervousness and assigned to rest a secondary rôle. Some of his followers have placed a greater emphasis on the rest element.
Laymen sometimes speak of the value of "relaxation" for
the relief of nervousness, and so may seek rest, change of
scene, physical culture or various pleasures. They understand the word as meaning diversion or recreation, and in
this sense several neurologists have written upon the topic.
Amazing to say, the word "relaxation" seldom appears in
works on the treatment of nervous disorders, from Beard and
Wier-Mitchell to Oppenheim and Dercum. Perhaps this explains why a direct and systematic study of the subject has
hitherto been largely neglected.

In addition to the "rest-cure," it is usually admitted that
a few remedies such as bromides and other sedatives, sug-

gestion, persuasion, hypnosis, and perhaps the analysis of Freud are useful in their place; but limitations and objections to each are common, and the gap in therapy is apparent. Accordingly, it has seemed logical to search in the direction of physiologic relaxation for a more direct and efficient means to bring quiet to the nervous system.

The use of relaxation in the practice of medicine hitherto has been for the most part scattered and occasional. In order to promote sleep and relaxation, prolonged tepid baths have been employed, particularly in a few sanitaria. General massage is sometimes added and occasionally also warm drinks. That quiet is important is generally but not always recognized: attendants sometimes show a strange neglect when they permit their patients to be disturbed at critical times. Every practitioner has had occasion to quiet some patient by removing irritants, or with soothing tones, reassurance, or the inspiration of confidence. Fresh air has often proved a simple but useful aid to rest. Anne Payson Call (1902), a Swedenborgian, helped nervous persons to cultivate poise and to use relaxing exercises. But her interests are practical rather than scientific, and she admits that an individual may remain nervous while relaxed by her measures, for she fails to study that extreme or finely drawn-out relaxation which is the essential point of the present method. Maloney (1918) applied relaxation, produced by passive movements (i.e., manipulations performed by the physician), in the treatment of tabes. Certain of his observations agree with my own. Occasionally calm has been induced in excitable patients by suggestion and persuasion. Hypnosis has been employed to the same end by a few German workers. As will later be emphasized, suggestive measures, which render the patient dependent upon the physician, must not be confused with the method of cultivating natural and spontaneous relaxation presently to be described. A diet low in proteins and without condiments or stimulants is commonly prescribed for arterial hypertension, but seldom, if ever, for purely nervous maladies. Where apoplexy seems imminent, avoidance of nervous irritation has been urged (Drysdale, 1924). Auto-condensation has been occasionally used to induce relaxation, an effect due to the current heating the muscles with profuse sweating (Tuttle and William, 1925). In most of the foregoing instances the aim for relaxation has not been systematically carried out; the measures have been usually more or less empirical in origin; and generally there has been no clear-cut purpose to get the patient relaxed. By far the most general and popular agents employed by physicians to induce nervous quiet are the various drug sedatives.

When an individual lies "relaxed," so-called in the popular sense or the sense of Wier-Mitchell or as brought about by any of the preceding methods, it can readily be shown, as will later be seen, that in many instances he still remains tense in certain parts. Various movements and reflexes often are present, as can be graphically shown, which disappear if relaxation deepens. Therefore, in the present volume something more will be meant than what in the past generally has been called "relaxation" by physicians or by the public.

CHAPTER II
GENERAL FEATURES OF NEUROMUSCULAR HYPERTENSION

NEUROMUSCULAR or nervous hypertension may be defined as a condition marked by reflex phenomena of hyperexcitation and hyperirritation. The terms "irritability" and "excitability" are not here used because they generally mean *possible* responses rather than *actual* and often involve theories about nerve substance. On the other hand, phenomena of hypertension can be directly observed and measured. They may be general throughout the body or may be localized in some part. Disorders due to nerve lesions or to structural nervous pathology, including myotonia, hemiplegia, spastic diplegia and others, have received much attention in the past; but the phenomena which occur in the absence of such organic changes and which have been relatively neglected are the ones we propose to study. In these are found all degrees of restless behavior or rigid posture of the individual. Muscles exhibit variable or persistent contractions, which may be graphically recorded, indicating where nerve impulses are present in exaggerated number. Obviously such states are the antithesis of habitual relaxation. Therefore it is necessary to understand the conditions under which they arise and to recognize the symptoms in order to apply appropriately the method of relaxation.

Symptoms of hypertension can be observed, not only in the neurotic and psychopathic, but with care also in persons who ordinarily would not be called "nervous." Here we are in agreement with Woodyatt (1927), who describes a patient with emotional glycosuria as "physically well built, vigorous, fond of outdoor sports, not abnormally egocentric, not a malingerer, but successful in business and in society." Accordingly the physician who is on the lookout for neuromuscular hypertension will broaden his horizon beyond the bounds of what is commonly branded "restless," "irritable" or "excited."

The causes of neuromuscular hypertension as encountered in medical practice are many and various. Acute conditions may occur after intense or prolonged pain or distress from whatever source, whether physical, as a trauma, angina or colic, or mental, as a fright, bereavement, quarrel or loss. Such excitations are commonly called "emotional disturbances." They may take place in the course of almost any kind of disease or surgical ailment. They are frequently seen during bacterial infections and after a great variety of physical and chemical sources such as surgical operations, burns, electric shock, sunstroke, and food and drug poisonings, including strychnine.

Chronic conditions of nervous hypertension, although not commonly known by that name, are nevertheless often met by the physician. Predisposing causes evidently include all agents that reduce the general resistance and vigor of the organism; for, as is well known, conditions of debility are likely to be attended by nervousness. Apart from such agents, how shall we explain the common experience that some individuals appear from birth more irritable and excitable than others? Shall we assume that there is some congenital defect of their nervous system, perhaps nutritive (Meynert, 1882) or chemical (Krafft-Ebing, 1895)? Such an assumption would lead us into the midst of theories advanced by various writers (cf. Löwenfeld, 1894; Binswanger, 1896; Goldscheider, 1898). Yet after such theories have been discussed and it has been seen that nothing definite has been

[7]

established, who can say of an irritable adult, how much of his nervous hypertension is due to faulty congenital factors and how much to general malnutrition or disease of his organism, or to habits and impulses resulting from past experience of pain and suffering? It seems evident that at the present stage of knowledge of this question we are in no position to be dogmatic: rather we may be encouraged to keep an open mind and to turn without prejudice to the study in such patients of the therapeutic results which follow the regulation of some of their nervous functions through relaxation.

Exciting causes of chronic neuromuscular hypertension include those which may lead to the acute condition, particularly prolonged pain and distress from whatever source. Accordingly a whole range of chronic infections, toxemias, painful or debilitating tumors, and irritative gastrointestinal disorders may be mentioned. Vitamine deficiency (water-soluble "B," McCarrison, 1919) is to be noted here. Focal infections perhaps often cause various nervous disturbances (Watson, 1925; Nixon, 1925; Ackland, 1925). This seems quite plausible if pain or distress is marked; but if not, as often occurs with apical abscesses, the matter remains open to doubt, at least in any particular instance. Such infections are widespread among individuals free from nervous afflictions, rendering it difficult to apply a crucial test. Therefore further investigation will be required.

As is well known, examination of many irritable and excitable individuals discovers no organic lesion or disease. Followers of psychoanalysis believe that evidence has been accumulated to prove that "unconscious" sex experiences are everywhere a factor in the production of nervous disorder, but this hypothesis cannot be discussed adequately here. It is generally agreed that prolonged worry and anxiety often cause nervous disorder, and likewise emotional shocks, trauma (leading to traumatic neurosis), and maladjustment to environment (where this occasions efforts and strains). Intellectual overwork is a frequent cause: Many students and others live in an intermittent or continual state of mild hypertension with frequent symptoms of insomnia or emotional irritation. Since this condition can scarcely be considered pathological, it is evident that in the field of nervous hypertension the line between the so-called normal and abnormal cannot be sharply drawn.

In contrast with nervous hypertension stands nervous hypotension. This is familiar to clinicians in conditions of coma, typhoid fever, general anesthesia, myxedema, cretinism, lethargic encephalitis, and in phlegmatic individuals. Combinations of hypotension in some parts with hypertension in others are frequent: for instance, in sunstroke, where the voluntary muscles appear highly relaxed, yet tremor and jactitation may occur.

What has above been called nervous hypertension would by many be called neurasthenia. This term is generally used in the sense of nervous exhaustion or a tendency thereto. There is an obvious rational basis for the treatment of states of exhaustion by a method of rest: we should reasonably expect conditions commonly called neurasthenic to be open to treatment by relaxation.

Nevertheless, the current conception of irritable and excitable nervous states seems to need revision. We may therefore ask, first, what is the meaning of the term neurasthenia as distinguished from the term nervous hypertension? and, second, is the term neurasthenia true to reality or is it misleading and speculative?

Neurasthenia was originally defined by Beard in 1869 as excessive fatigue, nervous exhaustion, irritable weakness. This definition is largely current today (Curschmann-

Rostock, 1924; Church and Peterson, 1923; Reynolds, 1923; Singer, 1925). Investigators have not as a rule clearly distinguished between neurasthenia and nervous hypertension, as defined in this volume. Oppenheim cites six objective symptoms of neurasthenia, some of which apparently imply nervous hypertension. Binswanger, however, makes the distinction definitely and concludes that nervous hypertensive symptoms are present in states of so-called "neurasthenia." The same conclusion can be inferred from analysis of the observations of other investigators.

There is a noteworthy difference between a state of exhaustion and the activity that leads to that exhaustion. Obviously the difference is that between cause and effect. Such is presumably the difference between nervous hypertension and neurasthenia. Excessive activity leading to exhaustion may cease when the latter state is established, or it may continue and so maintain this state. These considerations suggest that in the past the cart has been put before the horse: the attempt has been made to understand and treat "neurasthenia" without adequate emphasis on the rôle of the nervous hypertension or hyperactivity that produces or maintains the state of exhaustion.

This becomes more obvious when we recall the well-known fact that the amount of activity necessary to produce fatigue or marked changes in pulse, respiration, and other reflex activities varies from time to time with each individual, owing to many possible influences. At times the result may be produced by an increment of activity which ordinarily would be considered slight. Such an increment then would very properly be called "excessive activity" *relative* to the occasion on which it produces fatigue or other disorder. A simple arithmetical exercise may prove severely trying to a patient with fever. We conclude that the general state of the organism at a particular time, as well as the character of the activity, determines whether the latter should be considered excessive. During fatigue or high fever, when complete rest is appropriate, all unrequired activities, voluntary or involuntary, evidently are fittingly termed excessive. Clinical investigation of individuals during fatigued and nervous states, as will later appear, discloses the presence of persistent or frequently recurring reflex and voluntary activities which evidently interfere with complete rest and so maintain the disordered states. The importance of such activities in causing and maintaining fatigue has sometimes been overlooked. Upon complete relaxation, frequently repeated, the fatigue generally disappears.

Once induced, it is probably true that fatigue in its turn favors the occurrence of hypertensive phenomena (Janet, 1905) and that this influence is reciprocal. There is evidence that fatigue alters the physicochemical state of neuromuscular structures (Lee, 1907 *a, b*, and many others) and delays relaxation, at least in isometric muscular contractions (Fulton, 1926, p. 111). Possibly also, as Binswanger implies, fatigue diminishes voluntary inhibition, leading likewise toward excessive reaction. Such considerations should not be neglected. In order to take them into account, we recall that nervous hypertension is a descriptive term for phenomena sometimes attended by fatigue and sometimes not, but which tend, if persistent, to result in fatigue. Accordingly it is clear that in so-called "neurasthenic" conditions, fatigue alone is not the complete source of symptoms, nor is it necessary to resort to a hypothetical "irritable weakness" in order to explain them.

There are various objections to the use of the term "neurasthenia." Frequently, it is charged, this diagnosis has been carelessly made, overlooking focal infections and other organic pathology. But there is a more serious difficulty. "Weakness" of the nervous system of afflicted individuals is a mere word or conjecture; it has never been physiologically proved. Even exhaustion has not been generally verified: Beard admitted that the neurasthenic may be plethoric or rubicund and that even prolonged muscular labor may cause little or no fatigue; he even adds that "neurasthenia does not always mean mental and physical exhaustion." Admissions of the same sort are to be found in recent works which follow the views of Beard (e.g., Singer). Certain investigators have succeeded in showing that their neurasthenic patients were more readily fatigued by arithmetical exercises than were the normal controls and that there are changed vascular reactions (arm-volume, Weber, 1910); but this is not satisfactory proof of exhaustion of nerve-substance, since so-called "neurasthenics" admittedly show continued nervous and mental activity, which would account for fatigue and early exhaustion without requiring us to assume that they have a pathologic state of the nerve-substance. Several recent writers prefer to regard neurasthenia not as a single disease but as a group of diseases or a class of nervous reactions (Curschmann-Rostock, 1924; Peterson, 1925). Apparently, current views on the subject of neurasthenia are inconsistent and often widely divergent. In short, the absence of direct evidence makes it unsafe to assume that individuals with "neurasthenic" or hypertensive symptoms have a congenital or acquired weakness or exhaustion of their nerve-substance.

It is therefore suggested that the term neuromuscular hypertension or the briefer form nervous hypertension should largely replace the term neurasthenia, excepting perhaps in a relatively few instances where exhaustion can actually be demonstrated. The former term is intended as descriptive, while the latter curiously mingles symptoms and theories. To be sure, neuromuscular hypertension can be noted in many disease conditions which no one would call neurasthenic.

Nervous hypertension is a classification that reminds us of arterial hypertension. It deserves to have the same popularity. These two diagnostic terms are alike in not denying possible underlying pathology. Arterial hypertension may follow certain heart and kidney diseases. Nervous hypertension may follow various pathological conditions above mentioned. Likewise, both arterial and nervous hypertension may arise from functional conditions in the absence of demonstrable organic disease. Both terms represent a systemic condition but not a complete diagnosis, and both types of hypertension frequently appear together.

Clinical evidence of nervous hypertension is seen in the form of increased reaction or reflex hyperexcitation. The increase or exaggeration may be in the speed, extent, duration, or number of reflex and voluntary activities. In this connection it would be important to determine normal curves of knee-jerk extent (chap. ix) and of other reflexes, founded upon many observations under set conditions. We could then say that a particular curve of knee-jerk under the conditions of test fell below, within, or above the normal range during a period of examination. As it is, clinicians customarily resort to a sort of empirical standard with indeterminate limits. They recognize the knee-jerk under certain conditions as exaggerated but never stop to calculate what this would mean quantitatively. Among the evidences of nervous hypertension are (1) increase of tendon reflexes; (2) increase of mechanical muscle excitability and less often of mechanical nerve excitability; (3) spastic conditions of smooth muscle; (4) abnormal excitability of the heart and respiratory apparatus; (5) tremor; (6) restlessness, volubility, and other apparent overactivity involving the skeletal muscles.

We must recognize that exaggerated reflexes can and often do result either from increased excitatory impulses or from decreased inhibition. The latter course, according to Binswanger (1896), was not realized by the early neurologists who explained the exaggeration in terms of "nervousness" and "irritability." Their views, failing to explain some of the facts, had to yield to those of Beard. Binswanger argues that in neurasthenia the increased reactivity is more largely due to decreased inhibition than to increased excitatory processes; but the evidence is not conclusive, and we shall not attempt to settle the question. (For further discussion, see chap. xiv.)

The clinical symptoms of nervous hypertension may be classified according to the portion of the nervous system involved. An attempt has been made to distinguish between hyperactivity of the sympathetic and parasympathetic systems. Certain signs were described by Eppinger and Hess (1915) as appearing in vagotonia. Chief among these were myosis, accommodation spasm, widened palpebral fissure, increased salivary and lacrimal flow, vasodilatation of the skin of the face and head, slowing of the pulse, diminished strength of heart-beat with increased diastole, respiratory arhythmia, contraction of the bronchial muscles, eosinophilia, hypersecretion with or without gastric hyperacidity, increased gastric tone and peristalsis, pylorospasm, increased peristalsis or spasm of the small and large intestines, spasm of the sphincter recti, as well as increased excitability of the urinary bladder and sexual organs. It has been found increasingly difficult by subsequent observers to draw the line between "vagotonia" and "sympatheticotonia" (Carlson, 1922). A further difficulty is that many of the foregoing symptoms can be produced through diminished activity of sympathetic efferent nerves as well as by increased activity of the para-sympathetic. Therefore, in any particular instance the origin of autonomic hypertensive symptoms often cannot be fully ascribed to either the sympathetic or the parasympathetic system alone; but symptoms may appear to result chiefly from overactivity of either system (see Kiss, 1932). Analysis may be aided by tests of pharmacological reactions and of the oculocardiac reflex.

Most important are the results of clinical and laboratory examinations, along with a detailed medical history of the patient, in indicating the presence of hypertension in specific visceral systems. The vascular system is most readily checked in this respect by repeated blood-pressure readings, electrocardiograms, examination of retinal and other periph-eral arteries and other tests. The alimentary tract can be ob-served for spasticity and hyperperistalsis in its several por-tions by roentgenological methods, proctoscopy and stool ex-aminations. Hyperactivity of respiratory muscles is dis-played through well-known symptoms, as in bronchial asth-ma. Excessive muscular activity in the genital urinary tract may reveal itself in frequent urination or in painful spasm of the ureter or other parts; the symptoms may or may not seem sufficiently accounted for on the basis of discoverable pathology.

In accounting for symptoms resulting from muscle spasm in any organ, local pathology, if any, must, of course, be con-sidered; but very often this is not the whole story. For the genesis even of localized symptoms may commonly depend upon the nervous and muscular reactions characteristic of the individual in response to his environment. This view-point should help to bring to the doctor an understanding and a practical handling, not just of organs, but rather of the pa-tient as an individual.

CHAPTER III
NEUROMUSCULAR HYPERTENSION IN
VARIOUS DISEASES

ONCE attention has been drawn to the matter, one can with care detect nervous hypertension in a whole range of diseases. It precedes some diseases as the whole or part of the cause, either predisposing or exciting. It is manifested during the course of other diseases or afterward. Its symptoms then may be the direct result of the action on the nervous system of toxic, bacterial, nutritional, or other factors responsible for the disease; but far more often they are reflex reactions to sources of pain, discomfort and anxiety. Doubtless nervous hypertension frequently arises in a given disease from more than one of the foregoing sources. In many instances further investigation is needed.

We may begin with specific infectious diseases and consider first those of bacterial origin. During severe acute conditions, nervous hypertension is commonly seen in the form of restlessness, delirium and insomnia. Many patients become talkative or perhaps irrational when there is even a moderate rise of temperature. Apparent inability of the patient to lie quiet sometimes seems to prevent adequate recuperation. The deep reflexes are commonly increased, until with advanced toxemia a diminution may follow. Individuals who during ordinary life are deemed "nervous" are particularly likely to show hyperactive symptoms. Various objective symptoms of disease apparently may be increased in those inclined to worry: Cough, if present, may occur upon slight provocation, often resulting in paroxysms with little production of phlegm. Sweating may become marked. Palpitation, or increased pulse-rate disproportionate to the fever, is often observed. Every clinician has seen instances where exciting a patient has apparently produced an increase of fever, which later subsides as the patient becomes quiet. But this has not yet been tested with precise methods.

Among further signs of nervous hypertension during bacterial as well as other diseases are groaning, continual complaining, frequent shifting of the eyes and repeated micturition (when not due to inflammations, stones and other local disturbances of the urinary system). Not seldom a heightened pitch of voice (familiar to singers) proceeds from hypertension of laryngeal and pharyngeal muscles. Often noticed also is the nervous start upon sudden noises, pains, and other external and internal stimuli. Perhaps as the patient falls asleep, his trunk and limbs give a convulsive start which usually awakens him. The significance of this "predormescent start" awaits explanation. Years of observation of patients during advancing relaxation have thrown light on its occurrence. Apparently it takes place in individuals who have been hypertense during the preceding hours, that is, during the day's activities. It tends to disappear after a restful day but to appear again after exciting experiences. The further explanation is not known with certainty. Observation suggests that it arises from a startling experience in the dream state of the individual, which tends to be rapidly forgotten as he awakes. He dreams, for instance, that he is walking, steps precariously, slips, and awakes with a start. Presumably the physiologic mechanism is otherwise similar to that of the nervous start described in chapter vii.

Tremor, convulsions, dilated pupils, sobbing and various other signs of increased emotion are familiar to every general practitioner. The nervous patient is likely to report any subjective distress in increased degree. From the detail and vividness of such reports, the impression is gained that nervous hypertension may be responsible for an actual, not merely an imaginary, increase of pain. This impression is strengthened by the conclusions drawn from the experiments of chapter viii.

In convalescents, when the physician's attention is no longer engrossed with the bacterial phases of the disease, the nervous after-effects become more conspicuous. Then restlessness, complaining and insomnia may become troublesome and seem to interfere with recovery. Then also gastrointestinal symptoms, nausea, dizziness, anorexia, belching, flatus, even vomiting or diarrhoea may arise of nervous origin.

Tetanus is an instance in which virus acts upon nervous elements in the cord or directly upon muscle tissue, causing contractions which characterize the disease (Fröhlich and Meyer, 1912; Liljestrand and Magnus, 1919; Fulton, 1926, quoting unpublished observations of Sherrington).

In chronic bacterial diseases, various of the above-mentioned symptoms of hypertension often become prominent. Tuberculosis is a good illustration. Patients are now being trained in certain sanitaria to diminish unnecessary coughing and are frequently requested to rest (Pottenger, 1924). What is being done by simple requests might perhaps be better accomplished through systematic practice by the patient along directed lines according to the method of progressive relaxation.

Any nervous hypertensive symptoms during syphilis generally are reactions to pain and distress or are an expression of the anxiety of the patient over his condition; but if tertiary nervous lesions exist, a more or less excited condition occasionally is noted, sometimes to the point of delirium. Excitement in advanced general paralysis is sometimes noisy and violent; however, in early cases the complaint of insomnia and restlessness may present a picture like "neurasthenia."

From the foregoing instances it is evident that in the various diseases of bacterial origin, nervous hypertension commonly *results* from toxemia or reflex reactions to distress or anxiety; but it also plays the rôle of *causing* symptoms which may modify or prolong the course of the disease.

Whether nervousness may directly give rise to increase of temperature is not settled (cf. Gottlieb, 1891; Schultze, 1899; Aronsohn, 1902; Freund and Strasmann, 1912; Oppenheim, 1923). Upon admittance to the hospital, many patients exhibit a transient fever which seems due to excitement. Cer. tain recent writers (Mora, Amtman and Hoffman, 1926) offer blood-counts from dogs during excitement and from mei with non-inflammatory conditions such as hernia, varicocel and hydrocele in evidence that nervousness may cause leu cocytosis; their work should be repeated by others, for it is o' obvious importance. That the autonomic nervous system can excite leucocytosis is also affirmed by Mueller (1926) who describes two instances of insulin shock with increases to 19,000 and 28,000, respectively, in the brief time of 15 minutes.

Diseases due to physical agents are likely to leave a period of nervous hypertension in their wake. After accidents, patients sometimes exhibit tremor, restlessness, diminished "concentration," anxiety states, insomnia and other manifestations of irritability and excitement for years. Such symptoms are well known in traumatic neurosis. During sunstroke, relaxation of the skeletal muscles may become very marked, indicating the presence of nervous hypotension in certain centers. However, the occurrence of twitchings, jactitation, or very rarely convulsions points to hypertension in other centers.

[19]

Intoxications, such as alcoholism and morphinism, are characterized by well-known symptoms of functional disturbance, including muscular tremor and nervous irritability. But before taking his drink in the early morning hours, the chronic alcoholic may exhibit hypotension in the form of dull mental processes with inability to transact business. All severe acute forms of chronic intoxication, including food poisonings, may be followed by one set or another of hypertensive symptoms. Divers nervous symptoms due to poisoning by drugs, such as strychnine and numerous others, are familiar in textbooks on pharmacology.

In the field of metabolic disorders, worry or sudden mental shock is said to be among the directly exciting causes of an attack of gout (Osler and McCrae, 1926). No observations have been made that enable us to say whether the headache, migraine, neuralgia, sciatica, and paresthesias of gout are due entirely to the chemical change or whether nervous hypertension plays a reinforcing rôle.

Diabetic patients not seldom display an extraordinary degree of restlessness and anxiety. They are sometimes morose or hypochondriacal, showing a neuropathic disposition (Naunyn, 1906). Neilson (1927) suggests that the onset of coma frequently can be explained as due to the occurrence of nervous excitement. According to common experience (Osler and McCrae) there is no doubt that, once established, diabetes mellitus is influenced by the general state of the nervous system. Worry, mental strain, anxiety, etc., may have a marked influence. Severe nervous strain or excitement may cause temporary glycosuria (chap. xii, p. 204; also Woodyatt, 1927). In conformity with this, rest is of acknowledged importance in the treatment of ketosis (Campbell, 1925).

Tremor, restlessness, irritability and excitement are familiar symptoms of nervous hypertension in exophthalmic goiter. It is well known (Neilson, 1927) that emotional stress and psychic upsets often seem to initiate this disease; but we cannot be certain that such instances are not latent, in the sense of constitutional or potential susceptibility. Symptoms similar to those of Graves' disease, but not quite identical, have been produced in dogs by feeding thyroid substance (Kunde, 1927).

In various diseases of the digestive system neuromuscular hypertension appears to play a conspicuous rôle. One of the most common ailments in this class is the mildly spastic esophagus (chap. xvi); another is the spastic colon (chap. xvi). McLester (1927) in a series of approximately 1,600 patients who complained chiefly of digestive disorders was unable to find organic disease in 32.6 per cent and classed them as psychoneurotics. He estimates that one-third of the patients who come to the consultant because of digestive complaints are of this type. But in addition we must take into account that organic disease can reflexly provoke increased nervous irritability and excitement. Symptoms arising or increasing through this route come under the caption of nervous hypertension, although organic disease also is present.

Among diseases of the respiratory system, hay-fever and bronchial asthma have always had the reputation of occurring in individuals with unstable nervous systems. Dunbar's researches (1905) placed the etiology of the disease on a scientific basis, evidencing that in many instances the reactions were due to foreign proteins. Treatment by desensitization became highly developed and widely used. However, of late a reaction has set in (Kahn, 1927): it is doubted that the anaphylactic method of treatment has proved to be the success that was earlier anticipated. A recent article in Germany's most-read medical journal (Moos, 1928) claims cures by psychoanalysis in 100 per cent of a series of eighteen cases of bronchial asthma with disappearance of spirals, crystals, emphysema and excess eosinophils. Unfortunately, as too often occurs in psychoanalytic articles, the data lack the critical presentation required by the careful scientist and therefore have at most a suggestive value. Realizing that the problems of etiology and treatment of bronchial asthma are not yet completely solved, it seems likely that study of the autonomic nervous hypertensive symptoms in these diseases may some day yield important information.

During attacks of asthma, whether bronchial, cardiac or renal, various hypertensive symptoms can be noted. Accompanying the violent respiratory efforts, the forehead and brow are visibly tense and with the shifting eyes create an anxious expression. The jaw may be held tensely open or closed, while the limbs occasionally move with restless jerks, expressive of distress. These various muscular reactions of distress apparently are reflex results of dyspnea; but it should not be forgotten that muscular activities under such conditions tend to increase the very dyspnea that produces them.

Chronic bronchitis may also lead to hypertensive symptoms. Repeated stimulation from severe paroxysmal cough, due to an irritated mucosa, eventually has effects upon the nervous system. The general muscles of the face, limbs and trunk become tense during and following the paroxysm, giving the patient an expression of irritability. At such times the deep reflexes are increased. In consequence of this excess tension a vicious circle is created, with further increase of cough.

Movable kidney is well known for its connection with symptoms of so-called "neurasthenia with dyspepsia." Albuminuria is sometimes said to result from functional nervous conditions. Whether there is a connection between overactivity of nerve tissues and excess excretion of phosphoric acid still remains, I believe, an unsettled question. Hypertensive symptoms in uremia are well known. They include early irritability and excitement, with development in severe cases into convulsion and mania. Every physician has witnessed the nervous symptoms that accompany or follow disturbance caused by renal calculus.

Chronic nephritis is likewise often accompanied by familiar symptoms of nervous hypertension. As vascular hypertension increases, the nervous symptoms also are likely to become more conspicuous. Impatience, irritability and restlessness are common. The patient, however, may not be known to his fellows as nervous or irritated. Muscular hypertension may be manifested in the form of a certain rigidity of posture, particularly in the jaws. Possibly this is the feature, along with rubicund face, that occasionally suggests to experienced clinicians at a glance that a patient will show high blood-pressure. That nervous irritability may become marked during restless sleep at night, increasing the blood-pressure and resulting in cerebral hemorrhage, has been emphasized by Drysdale (1924).

Anemia is characterized by increased respiration and pulse-rate upon relatively small exertion or emotion. These phenomena often are marked in the pernicious type. As a factor in such symptoms, the nervous element may be quite apparent. It seems worthy of investigation whether fatigue from persistent excess of neuromuscular activity might not in some conditions increase the degree of secondary anemia present.

Nervous symptoms are generally not looked for in endocarditis and pericarditis. Yet restlessness and insomnia may appear as disturbing factors and perhaps even turn the course of the disease by interfering with recuperation. The "sleep-start" may occur repeatedly: there may be groans and fre-

quent cries as the patient falls asleep with shifting and fright-
ful dreams. Likewise in chronic valvular or in myocardial
disease the sleep-start may be frequent, and nervous overactivity during the day may persist at night in the form of restless and broken sleep which fails to bring adequate refresh-
ment. Foster (1927) calls attention to sleep disturbances in
cardiac insufficiency and to other symptoms of daytime restlessness, with finally the terminal delirium. He reminds us
also that hypochondria can produce distress simulating organic disease. Jaquet (1922) quotes data to the effect that
only 25–30 per cent of all confirmed cardiac patients are free
from neurosis; and Neuhof (1922) cites five patients with
cardiac lesions who had been improving until apparently
made markedly worse by the occurrence of psychic disturbances.

Arteriosclerosis may or may not be accompanied by apparent symptoms of nervous hypertension. The latter sometimes has been considered to be the causal factor (Watermann
and Baum, 1906). Relationship of symptoms and signs in
chronic arthritis with those of nervousness was early mentioned by Vigoureux (1893) and recently by Pemberton.
Ophthalmologists (Lunt and Riggs, 1924) have estimated
that there is a neurotic element in 10–20 per cent of their
patients, and that perhaps 5–10 per cent have ocular symptoms explainable by their neuroses. Cutaneous manifestations due to neuroses have recently been discussed by Klauder (1925). I have seen in one case a sharply demarcated
erythema on the neck and groin, in another a diffuse erythema with urticarial-like plaques limited to the neck appear,
disappear and reappear concurrently with symptoms of
nervous hypertension.

In "functional nervous disorders," the psychic phases
have received most attention. The patient's complaints have

led the physician to give heed to his anxieties, fears, griefs, depressions, obsessions, abnormal feelings, emotions and other subjective disturbances, while physiologic signs of hypertension have generally suffered neglect. Yet it has been my experience that wherever there is psychic disturbance, trained observation will reveal corresponding signs of neuromuscular hyperactivity or hypo-activity. There is an old principle, long accepted in experimental psychology, that every psychic process has a neural correlate. This is doubtless true so far as it goes, but physicians and others have been prone to assume that the "neural correlate" is largely limited to activity within the nervous system and therefore have neglected to look beyond. I shall later present evidence and give reasons why this narrow conception must be abandoned; why we should look during the moment of psychic disturbance, not solely for cerebral activity, but particularly for the efferent neuromuscular changes, wherein we can use graphic, fluoroscopic and other precise measures. When this is done, according to my experience, no functional nervous disorder fails to reveal a characteristic element of neuromuscular hypertension. But this statement needs to be clarified, since in states of grief and depression it is generally the limpness or relaxation that most strongly strikes attention. However, a study of individuals during such states quickly reveals the existence of mental processes of worry, fear, anxiety and despair and, corresponding to such processes, various physiologic phenomena of hypertensive character. That hypotension can exist in some neuromuscular localities although hypertension is present in others was witnessed in the foregoing example of sunstroke. Relaxation in some portions of the body musculature with concomitant tension in other portions will later be discussed (chap. vi). Such *differential relaxation* obviously has its counterpart in what might be called the *differential hypertension* found in states of grief and depression.

In obstetrics, hypertensive symptoms are familiar in pregnancy, labor and the puerperium. Following a complicated pregnancy or difficult first childbirth, many women give an extensive medical history in which hypertensive symptoms from the nervous system may be difficult to differentiate from symptoms due to other organs.

In surgical practice, nervous hypertensive symptoms must be kept in mind. Mild abdominal pain in a patient who is so tense that symptoms and signs become exaggerated may readily lead to a mistaken diagnosis resulting in surgery. Furthermore, even if an organic lesion has been found and surgically removed, it must be remembered that the possibility of nervous hypertension is not excluded. Symptoms sometimes continue, due to the hypertensive habits of the individual; and neglect of such factors may lead to an unsatisfactory result.

An increased measure of symptoms of nervous hypertension is commonly observed in emaciated states; marked reduction of diet is said to increase the irritability of some patients; but there is no evidence of a necessary connection between nervousness and the state of nutrition. Beard's method of treating nervousness by nutritive methods has long since been abandoned. Persons well nourished or gaining in weight may nevertheless show increasing symptoms of worry, anxiety or other forms of irritability and excitement; in this, the author's experience confirms that of many neurologists (Beard, Singer and others).

As is well known among diagnosticians, careful search for etiological factors frequently fails to reveal any noteworthy focal infection or other pathological change to explain the symptoms of nervous hypertension.

NEUROMUSCULAR HYPERTENSION

The preceding fragmentary summary of the appearance of phenomena of nervous hypertension in various diseases and disorders indicates that these factors occur in the guise of symptoms, causes or effects almost throughout the whole range of the practice of medicine, surgery, and the specialties. Accordingly the opportunity for a wide and varied application of a method of relaxation is suggested.

CHAPTER IV
GENERAL FEATURES OF PROGRESSIVE RELAXATION

I N THE course of laboratory studies presently to be described, a method to produce an extreme degree of neuromuscular relaxation was gradually developed. What
is customarily called "relaxation" was found in many instances to be inadequate and undependable for our purposes,
both investigative and clinical. I found, as others had found
previously, that an individual might lie on a couch apparently quiet for hours, yet remain sleepless and nervously restless. Even as he lay there, he might continue to betray signs
of mental activity, organic excitement, anxiety or other emotional disturbance; he might breathe irregularly, fidget and
start, with restless movements of the eyes, fingers or other
parts, or perhaps with an impulse to unnecessary speech.
These signs might occur occasionally or frequently and might
be quite obvious to the observer or require close inspection.
When attention is once called to the matter, it is evident
that such rest at best is not complete; following it, the patient often fails to seem refreshed, retaining his symptoms
and complaints of nervousness, fatigue or other ills. Accordingly I was led to inquire whether the foregoing phenomena would not diminish or disappear if a greater degree and
bodily extent of relaxation were cultivated.

It was evident that an extreme degree of relaxation was
required, and for convenience this was called "progressive
relaxation." As applied to patients, as will later be shown,
this developed into a physiological treatment for nervous hypertension. The plan was to test whether excitement which

[28]

has stubbornly persisted will tend under conditions favorable to progressive relaxation to give way to sleep, and whether spells of worry or rage or other emotional disturbance will tend to pass off. A further development of method (Jacobson, 1920, chap. vi) aimed to produce a certain measure of these quieting effects, operative while the individual continued at work or other activity. In substance, the hypothesis was that a method could be evolved to bring under varying conditions absolute or relative rest to the neuromuscular system, including the mind.

When the unpracticed person lies on a couch, as quietly as he can, external signs and tests generally reveal that the relaxation is not perfect. There remains what may conveniently be called *residual tension*. This may also be inwardly observed through the muscle sense. Years of observation on myself suggested in 1910 that insomnia is always accompanied by a sense of residual tension and can always be overcome when one successfully ceases to contract the parts in this slight measure. Residual tension, accordingly, appears to be a fine tonic contraction along with slight movements or reflexes. Often it is reflexly stimulated, as by distress or pain; yet even under these conditions relaxation is to be sought.

Doing away with residual tension is, then, the essential feature of the present method. This does not happen in a moment, even in the practiced person. Frequently the tension only gradually disappears; it may take 15 minutes progressively to relax a single part, such as the right arm. The desired relaxation begins only at the moment when the individual might appear to an inexperienced observer to be very well relaxed.

When an individual lies "relaxed" in the ordinary sense, the following clinical signs reveal the presence of residual tension: respiration is slightly irregular in time or force; the pulse-rate, although often normal, is in some instances moderately increased as compared with later tests; voluntary or local reflex activities are revealed in such slight marks as wrinkling of the forehead, frowning, movements of the eyeballs, tenseness of muscles about the eyes, frequent or rapid winking, restless shifting of the head, a limb or even a finger; the knee-jerk and other deep reflexes can be elicited (if there is no local nerve-lesion); a reflex start generally follows any sudden unexpected noise; in the studies undertaken up to date, if the esophagus or colon is spastic, it continues in this excited state; finally, the mind continues to be active, and once started, worry or oppressive emotion will persist. It is amazing what a faint degree of tension can be responsible for all this. The additional relaxation necessary to overcome residual tension is slight indeed. Yet this slight advance is precisely what is needed. Perhaps this again explains why the present method has hitherto been overlooked. As relaxation advances past the stage of residual tension (chaps. vii, viii, ix, x, xvi), respiration loses the slight irregularities, the pulse-rate may decline to normal, the knee-jerk diminishes or disappears along with the pharyngeal and flexion reflexes and the nervous start, the esophagus (assuming that the three instances studied are characteristic) relaxes in all its parts, while mental and emotional activity dwindle or disappear for brief periods. The individual then lies quietly with flaccid limbs and no trace of stiffness anywhere visible, with no reflex swallowing, while for the first time the eyelids become quite motionless and attain a peculiar toneless appearance. Tremor, if previously present, is diminished or absent and slight shifts of the trunk or a limb or even of a finger now cease to take place. Subjects independently agree in reporting that this resulting condition is pleasant and restful. If persistent, it becomes the most restful form of natural sleep. No university subject and no patient has ever considered it a suggested or hypnoidal or trance state or anything but a perfectly natural condition. It is only the person who has read a description without witnessing the actual procedure who might question this point.

The symptoms and signs of overactivity of the cerebrospinal system perhaps accompanied by those of the vegetative system have been described in the previous pages. Overactivity or nervous hypertension due to the central nervous system, as shown in tenseness or exaggerated or excessive movements of striated muscles, presumably is subject to voluntary control. Every individual at least in some measure relaxes his muscles when he goes to rest. It would seem strange, therefore, if this natural function could not be specially cultivated to counteract an excess of activity and bring quiet to the nervous system. Such is the aim of the present method. As may be readily noted, the neurotic individual has partly lost the natural habit or ability to relax. Usually he does not know what muscles are tense, cannot judge accurately whether he is relaxed, does not clearly realize that he should relax and does not know how. These capacities must be cultivated or re-acquired. Accordingly, it is usually futile to tell the sufferer to relax or to have him take exercises to this end in gymnasiums. Furthermore, relaxation as heretofore understood is not adequate as a clinical method. Following previous standards, a patient may be apparently "relaxed" in bed for hours or days, yet be worried, fearful or otherwise excited. In this way at times the patient has been wrongly considered at rest, overlooking voluntary or local reflex activities described above. The detection of such signs is useful in diagnosis and in directing the patient or subject so as to bring about nervous and mental quiet.

[31]

According to the present clinical and experimental experiences up to date, if the patient is shown how to relax the voluntary system, there later tends to follow a similar quiescence of the vegetative apparatus. Emotions tend to subside as he relaxes. To be sure, there may be a vicious circle: vegetative nervous overactivity seems to stir up local or general cerebrospinal reflexes (Carlson, 1912; Daniélopolu and Carniol, 1922; Johnson and Carlson, 1927), and these again vegetative reflexes (see chap. xvi). The one system must become quiet before the other system can become quiet. So in certain chronic cases, relaxation becomes a gradual progress —a matter of habit formation that may require months. Various stimuli that occur during pain, inflammation, or disordered glandular secretion, such as hyperthyroidism, may give rise to smooth muscle spasm and therefore hinder relaxation. Under these difficult conditions, it is traditional to assume—and patients often assert—that an individual "cannot" relax. Yet such inability would be difficult to prove. The presence of a reflex response to pain or other stimulus, as will later be shown (chap. viii), is not in itself proof that the reflex "could not" have been relaxed. That is precisely what needs to be investigated, for the patient's subjective views as well as the physician's a priori conceptions should not take the place of laboratory and clinical tests.

After emphasizing the difference between "cultivated" and "ordinary" relaxation, it is equally important to emphasize their fundamental identity. Under favorable conditions untrained individuals relax, as shown by the knee-jerk and other tests, although perhaps not so fully as after training. So-called phlegmatic individuals are particularly likely to succeed. It may be assumed that whatever the natural propensities of an individual toward relaxation, there is always considerably more that he can be taught; just as anyone with a naturally good voice nevertheless improves greatly with proper training. In my experience persons who have not been trained to relax are less likely at times of emotional disturbance to resort to voluntary relaxation: they fail to apply the ability even if they have it; yet the process of relaxation, whether natural or cultivated, is essentially the same.

Before training any patient, a detailed history should of course be taken, followed by thorough physical, laboratory and roentgenological examinations leading to a correct diagnosis. Treatment, when required, in the form of surgery or drugs or hygienic measures should accompany progressive relaxation. It is of course important to remove so far as possible both physical and mental sources of excitement. Since this ideal often cannot be realized, the method of relaxation will seek to lower the nervous reaction even where those sources remain unavoidably active. If the purpose in employing relaxation is investigation rather than practical results, other measures of treatment must be excluded so far as possible. However, even considering only the practical interests of the patient, it is best in many instances to exclude additional measures of therapy until the effects of relaxation have been thoroughly tested; for otherwise, if the condition should become improved, doubt would arise as to what agent were responsible for the result, and in consequence the wrong one might be selected for continued use.

The present method consists of voluntary continued reduction of contraction or tonus or activity of muscle-groups and of motor or associated portions of the nervous system. When the relaxation is limited to a particular muscle-group or to a part, as a limb, it will be called *local;* when it includes practically the entire body, lying down, it will be called *general.*

[33]

General relaxation, if sufficiently advanced, as will be shown, is characterized by reduction of reflex irritability. In the present method, general relaxation is progressive in three respects: (1) The patient relaxes a group, for instance, the biceps-brachial of the right arm, further and further each minute. (2) He learns consecutively to relax the principal muscle-groups of his body. With each new group he simultaneously relaxes such parts as have previously received practice. (3) As he practices from day to day, according to my experience, he progresses toward a habit of repose, tends toward a state in which quiet is automatically maintained. In contrast with this, experience indicates that the individual who indulges in unrestrained excitement renders himself susceptible to further increase of excitement, as a consideration of the familiar phenomena of augmentation or summation (chap. xiv) would lead us to expect.

By *differential relaxation* will be meant the absence of an undue degree of contraction in the muscles employed for an act, while other muscles, not so needed, remain flaccid (chap. vi). Clinical relaxation may as a rule be called *systematic, habitual* or *cultivated* and the meaning of these names will be obvious; but they will not apply in many cases where relaxation is briefly used to aid in overcoming some acute or transitory disturbance. Relaxation, whether general or local, is defined as *complete* if it proceeds to the zero point of tonus for the part or parts involved, and *incomplete* if it falls short of this.

Current usage applies the term "relaxation" to nervous as well as to muscular phenomena; I have followed this because of the difficulty as a rule of separating or distinguishing the manifestations and effects of the two types in the intact organism. The reader may, if he choose, reserve some other term such as "inaction" or "progressive inactivation" for the nervous system, and apply the term "relaxation" only to muscles.

When we say that a muscle is "tense," we mean that it is contracting, its fibers are shortened. Such tensions make up much of the warp and woof of living. Walking, talking, breathing and all of our activities involve a series of complicated and finely shaded tensions of various muscles. To do away with all such tensions permanently would be to do away with living. This is not our purpose, but at times we need to control them, and relaxation is a form of such control.

It is customary to apply the term "tenseness" not only to a muscle but also to the individual as a whole; then we mean that he acts as if his muscles were unduly contracted; he is overalert, in popular terms, or "high-strung." In addition, a third meaning of the term "tense" is used in this volume; the individual who contracts a muscle has *an experience* which it is agreed to call "tenseness." He notes the local feeling so as to recognize it later wherever it occurs in his body. It might be called by some other name, such as the "*X*-sensation," and it is often well to have the subject use this term in order to avoid the belief that he is to report whether his muscle is contracting. The latter is evidently a matter for a graphic record rather than for a subjective report. These three meanings of the term "tense" should be clearly distinguished by the patient; but wherever the term is used in the present volume, it is believed, clarity of context will leave no ambiguity as to which meaning is intended.

There is evidence that the average person does not know when he is tense. Psychologic experiments have indicated that sensations of tenseness or contraction constitute some of the very essence of the processes of attention and thinking (Wundt, Pillsbury, Ach, and many others), but this does not mean that the thinker is aware of the sensations as such. In the present method the patient is made aware of sensations of contraction, and later may observe them taking place while he thinks or is excited. He may note how tense he becomes; and his attention, thus drawn to the point, shows him what to relax; or again, when he becomes tense, the habits he has newly acquired may lead him automatically to repose.

In my experience, the individual profits by knowing what muscle-groups are tense in order to know where to relax. Therefore I have cultivated the "muscle-sense." This term is here used to mean the sensations that arise within contracting muscles. The term "muscle-sense" was first used by Charles Bell (1830). He found that sensations of movement were not cutaneous and it seemed to him that they came from muscles. Duchenne (1855), on the contrary, suggested that they came from joints. He based his opinion upon certain instances where muscular sensibility was lacking upon pressure or electric stimulation, yet active and passive movements were clearly felt coming from the joints. Lewinski (1879) and Goldscheider (1889) came to the same conclusion. Sherrington (1894) discussed the sensory function of the muscle spindles. In addition to these there are various sensory organs connected with the muscles—Pacinian corpuscles, corpuscles of Golgi-Mazzoni and the terminal cylindrical and spherical bulbs of Kraus. Woodworth concluded in 1903 that it was as yet impossible to decide the source of movement sensations. He observed that sensations of rotation evidently came from the region of the joints, sensations of cramp from the muscles, and sensations of fatigue from both sources. Winter (1912), on evidence secured by anesthetizing certain joints through electrical measures, concluded that muscles are the chief sources of sensations of movement. Since Woodworth's time there has been considerable discussion as to what sensations are used in the comparison and judgment of weights placed upon the outstretched hand. Such discussions have no interest for our present purposes except as they furnish a starting-point for our conceptions.

Sensations from contracting muscles can with care be distinguished from those arising in joints. The individual finds that he can relax away the former but not the latter. If the physician raises the *limp* arm of the practiced individual and pulls the hand in any direction so that the arm is in marked extension, sensations of *strain* (but not of muscular contraction) arise in the shoulder joint which obviously cannot be relaxed away. Frequent repetitions tend to make the individual familiar with the peculiar qualities of the experience of joint sensations in contrast with those arising in contracting muscles, which can be voluntarily relaxed.

Training in the present method thus tends to make the individual increasingly observant of muscular contractions in various parts of the body. Many lay subjects, especially musicians and athletes, evidently have made such observations previously. It would not seem quite correct to state that anyone meets the sensation for the first time, or that the training brings these sensations to consciousness. We must remember that the layman is prone to call an experience "unconscious" when he should more accurately say that he is "unaware of it." There is an important difference, which even some psychologists overlook, between having a sensation and observing it. The same is true of objects: for example, some of the most familiar things in our home escape our notice and observation until perhaps somebody brings them to our attention. Every normal individual is familiar (although not necessarily observant) with sensations from muscular contraction and from joints and tendons, for a fine play of these is often required in order to maintain balance or co-ordination. We conclude that these sensations are conscious although not generally noticed as such by the possessor, because trained and often untrained individuals can recall and describe them, and because it is evident that the individual who balances himself or co-ordinates delicately must give his attention to it. The use of muscular sensations was isolated from that of joint, tendon, and skin sensations in the experiments of Allers and Borak (1920). Their subjects compared weights applied on muscle masses left after amputation. Changes in the skin probably did not figure in the judgments, because the position of the skin had been altered by the operation. Sensations from the joints and tendons were of course excluded. These workers found that their subjects judged weights most accurately by means of active contraction of muscles. They concluded that the sensations normally used in judgments of weight and of extent of muscular contraction as well as of position of limbs are mostly of muscular origin, while sensations from joints and tendons are not absolutely required. Obviously the subjects in these experiments must have had conscious sensations in order to judge the weights. It seems safe to infer that such judgments by the average individual are made consciously. Evidently in the same sense conscious muscle sensations play an important part in every co-ordinated bodily act. The average individual, however, needs to be trained before he is definitely aware and able to describe his sensations.

While such training is helpful, it is not invariably needed, for relaxation often proceeds reflexly without conscious assistance, and this is ever to be encouraged. If performed unskilfully or at a wrong moment, conscious observation may even interfere with relaxation. Furthermore, the individual must not continue to watch his muscles intently, else he may fail to relax. On the other hand, it is my common experience that some muscular regions fail to relax until he locates them. A happy medium is reached when, with a minimum of attention, the disturbance is located and then relaxed. Moments of attention to muscles while trying to relax become increasingly unnecessary in the course of months when relaxation has been made habitual. Here the matter is like any other learning process, requiring less attention as time proceeds. After relaxation has been cultivated, it proceeds, at its best, automatically with little or no clearly conscious direction.

CHAPTER V

THE TECHNIC OF PROGRESSIVE RELAXATION— GENERAL FORM

LIKE any new procedure, for instance a surgical operation, the method of progressive relaxation presented here is best learned by the physician who sees
what is done rather than from a written discussion. He then
avoids any preconception that suggestion is used in the present method; for the falseness of this idea becomes patent
during the first period of relaxation under instruction. The
aim is to train the patient to use *his own initiative*. He learns
to localize tensions when they occur during nervous irritability and excitement and to relax them away. It is a matter of
nervous re-education.

The method of progressive relaxation has many forms,
depending upon the condition to be met, whether acute or
chronic, its previous duration, and the ability of the patient
to follow directions. For purely clinical purposes in acute
conditions, treatments may be made very brief and few in
number: occasionally two or three periods may seem to serve
the purpose. Results may then have a practical value but
will have less interest from a scientific standpoint. In conditions that have persisted for years, the period of treatment
varies greatly. Some require months or a year or longer. In
others, where the individual shows aptitude, a practical
therapeutic purpose is sometimes accomplished with perhaps
a dozen sessions, followed by diligent practice. Where thorough work is to be done, periods of treatment or instruction
last about $\frac{1}{2}$ hour to 1 hour and take place about three or

four times a week or daily. In addition, the individual practices by himself each day for an hour or two or more.

Securing Co-operation of Patient or Subject

If it is certain that the individual will strive to follow directions, no preliminary statement at all need be made. He simply lies down and the work begins. If desired, he may be informed that he is to learn to relax to an extreme degree.

If co-operation is a little doubtful, some preliminary explanation may be required. But prejudice will be avoided by saying as little as possible in advance. Perhaps as he chats, his hand fidgets nervously, or his brow becomes anxiously furrowed or his legs shift restlessly; if so, these tensions may be pointed out to him as symptoms of nervousness that are to be relaxed. Not seldom the patient himself volunteers that he is habitually tense, often pointing to the apparent seat at the back of the neck or across the brows, or stating that he frequently feels "bound up in a knot and cannot relax." This likewise opens a ready avenue of approach. Many, however, have never observed the connection between tenseness and nervous excitement, or between relaxation and nervous calm. To such a one, practice at relaxing a muscle may seem absurd and out of place so far as his nervous troubles are concerned. If he will co-operate, it is well to let his skepticism remain, for it will gradually yield when his observations and insight progress.

To tell in advance of the complete purpose of the work and of the benefits that may be expected or that others have obtained is unscientific. It must not be done if the record is to be of value for investigation. Evidently this involves a certain element of practical disadvantage, for the patient who is in the dark as to what is being done may become discouraged and may even abandon treatment. If the record is not desired for scientific study, but practical improvement is the only purpose, it is possible to explain fully in advance. But unless this is done with extreme caution, the explanation may be received in a suggestive manner.

The present method may be deliberately used along with suggestion, or along with bromides or other drug sedatives; but the conditions of scientific study require that these accessory agencies be avoided in order that we may know what has produced our results.

GENERAL CONDITIONS

The patient or subject lies comfortably on his back with his arms at his sides and with legs not crossed. A collapsible canvas couch without sides is most convenient, but a bed will serve the purpose. Occasionally a chair with the sitting posture is used from the outset. The room is kept fairly quiet, at least during the first periods of practice. As many as eight patients may be handled at the same time, but in different rooms.

OUTLINE OF PROCEDURE

The individual is to be taught to recognize the presence of muscular contraction, no matter how slight. Such ability, as previously said, is found to be highly useful, although fair relaxation can be acquired without it. After he reports that he recognizes contraction in a muscle-group, he is shown how to relax it extremely. This is to be done with almost all of the noteworthy muscle-groups of the entire body. He learns to recognize contraction in the various parts in a certain order. The large muscle-groups are studied first, because the sensation therefrom is most conspicuous. As he relaxes a given part he *simultaneously* relaxes all parts that have previously received practice. Most often I have used the following order (the muscle-group in some instances is designated by its chief muscle or by its function): left biceps, l. triceps, l. hand flexors, l. hand extensors, right arm similarly, l. foot flexors, l. foot extensors, l. leg flexors, l. leg extensors, l. thigh flexors, l. thigh extensors, right lower limb similarly, abdominal muscles, respiratory muscles, erectores spinae, l. pectoral group (forward and inward extension of arm), l. interscapular group (backward movement of shoulder), r. pectoral group, r. interscapular group, elevators of shoulders in shrugging, bending head to right, to left, forward, back, holding it up stiffly, wrinkling the brow, frowning, closing eyelids tightly, with lightly closed lids turning eyes to look toward right, left, up, down, straight forward. Further eye-work at this stage will be described later. Smiling, rounding lips to say "O," protruding the tongue, retracting the tongue, closing the jaws tightly, opening them, counting one to ten, and swallowing, complete the usual list. This list can of course be extended for any particular patient to include any other muscle-groups in which undue tenseness appears. The foregoing indicated contractions are performed solely to make the individual acquainted with the experience of tenseness, that is, to show him what not to do. They are occasionally repeated in the presence of the physician, but as a rule not otherwise. *They are not designed to be an aid to relaxation and should not be repeated immediately before or during attempted relaxation.*

The Cultivation of the Muscle-Sense

When the untrained person is requested to relax, he may fail because he does not know where his tensions lie. Therefore, as an aid toward the control of relaxation, I have cultivated the muscle-sense (chap. iv).

It is the sensation of muscular contraction that is of prime interest rather than that from the joints or tendons. The former arises when the subject contracts voluntarily, and is

[43]

called to his attention during the procedure. Simultaneously with the muscular contraction other sensations arise from contact with the clothes or couch, but particularly also from the joints and tendons. These other sensations are more conspicuous to the beginner than are the sensations from muscular contraction, and he should learn to make the discrimination. Because the sensation from muscular contraction seems relatively vague at first, he will be likely to report that his sensations of tenseness lie in the region of the joints and tendons. Practice tends to make the distinction familiar, just as the average person can learn to distinguish a delicate shade of color if he sees it often enough. This distinction will be discussed below in connection with the shoulder joint. Little can be expected at first in the way of such observations from disturbed patients or from those of the distracted or dependent types. For them, at least in the early stages, it may prove better to omit any considerable cultivation of the muscle-sense. Subjects are often aided in their observations if they are requested to describe their sensations. The sensation from muscular contraction is described by our subjects as a rule as dull or faint, readily obscured by other experiences, fairly localized but diffuse and ill outlined, neither agreeable nor disagreeable, but particularly indistinct and characterless. It is to be distinguished also from other somatic sensations due to touch, warmth, cold, tickle, and pain. These distinctions are likewise familiar to the student who has had an adequate elementary course in experimental psychology.

In order to bring out the sensation from a particular muscle-group most clearly, the individual should contract the part steadily while, if applicable, the hand of the physician retards the movement. A few patients queerly fail to understand and carry out this simple direction, even after many repetitions. For example, they fail to flex the forearm steadily, but flex for perhaps a second, then extend, then flex again, and so on in an irregular manner. It is useless to proceed with them until they have gained a better idea of the simple procedure that is desired. Unexpected awkwardness may be encountered: one patient with general high intelligence quite failed (in connection with relaxation of the eyes) to look off in the distance in the manner ordinarily requested by an oculist. Such simple directions can nevertheless be followed by the average child above six years of age. A feeble-minded girl of twelve met no difficulty on such points.

In cultivating the muscle-sense, only one muscle-group is contracted at a time, so as to bring out the experience alone, while other groups remain so far as possible inactive. Accordingly, when the forearm is flexed, the upper arm should rest upon the couch so that the shoulder muscles are not in play, while the fingers and hand should be limp. Analogous conditions should prevail for the triceps. Again, when the leg is flexed at the knee to bring out the sensation from the hamstrings, it is best to have the heel drawn along the couch without being raised, for this would cause the flexors of the thigh also to come into action. The foregoing rule can readily be applied to other muscle-groups. However, it is not necessary to follow it strictly if the individual is fairly observant and if it is evident that the distraction is of no great consequence. Closing the eyes sometimes assists the subject to note a faint sensation, as he would note a faint odor; but usually the eyes are kept open. As a rule the individual who is genuinely observing his sensations has a characteristic appearance: the eyes deviate to one side in a steady manner in contrast with the continually wavering movements in the individual who is just guessing.

By asking questions at the proper time, the physician can

satisfy himself whether the individual is really observing. Light will be thrown on this matter by the record as taken below. The patient who generally localizes the sensation on his first or second attempt (-1S or $-1+1$S) is probably observing. This is the more probable if he has not studied anatomy. Excellent tests in this respect are flexion of the thigh and shrugging of the shoulders, where the location of the contracting muscles is particularly unlikely to be guessed.

The patient is not to learn and memorize what muscles are responsible for various movements. *He need not and should not concern himself with whether his muscles are contracting, nor should he palpate his own muscles or watch his own movements.* Not the patient but the physician needs to know physiology. Observation at a given moment and localization of the experience of tenseness is all that is required of the patient. We are reminded of the discussions of Martin and Müller, Titchener, and others on the "stimulus error," which will be recounted in chapter xi.

It has been stated that the physician offers resistance to the contracting limb in order to aid the patient's observation by intensifying the sensation. As the patient acquires skill in observation, such assistance becomes unnecessary. There should be frequent review with each muscle-group, and the extent of contraction should be lessened until the sensation is reported as clearly perceived even with very slight contractions.

Cultivation of the muscle-sense is not successful with some patients. These generally fail to attain the fine control of relaxation gained by those more clever. Their lesser skill at relaxation is an argument for the advantage of retaining the cultivation of the muscle-sense in the present method. However, as previously said, the ability to localize tensions is not indispensable. Even without this ability a fair if not a complete measure of successful relaxation may be attained.

The Record To Be Kept

All instructions, no matter how casual (and during investigations, even the substance of all remarks), are to be jotted down. To keep data in order, a record sheet is required; and abbreviations or symbols may stand for frequent procedures (Chart I).

Initial Instructions

A minimum of words is to pass between the physician and the patient, who is to learn from concrete experience rather than from discussion. While lying down, the individual flexes his left arm steadily, avoiding unnecessary contraction of other groups as previously stated (Fig. 1), and reports whether he notes the sensation from contraction of the biceps group. To strengthen the sensation, the physician offers passive resistance. Some patients perceive at once, while others require many repetitions, particularly the agitated and unobserving types. The experience from muscular contraction, as illustrated both by a maintained flexion of the forearm and by a quick flexion, is by agreement between physician and patient to be called "tenseness." This applies whenever this type of experience appears anywhere in the body. But the physician clearly explains, at this or some other more favorable moment, that "tenseness" will not mean the same as physical tautness occurring when a tissue is stretched during some movement or otherwise. As previously said, the patient is to understand (if his capacity permits) that *he is not called upon to report whether his muscles are or are not contracting, but only whether he has a certain kind of experience.*

When the sensation is clearly perceived while the indi-

[47]

PROGRESSIVE RELAXATION

CHART I
RECORD OF PROGRESSIVE RELAXATION

PATIENT'S NAME _____

    TYPE OF RELAXATION_____

        DAILY OCCUPATION _____

            PHYSICAL CONDITION_____

Date	No.	General Condition	Instructions	Old MM.	New MM.	Report	Relaxation	Restlessness and Difficulties	Observations and Remarks	Pulse B. P. and Bas. Met.

Type of record sheet reduced from full size, 8×11 inches (width). The headings of the columns are almost all self-explanatory. In the second column is the number of the period. The sixth column contains the abbreviated name of the muscle-group, such as "l. b.," which first receives practice opposite a given date. Should l. b. receive practice a second time, it will be found in the fifth column. Subjective reports are in the seventh column, while the eighth and ninth are reserved for objective data. Any data on pulse, blood-pressure, and basal metabolism are in the last column. If the description requires more space, the back of the sheet is used. During a particular period events recorded in various columns may take place in a certain temporal order; to designate this the letters of the alphabet are used before each notation. Then "*a*" represents the first event, "*b*" the second, and so on.

A record is kept of the individual's reports on the recognition of the sensation of muscular contraction in each locality. Success is indicated by the sign +, failure by −; digits represent the number of reports given, "S" means the spontaneous or unassisted report of the subject, but "I" means that he was instructed as to the locality. For instance when he contracts the right triceps, he is asked, "What do you experience?" If he replies, "A sensation from contraction there," and designates the place, it is recorded as +1S. If the notation were −2S−1+1I, it would mean that twice he reported incorrectly without assistance as to the locality of the sensation, once he reported failure to perceive the sensation although the location was pointed out to him, but finally he gave a positive report. (Sometimes the subject pauses before he reports, contracting the part a number of times; this number is not recorded.)

[48]

vidual is flexing the biceps, his attention may be sharply
called to the issue by saying to him, "This is *you* doing!
What we wish is simply the reverse of this, namely, not do-

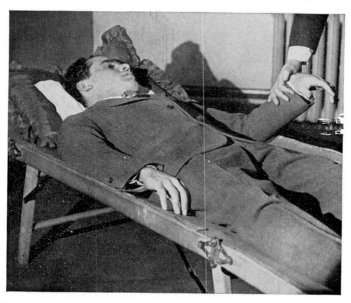

FIG. 1.—The patient flexes his left forearm in order to experience the sensation
produced by contraction in the biceps-brachial muscle-group. To increase the
strength of this sensation, resistance is offered by the hand of the physician. After
the subject reports that he has recognized the sensation, he is directed to cease what
he has been doing, that is, to relax. It should not be inferred that the patient is customarily to contract before he relaxes. This would be poor practice. Contraction in
this and the following illustrations is only designed to make him acquainted with the
experience therefrom in order that he may be guided what *not* to do.

ing!" As he then relaxes, he begins to learn clearly what it is
not to do. He begins to realize that progressive relaxation is
not subjectively a positive something different from contraction, *but simply a negative*. After he has relaxed his arm for
several minutes to illustrate this point, he is requested to
contract it, and then to relax again. This time it is called
to his notice that *his act of relaxation involved no effort:* he

did not have to contract his arm or any other part in order to relax. These are important points to learn, for the untrained individual who fails to relax will contract various muscles in a vain effort to do so. Such effort is often disclosed by a furrowed brow or forehead, in which event he should be interrupted in his attempt to relax, and after in-

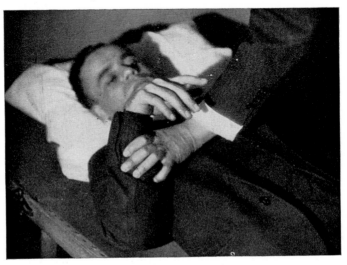

FIG. 2.—The patient extends his right forearm against resistance. Note that the hand falls relaxed in order to isolate so far as possible the action of the triceps group. When possible, he should spontaneously indicate to the physician the site of the muscle-sense experience.

struction on this point, requested to begin anew. It is important to make clear that relaxation is never "hard" and cannot be; it is either done or it is not done, and that is all.

If these matters evidently have been assimilated, the individual is made to contract the biceps again and then to let it go. He is requested to let the part go further and further every minute. *"Whatever it is that you do or do not do when you begin to relax, that you are to continue on and on, past the point where the part seems to you perfectly relaxed!"* This instruction, if clearly illustrated, conveys to him the meaning of progressive relaxation in terms of an immediate experience.

If he now lies quietly with eyes closed and seems to be set to relax, he may be left alone. The reader will recall that

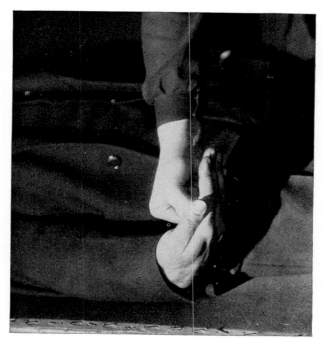

FIG. 3.—The patient flexes his right hand against resistance in order to experience the sensation of contraction in the flexor muscles.

after the individual has relaxed in the popular sense there remains as a rule a certain degree of tension called "residual tension." To undo residual tension in a part may proceed quickly or may require as much as 15 minutes. If he is restless and fidgety, his attempt at relaxation may be interrupted, the biceps again contracted, and relaxation again begun.

As will be seen, only instructions are given, just as in

learning how to play a musical instrument or billiards or how to drive a motor car. Suggestions, in the technical sense of the word, are always avoided. From the outset the learner

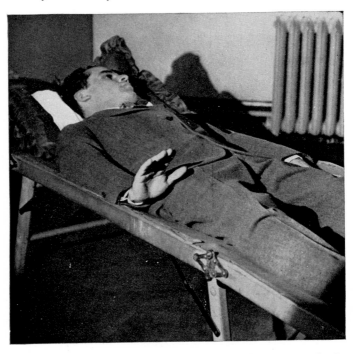

Fig. 4.—The patient extends his right hand, here against no external resistance, in order to observe the sensation of contraction in the extensor muscles. Note that he lies fairly relaxed so that the experience in question is not obscured by sensations arising from contraction in other muscle-groups.

does everything for himself. If he fails, he may be scolded and made to try again, which is entirely different from a suggestive procedure.

When it is evident that the patient is beginning to relax, the physician may quietly leave the room, carefully closing the door to attend to some other matter elsewhere. He returns in from 5 to 15 minutes or more, according to his judgment that the patient should be left alone for practice or needs further instruction and criticism.

The entire first period may be devoted to the biceps, or one or more of the three following groups may be added (Figs. 2, 3, 4). Occasionally it is well to vary the instruction with a phrase such as, "Just let the arm become limp as a rag!" or "Go in the negative direction!"

FURTHER EARLY PROCEDURES

After the individual has become familiar with the entire arm, the opportunity arises *to let him experience that a part need not be moved in order to be progressively relaxed.* He is directed to "stiffen the arm without moving it, more so, still more, and still more! Then not quite so much, a little less, still less, and so on and on past the point where it seems perfectly relaxed, and even further!" He is requested to note that this is what will be meant by "going in the negative direction." *This is then taken as the type-form of progressive relaxation for all subsequent work.* It is necessary to warn the patient that one part of the illustration is not to be followed in the future: he should not as a rule contract before he relaxes, but should begin to relax at whatever stage he finds himself. Graphic records have revealed that subjects sometimes fail to reach extreme relaxation because they contract at the moment they try to relax. The instruction to "go in the negative direction" has come to be the chief of those in present use.

If the individual has displayed a delicate sense of recognition, he may then be made familiar with what I shall for brevity call *diminishing tensions.* He flexes the arm with no resistance from the physician; and if he reports a sensation of tenseness, the flexion is repeated half-way, then half of this, and so on. A point is soon reached at which he is

[53]

requested to flex so that the physician can scarcely note the movement. If he still reports positively, he is requested to go as if to flex it, but a little less, so that no movement is discernible. My laboratory subjects as well as patients have found that the sensation or experience of muscular contrac-

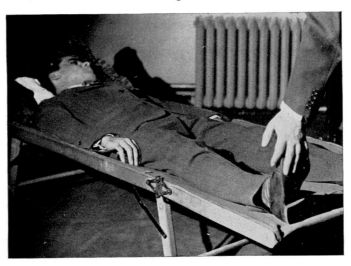

FIG. 5.—The patient extends one foot in order to observe the experience of contraction in the corresponding calf muscles. In this and other illustrations, he should perform a steady, persistent contraction. Intermittent contraction should not be permitted, particularly if he tends to flex and extend alternately, thus confusing the picture.

tion is again repeated but is considerably fainter than before. Accordingly, if the beginner still reports positively, he is requested to repeat again but still less than before. After several diminutions from this stage the experience vanishes.

No one can learn to control his relaxation who does not know the difference between what we call "tenseness" and "strain." This test arises, for instance, when the individual extends his right arm forward and inward and reports where he notes the tenseness. A common error is for him to point to the back of his shoulder region, for here the sensations are strongest, and to fail altogether to note the tenseness in the pectoral region. If he makes this error, he is requested to keep the arm relaxed while the physician pulls it in the same direction as previously. The alert observer then notes the same sensation as before in the posterior region, showing that

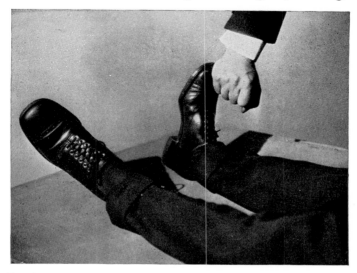

Fig. 6.—The patient flexes his right foot at the ankle joint in order to observe tenseness produced in the region of the tibial muscle.

this cannot be "tenseness." It is agreed to call sensation of this latter type "strain." This opens the opportunity for the individual to look again for the sensation of tenseness. When finally localized, perhaps with assistance from the director, the distinction becomes apparent. In this manner is illustrated that sensations of tenseness are readily overlooked because of their relative faintness. These sensations are sometimes "unconscious," in the sense that they are commonly overlooked. Without doubt, in this sense, "unconscious" experiences can be relaxed away.

Another distinction to be taught is between the experiences of *moving* and *static* tensions, that is, changing mus-

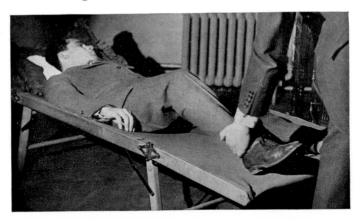

Fig. 7.—Flexion of the right leg is performed to excite the experience of contraction in the right hamstring muscle-group.

Fig. 8.—Extension of the leg against resistance aids the recognition of the experience of contraction in the quadriceps-extensor muscle-group. Resistance may be omitted in the contraction of this and other muscle-groups if the patient can recognize the sensation without this strengthening aid.

cular contractions versus tonic states. Flexion of the arm provides an instance of a moving tension, while holding it rigid illustrates a static tension.

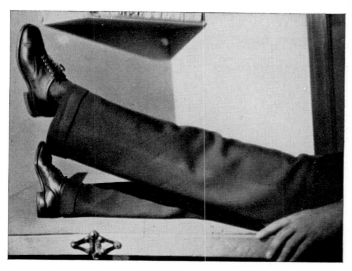

FIG. 9.—Forward extension of thigh arouses sensations in the psoas muscle-group, a region which most patients fail to point out, unless they are informed of the location.

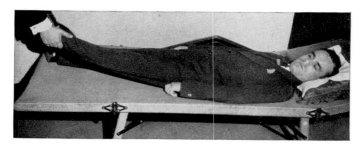

FIG. 10.—Here the left lower limb is being flexed backward toward the couch, against the resisting hand, involving contraction of the gluteal group.

Other muscle-groups receive practice in the preceding manner (Figs. 5, 6, 7, 8) with variations according to the

[57]

needs of the individual, but further description seems required about the conditions to be observed during the contraction of certain ones. Forward flexion of the thigh (Fig. 9) involves the psoas muscles deep in the abdomen, and spontaneous localization of this group is not expected of the average but only of the keen observer. Extension of the

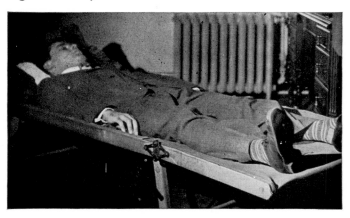

Fig. 11.—The patient draws in the abdominal muscles, producing a diffuse experience of contraction in this region, which he is to indicate as soon as he recognizes it.

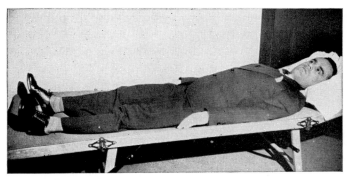

Fig. 12.—Upon taking a deep breath, tenseness should be noted faintly and diffusely over the entire thorax and in the diaphragm during inspiration alone. Subsequently, practice is taken in omitting such voluntary tenseness until breathing becomes quite automatic and uniform.

[58]

thigh is illustrated in Figure 10. To contract the abdominal muscles (Fig. 11) the individual is requested to draw in this part and report what happens, or he is requested to begin to sit up from his supine position. In both instances, he should indicate general tenseness of the abdominal muscles. Special practice is generally required for the thorax. The

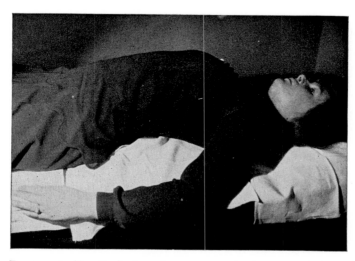

FIG. 13.—Arching the back produces tenseness in the *erectores spinae*, where the sensations are readily localized by most patients.

individual is requested to take a deep breath or to breathe irregularly (Fig. 12). It is counted an error if he reports tenseness just at the end of inspiration, for he is then mistaking the sensation from air pressure for the muscular sensation. If he is a chest-breather, he should report diffuse tenseness in the diaphragm and all over the chest during inspiration alone but no tenseness during expiration. An opportunity is given to observe what happens during ordinary natural respiration. His finding should be "faint tenseness during inspiration alone." Evidently now is the moment to point out

[59]

to him that complete relaxation of the chest would be impossible, since breathing goes on automatically. The instruction is worded, "Let the chest go just like your arm: let the breathing go entirely by itself!" "Controlled breathing" is not used as an aid to relaxation in the present method. Rather, the aim is to free the respiration from voluntary influence,

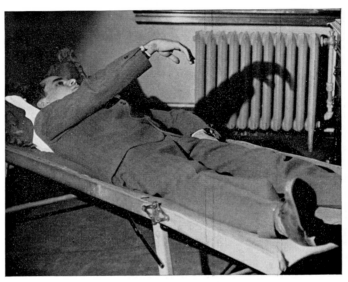

Fig. 14.—Forward and inward extension of the right arm produces an experience from contraction in the pectoral muscle-group, which should be distinguished from the more intense experience of strain at the rear of the shoulder.

leaving its regulation to the autonomic system. This results in uniformity and regularity of period, as shown by kymograph records. Illustrations of further steps will be found in Figures 13–17, inclusive.

THE OBJECTIVE ESTIMATION OF CONTRACTION AND RELAXATION

While the patient judges his progress by the diminution of the sense of contraction, the physician watches closely for external signs. Objective tests of advancing relaxation include: (1) palpation of the muscle-group; (2) passive motion of the part; (3) observation of the regularity and force of respiration; (4) visual observation of the flaccidity of the muscle-group or region; (5) the absence of movement or contraction, including speech and winking of the closed eyes;

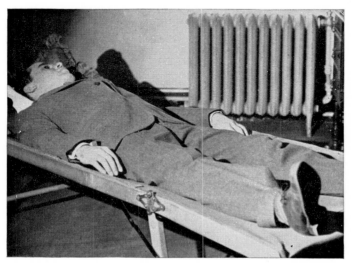

FIG. 15.—Drawing the shoulders backward and medianward produces an experience to be localized in the interscapular region.

(6) the presence of a sudden involuntary start or jerk, local to a part or generalized, which often marks the onset of advanced relaxation in an individual who has been previously hypertense (chap. vii); (7) increasingly slow responses to interruption, or failure to respond (chap. viii); (8) the sleepy-eyed appearance of the individual who arises after successful relaxation; (9) when the individual learns to relax the eyes while open, their vacuous appearance, with the facial musculature so relaxed that it is expressionless, is characteristic (Figs. 27, 28, 29); (10) graphic records of respiration, pulse,

eye movements, or of the tone of internal organs may be employed and in certain instances, fluoroscopy and Roentgen films have been used (chap. xvi); (11) diminution or absence of the knee-jerk during advanced relaxation is a valuable test (chap. ix; to avoid error the leg must be tested for per-

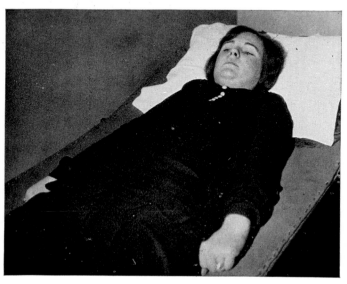

FIG. 16.—Shrugging the shoulders produces contraction which the patient un-aided should localize at the top of the shoulders and in the sides of the neck. If he proves backward at such recognition, in this and other instances, the site is indicated to him.

fect flaccidity by finding that the foot yields to any very gentle push to and fro; this will prove that the jerk is not absent because of contracted antagonistic muscles); (12) measurements of muscle tonus or contraction are being made with a string galvanometer and amplifier assembly (Jacobson, 1931). This will be further described in chapter xvi.

ADVANCED INSTRUCTIONS

The patient practices at home for an hour or two per day after he has once learned what to seek. *He relaxes but does not repeat the preliminary contractions when alone.* Many persons do better alone than at the physician's office. Generally they have questions to ask and difficulties which they wish to overcome. A common complaint is an unyielding tenseness of the neck muscles, even after this region has received

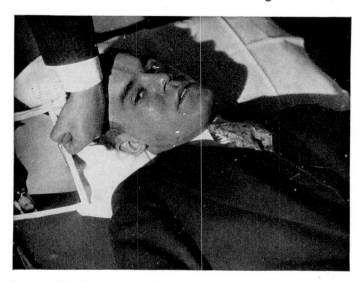

Fig. 17.—This illustrates one of various exercises to bring out the experience of sensation from contraction in muscles of the neck. The head is bent to the right against resistance, when the subject should indicate an experience of muscle tenseness on the right side of the neck but a different experience, namely of "strain," on the left side of the neck. After a period in which the patient is directed to cease all tendency to bend the head to the right—that is, to relax progressively—a similar proceeding is carried out in bending the head to the left. For further particulars see the text, since some of the proceedings are not shown in these illustrations.

practice. To aid in overcoming this, the patient is shown that it is he who is holding the neck stiff; he is directed to "hold it stiffer, more so, more so, extremely stiff—not quite so extremely—a little less, a little less," and so on and on. This illustrates progressive relaxation of the neck muscles and may need to be performed slowly and repeatedly.

[63]

PROGRESSIVE RELAXATION

Repetition is the keynote of the entire method of progressiv *relaxation.* It is the old story, "If at first you don't succeed, try, try, again!" The principles involved are very simple. That is just the trouble: they are too simple. Just the negative of doing is required, and everybody is capable of this, since whoever can raise his arm, for instance, is also capable of not raising it. But when requested to relax, the average individual does not do the simple thing: instead, he makes various efforts. *Instruction in relaxation largely consists of preventing the beginner from doing the wrong thing.*

Often the patient asserts that he cannot relax. It would be unscientific for the physician dogmatically to say to him that he can, for precisely this is to be proved in each case. But it is safe to point out that he has no proof that he could not relax and that it is better not to go beyond the facts: he did not relax. No assumption need be made either way, and it is sufficient if the patient, however skeptical of results, keeps an open mind and continues in his endeavors. Nevertheless, many patients persist in saying that they cannot, and so they fail to try when they should. It is well to show them that they can contract any muscle by having them do so, and then, while having them cease that same contraction progressively, demonstrate for at least this instance the absurdity of the claim that they *cannot* relax.

There are other matters which can be but touched upon here: Upon failure to recognize or localize a particular tension a preliminary period of relaxation may aid. Again, the patient as a rule continues at his affairs, but his life is to be reorganized so that rush, strain and worry are at a minimum. So far as possible he is to keep relaxed during his daily duties (chap. vi). He is not to make a task of learning to relax but is to have an attitude of not caring whether he succeeds in this or not. If he complains that he finds it hard to lie quietly at practice, he is evidently confused about what is wished. For he has never been instructed to hold still, since this is not relaxing. Let him stiffen his arm, holding it still, and it will become clear that this is a form of tenseness, not relaxation. It should be demonstrated to him that it is never "hard" to relax, for this word implies effort. Of course an untrained individual may stiffen when requested to relax, and so make a task of it, but this is not relaxation, for it is only an unsuccessful attempt. Correct speech is important for a clear understanding: every person will find that he does not readily relax on some occasions, and it is proper to say so. On some days he will have farther to go, more tenseness to undo, than on others. But the patient who complains of his practice, who asserts that "relaxing makes him nervous," gives a clear indication that he is doing something wrong: he is doing, in place of undoing. When he spontaneously announces that he is beginning to enjoy it, he has probably found the right track.

RESTLESSNESS AND DIFFICULTIES

Before a period of relaxation the patient does well to avoid discussion of any topic, since the mental activities so set up tend to persist, giving him extra tensions to relax. However, if discussion has taken place, he had better not close his eyes at once after lying down, but rather let activities die down for a few minutes before definitely beginning to relax.

Sometimes the fairly quiet patient who has just been directed to relax still further protestingly inquires, "What shall I do?" This calls for repetition, perhaps many times, of the illustration given above of going in the negative direction. Any set of muscle-groups previously covered may be used.

During attempted relaxation, the physician watches closely for signs of restlessness in the form of speech, opening of the eyes, or gross or slight contraction or movement of any part. Such contractions indicate that the patient is failing to relax the particular parts which are receiving practice. In this event, it is often best to interrupt the patient, telling him that he is not proceeding properly.

The nervous start in response to a sudden noise or other external stimulus, as when the door is opened, indicates poor or incomplete relaxation (see chap. vii); but the physician will not confuse this with the predormescent start (p. 17) which occurs with advancing successful relaxation in patients who have been hypertense during previous hours. If the patient complains about the latter type of start, he should be informed that he has been relaxing very well and should continue.

Sometimes the patient inquires whether he should keep his attention on the muscle he is to relax, or the physician learns that he is attempting to relax by this route. This must be corrected. Accordingly the beginner may be reminded that when he learned to waltz he marred the performance if he concentrated his attention upon his feet and that he danced best with a minimum of attention, although at first relatively more was required. A second reply is to have him flex his arm while having his attention elsewhere, proving to him that he need not be attentive in order to contract. This may then be repeated with relaxation. It may be well to explain that a slight and fleeting stroke of attention may be needed at or before the onset of relaxation, but the aim will be to minimize attention and to make the process habitual and automatic.

Not infrequently the presence of the physician proves somewhat disturbing to relaxation. To counteract this in some measure it is important to instruct the patient to *take*

the entrance of the physician as a signal to relax all the further.
Any other disturbance from without or even from within his
own organism is likewise to be taken as a signal to relax.
The beginner who opens his eyes, ceasing to relax whenever
the physician enters, needs to be reprimanded. Some patients continue to find the presence of the physician disquieting, but this serves as good practice for them to overcome. A
helpful function of the physician, like that of an instructor,
arises from his opportunity to inform the patient when an
error is made.

The untrained patient or even he who has had some experience at relaxation may nevertheless appear on some occasion in a highly emotional condition—perhaps weeping mis-
erably or displaying evident anxiety over some misfortune or
some approaching event. Tensions are to be noted and localized by him and then relaxed precisely as he has relaxed the
biceps. This calls for a test of technic of both physician and
patient if the former deliberately avoids solace and suggestion
and the latter is led to quiet in a mechanical manner.

It is consistent with the method of relaxation to remove
the patient from irritating environment, when practicable,
or to lead him by whatever means to alter his affairs in order
that emotional disturbance will be so far as possible avoided.
In many instances, however, it is good training for him to
stand his ground and to learn to avoid nervous irritability
and excitement through relaxation.

Backwardness in becoming relaxed commonly results (1)
from failure to understand how to relax a muscle-group or (2)
from failure to attend and to localize the experience of muscular contraction or (3) from lack of initiative to relax despite
the ability being present. In the first case, instructions must
be simply repeated along with illustrations until the difficulty is overcome. The second case often is ascribable to

[67]

pain or to distractions arising from personal affairs and anxieties. Where pain is unavoidable, the attempt to relax may
be made in spite of it, particularly if the patient has previously acquired skill; but if pain such as acute headache
chances to occur during one of a series of periods devoted to
learning to relax, it is better to postpone the matter or to
administer an analgesic drug before the period. Distraction
due to anxieties and to personal affairs often demands prolonged drill in the relaxation of the musculature of vision and
of speech. We must be continually on the lookout for the
patient who knows how to relax but fails to use his initiative
in this direction. Occasionally a characteristically tense (neurotic) individual with no preliminary instruction whatsoever
discloses the ability to relax in normal fashion as tested by
the knee-jerks (chap. ix). Evidently this does not conflict
with the view of the present investigation that neurosis is
characterized by the chronic failure to relax, but only indicates that such neurotics fail to take advantage of their natural abilities. Likewise, some patients who have not yet completed their course of training repeatedly fail to relax upon
occasions. One, for instance, upon arising after relaxation,
wrinkled his forehead, rubbed his face, and engaged in various other tensions. When his attention was called to this
failing, he argued that he "had to do so in order to get awake
and active," but soon he abandoned this view and sought to
keep relaxed upon arising. Nevertheless he continued to display the same fault in lesser degrees, and in consequence the
instruction had to be repeated many times. Considerable difficulty is sometimes encountered with such patients, who,
having every intention to relax, nevertheless fail to do so
spontaneously. Frequently they will resort to excuses such
as the pressure of business or domestic duties. It is well to
tell them that they can always find a reason to be tense and

that if relaxation is to be acquired it must be by their own choice. In such instances initiative to relax is cultivated by practice when alone, by frequent reminders and by pointing out moments of failure.

Fig. 18.—Wrinkling the forehead produces a regional tenseness which is reported correctly by most patients at once.

RELAXATION OF THE EYES

The ability to relax the eyes, including the brow and lids, is a crucial test of skill. It is easy for most persons to distinguish tenseness in wrinkling the forehead (Fig. 18). This region is then permitted to flatten out for a period of 10 min-

Fig. 19.—The patient frowns steadily in order to learn to recognize when he is contracting "unconsciously" in this manner.

utes or more until it is successfully done. After frowning (Fig. 19), the brow likewise is gradually unfurrowed. The physician watches closely to see that the patient does not wrinkle his forehead in order to smooth his brow. If this error takes place, he must try again, for he must learn that *it is not necessary to move in order to relax*. Next in order, the lids

Fig. 20.—Closing the eyelids tightly involves the *orbicularis oculi*. This illustrates to the patient what he is not to do in the slightest degree when he relaxes. He is never directly instructed to cease winking, but winking of closed eyelids gradually ceases during extreme progressive relaxation.

of the closed eyes are shut still more tightly and the tenseness noted (*orbicularis oculi*, Fig. 20). These are gradually let go, until the physician notes that winking of the closed lids has become less frequent. After a period of relaxation, the individual looks to the right without moving the head and

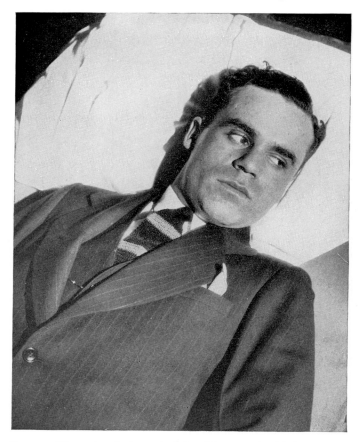

FIG. 21.—The patient looks to the left, without moving his head, and should report muscular sensations in the region of the eyeballs. A more precise localization is not required. As a rule this exercise is carried out with eyelids closed rather than open as here shown. Other eye movements are performed but will not be illustrated.

notes the ocular tension, then left (Fig. 21), up, and down, each time reporting his experience. He looks straight forward, noting the static tension of convergence. Each movement or act is repeated until the experience is clear. Then he is directed to let the eyes go completely, *just as he let the arm go.* He is not to try to look in any direction. If he

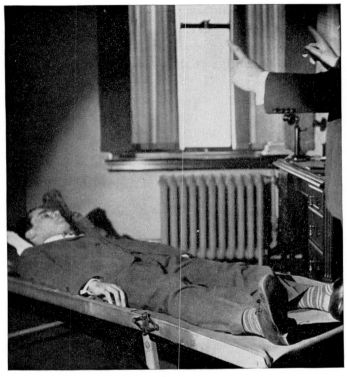

FIG. 22.—The patient looks from finger to finger held the largest distance (about 3 feet) apart, noting the sensations from muscular contraction as the eyes move. For description of other exercises in this class, see text.

fails, he is made to contract the arm, then gradually to let the arm and eyes relax together. As a rule many repetitions are needed.

After a fair measure of success the eyes are opened. The physician, standing at the foot of the couch, holds his index fingers horizontally about 3 feet apart to permit the patient to look from one finger to the other and report his experience (Fig. 22). This is repeated with half the distance, then about 2 inches apart. Next a single finger is held up for some seconds. One finger is held above the other for eye-movements up and down. The subject should report, in each test, without any help from the physician, that he looked from finger to finger and noticed either tenseness of eye movement or static tension. Often he is not successful at first. If he uses the term "eye-strain," he is required to explain what he means. It has been agreed to call a certain sensation "tenseness." *Whenever doubt arises, the biceps should be contracted again in order to illustrate the experience.* In case of failure to perceive tenseness in convergence upon the single finger, that member may be held close to the eyes, for this brings out the sensation more distinctly. When the individual has become adept at noting tensions of the open eyes, *the physician shows him by example how to relax the eyes while open* (Figs. 27, 28, 29); for learning to relax, like learning any other new art, is often most readily done by imitation.

THE METHOD OF RELAXATION OF MENTAL ACTIVITIES

Patients often continue to complain that the "mind keeps on working" after they lie down to relax, and perhaps this keeps them from sleeping. Some even inquire in advance whether muscular relaxation will quiet the mind. In place of answering such complaints or queries directly, the matter should be left to the individual's own observation. Thus prejudice is avoided. *At no time is the instruction given to stop thinking or to make the mind a blank. The instruction to relax muscles progressively pervades the entire course, and whatever is said always points in this one direction.*

An experimental discussion of the influence of relaxation

on mental activities will follow in chapter x.[1] Here a brief description of method is in order: With the eyes closed, the individual is requested to imagine a motor car passing and to report his experience. Often the physician can note a quick movement of the eyeballs beneath the lids. The patient may report that he saw the car in imagination and felt tenseness in the movement of the eyes to follow the car. This is repeated with other simple objects, moving or stationary, such as a train passing, a bird flying, a flower fluttering in the wind, a tall tree or tower, a ball rolling on the ground, a triangle, square, circle, and point, a blade of grass, a sailboat in the distance. As skill is gained by the patient in noting slight ocular tensions, the experience may be made somewhat more complex, and he may report his experience after recalling the morning newspaper or after a simple problem in arithmetic or after thinking of some social or business matter. In each case, as a rule, following adequate practice at observing, most persons report that their visual images were accompanied by experiences as of tenseness in the eye muscles from looking at the object imagined.

Of course the physician must not even hint to the patient that he is to look for tensions during visual imagination; for if the investigation is to have scientific value, the patient must observe for himself. Even if only therapeutic results are desired, it is better to have the patient rely on his own observations; but if time must be saved, he may for merely practical purposes be told what he is to observe, since this simplifies the task.

The patient can find for himself—with no hint from the physician—that when he has learned to relax such slight ocu-

[1] The importance of objective proofs of the presence of neuromuscular tensions during acts of imagination and recollection is obvious. These tests are now being completed by electrophysiologic methods as described in chapter xvii.

lar tensions as take place during the activity of imagining, the mind ceases to be active. This might not apply to individuals who use little or no visual imagery, for some do not (Fernald, 1912).

Individuals who are mentally hyperactive may succeed very well in discerning their tensions yet persistently continue to evince slight eye movements. After moments when these have appeared, questioning generally elicits the admission that reflection took place, while persistent flaccidity of the eye muscles and surrounding regions often brings the contrary report (chap. x). Such individuals often require special drill before they learn to relax their "reverie." With eyes closed they may be requested to look slightly in each direction, then straight forward; to imagine some object slightly off from the center in each direction, then directly ahead; repeating these exercises in slight voluntary movements of the eyes just as exercises in slight voluntary movements of the fingers have to be frequently repeated by the pianist who would learn to exercise delicate control. If such exercises with the ocular muscles are not frequently performed, it may be found that the individual lost in reverie fails to know that his eyes have been moving even when this has been quite conspicuous to the external observer. The instruction is to cease to move the eyes in any direction or to look forward, yet not to hold them still: to relax them like the biceps.

After serial practice with the musculature of the face, jaws, and tongue, the patient counts to ten, noting the various tensions in the speech apparatus. He should report or indicate tensions in the cheek muscles, lips, tongue, floor of the mouth, jaw muscles including the temporal regions, throat, chest, diaphragmatic and perhaps abdominal regions, due to altered breathing during speech. A complete report is not to be expected without many repetitions. The method of "diminishing tensions" is again used. He counts half as loud, notes the processes, then scarcely perceptibly, then not quite perceptibly, then less than this, and again less. According to reports, this becomes the same as imagining from one to ten in verbal terms. Upon relaxing the speech apparatus, such imagination of course vanishes.

Thereupon he may imagine or recall recent speaking in various ways, such as telling a waiter to bring his dinner, or requesting a conductor to let him off the car. The observing patient reports that he has slight tensions in the tongue, lips or throat as he speaks in imagination (sometimes also in the muscles of the jaws and floor of the mouth), and he should not fail to mention the thoracic or abdominal tensions from breathing which varies in its periods and pauses according to the character of the inner speech. Imagining sounds is also accompanied by tensions, usually ocular (as the source of sound is visualized), also possibly in the *tensor tympani* and muscles about the ear, although this has not yet been established.

If desired, illustrations of other types of imagery that take place during thought-processes may be made familiar. The patient is requested to imagine (not, however, to visualize) lifting a heavy weight with his right arm, or raising his left leg to take a step, or bending backward: generally he reports a subjective sense, as of slight tenseness in the muscles involved in each act. When he imagines other persons engaged in physical activity, such as a prize fight, he is likely to report a complicated series of slight tensions in various muscles corresponding to what is imagined. Occasionally neurotic patients spontaneously assert that when they witness athletic activities or attend the opera there occurs a play of mimetic tensions on their own part, with resultant fatigue.

[77]

Practice may be given at reporting what takes place during a few seconds of thinking of various matters, some suggested by the physician or experimeter and others selected by the subject. The beginner is likely to be vague and to speculate or state conclusions instead of making the simple report that is desired. This tendency must be corrected, but it is better to do so by a teaching method which encourages observation step by step than by a full explanation in advance. *Gradually the individual learns to observe his sensations and images and to report these first, adding thereafter what meaning they carried for him.*

It is good practice for the individual to note and report his tensions during such common acts as speaking, walking, reading or writing, stating the function to correspond with each set of tensions. For example, tensions are reported in certain localities of the leg in order to lift the member to walk. In passing to reports of fainter and vaguer tensions, like those present during imagination and thinking, statements may again be forthcoming about the functions of the reported tensions. Again, tensions are reported in the eyes, as if to look to see an imagined object at the right; or tensions are reported in the tongue and lips as if to utter certain things. When the individual reports muscular tenseness, but fails to state the function, questioning generally elicits one. *This leads to the working hypothesis that any report of the experience of muscular tenseness is incomplete until a function is stated.* The subject is simply asked, if necessary, "A tension to do what?" If the answer is not ready, the matter should not be pressed and he should not be encouraged to speculate or conjecture if he does not know. In this event, repetitions of the experience should be given. Sherrington (1915) goes so far as to believe that tonus is always to be understood in the light of its aim or function. He states,

[78]

Every reflex can, therefore, be regarded from the point of view of what may be called its "aim." To glimpse at the aim of a reflex is to gain hints for future experimentation on it. Such a clue to purpose is often difficult to get; and attribution of a wrong meaning may be worse than absence of all clue.

In nervous hypertension, the trained individual reports numerous gross and vague tensions in various parts of the body, sometimes fragmentary and elusive, abortive acts to

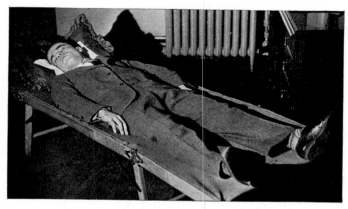

Fig. 23.—The patient is shown during general relaxation. In practice at home, when the physician is not present to direct him, he should not engage in preliminary contractions but should proceed at once to relax. However, he should be encouraged to observe the experiences of muscular contractions during his daily activities so that he may eventually become familiar with various localities as an aid to general or differential relaxation.

do now one thing, now another, often without harmony or adequate co-ordination. Thus in connection with objective observations of shifting, restlessness, grimaces, tics, and other manifestations, we are enabled to get an added insight of the corresponding phases of the inner life of the neurotic. We are led to the view that what the patient calls the "feeling of nervousness" consists of the varied sensations from the disorderly muscular tensions, voluntary and involuntary, that mark his responses to environment. From time to time, therefore, when he relates his anxieties and difficulties, we gain further knowledge about his tensions and are aided in directing him to relax.

It requires extreme progressive relaxation of the muscles of the eyes and speech apparatus to diminish mental activities. The individual simply lets these muscles go extremely in the same manner as in relaxing the muscles of the arm. Practice is required before this is accomplished, and the patient may continue to gain skill after he has been discharged. General relaxation is illustrated in Figure 23.

Simpler and Briefer Methods of Relaxation

The method of relaxation may be simplified or shortened in one or more of the following respects: (1) Training may be limited to a few muscle-groups. So it may include the biceps group but none other in the arm and none but the calf muscles in the leg. In this manner the patient may learn to relax the whole limb fairly well as he relaxes the part. (2) The patient may contract right and left groups simultaneously, noting the tensions; then relax them together. (3) Adjacent muscle-groups may be contracted or relaxed as units from the outset. For example, the whole right arm is progressively stiffened and later relaxed. (4) Cultivation of the muscle-sense may be omitted. (5) In its simplest form, the method of relaxation consists of directing the patient to relax whatever parts appear to the physician to be tense with no previous training as to muscle-groups and sensations.

The method last mentioned has many applications. First developed in 1908 for laboratory purposes, it is useful in general practice, particularly when the patient is seen for the first time in an excited and perhaps a weakened state, as during an acute illness or an emotional upset.

CHAPTER VI

THE NATURE AND TECHNIC OF DIFFEREN-
TIAL RELAXATION

THE use of relaxation in the recumbent position (chap. v) obviously is limited in its possibilities if, when the patient is up and about for the day, we make no attempt to avoid continual symptoms of nervous hypertension. But how can this be accomplished? I faced the question, "Is it possible for the ambulant patient to remain at his daily duties and affairs while at the same time something is being done to replace irritability and excitement with nervous quiet?" If so, perhaps many patients might be spared the worry and loss of giving up business during such disorders, as commonly occurred when they were ordered to take the old-time "rest-cure" or to "leave town for a change."

Evidently this problem, or at least one aspect of it, can be translated into terms of tension and relaxation. We might ask whether there is any way for the individual to become partially relaxed while engaged in such activities as are required for the conduct of his affairs. Can he learn to do essentials, yet omit non-essentials? make necessary movements, yet omit those which reveal irritability and excitement? Light is thrown on this possibility if we turn to the methods of schools and university departments of physical art education. Here we gain suggestions that may perhaps be turned to practical and scientific account. It will be convenient to postpone the discussion of laboratory investigations which attempt in part to answer the foregoing questions (chap. ix). We shall here assume that, as tested by the knee-jerk, these investigations justify the belief that some degree of differential relaxation commonly takes place during reading, writing, and other customary activities in normal persons and can be specially cultivated, if desired, in patients. In the present chapter, after a few introductory comments in the field of art education, the method of differential relaxation will be described, leaving applications and evidence to later consideration.

Differential relaxation, although not so named, is well known in the arts. Teachers of vocal culture, including the operatic type, devote much time to the relaxation of the muscles of the throat, larynx and respiration. Singers early learn that a loud tone is not required in order to be heard in the back row of an auditorium with good acoustics. "Carrying-power" of a voice, as it is called, increases not alone with loudness but particularly also when the voice is properly "placed." Even a whisper should be so uttered as to carry to the last row. It is commonly understood that voice placement depends largely upon proper relaxation. Generally the student is taught to locate his breathing expansion in the lower portion of the thorax. When the lips, tongue and jaw are in the proper position for utterance, a minimum requisite of breath is expressed from the lower thorax. He is particularly taught to sing with the throat and jaw muscles so far as possible relaxed. He is not to sing or speak "from the mouth," else carrying power will be lacking. Likewise, the timbre of the voice somewhat depends upon proper relaxation. The so-called "throaty" tone, which may mar a performance, is due to excess tension in throat and laryngeal muscles. Unfortunately, vocal teachers commonly lack an adequate knowledge of anatomy and physiology, which doubtless would greatly expedite their work.

In aesthetic and ballet dancing, relaxation plays a conspicuous rôle. The individual who holds himself rigid in these arts fails in his effects. A particular exercise is repeated until grace is attained. This means that those muscles alone are used which are needed for the act, and that no excess tension appears in them. Delsarte undertook to prove that relaxation underlay the art of sculpture, indeed all the physical arts, and developed his so-called "decomposing exercises" in order to secure his end. Certain philosophical works on aesthetics seem somewhat to realize these points, but they fall short of clear, well-defined statements.

Differential relaxation accordingly means a minimum of tensions in the muscles requisite for an act along with the relaxation of other muscles. A large variety of instances can be found in daily life. The speaker with trained voice does not tire even after prolonged effort if he keeps his throat differentially relaxed. The billiard player spoils the delicate shot if he is generally too tense. The golf or tennis player must learn to mingle a certain relaxation with the strokes in order to be successful. The restless or emotional student finds it difficult to concentrate. The excited salesman fails to impress his prospective client. The clever acrobat produces an impression of grace and of ease by relaxing such muscles as he does not require. The comedian often makes his laughable effects depend upon extreme relaxation of certain parts while others are active or held rigid. It seems safe to say that every learning-process depends upon the acquisition of certain tensions with concomitant relaxations. Psychological textbooks commonly illustrate the early learning-process with the child at the piano who squirms and shifts, perhaps even protruding the tongue, as the notes are first studied. As skill is acquired, these tensions disappear: in the present terms, a certain degree of differential relaxation sets in.

One can with care observe excess tension in daily life on

all sides. Individuals can readily be noted who gesticulate unnecessarily, speak rapidly or with shrill pitch, shift or turn about excessively, wrinkle their foreheads or frown too often, move their eyes unduly, or show other signs of overactivity or excitement. On the visual side, interesting imitations of hypertension in normal individuals under conditions of excitement can be found in almost every current exhibition of moving-pictures. Obviously stage-craft is successful in its imitations of human activity in proportion as the total muscular patterns are duplicated, including both tensions and relaxations.

The Technic of Differential Relaxation

Differential relaxation can be taught independently of general relaxation or prior to it. It has usually seemed most simple and convenient, however, to have the differential follow the general form. In therapeutics, both types of training have appeared necessary for stubborn chronic cases, because the individual who remains excited during his daily activities does not readily relax when lying down. The tensions appear to be cumulative in their effects. For instance, according to present experience, the individual who has had insomnia for many years needs to be shown not alone how to avoid restlessness at night but likewise how to avoid undue excitement during the day. Conversely, the nervous, excitable individual needs to be shown not only how to be relaxed while at his activities but also how to avoid restlessness at night if sleep is to be made profound and restorative.

A convenient way to introduce differential relaxation after a course of general relaxation is to have the individual arise slowly and limply from the couch on which he has been resting and sit in a nearby chair. He then relaxes all his parts in this new posture as well as he knows how. He is instructed to maintain sufficient stiffness of the back to prevent him from falling, but no more.

In the sitting posture a review of the same procedure as that while lying down may now be made. Depending upon the type of disorder, its severity, the time available, and the skill and other characteristics of the individual, the muscle-groups may be reviewed thoroughly or briefly. If necessary, an entire period may be devoted to the recognition of tenseness and the relaxation of each muscle-group, and the same order may be followed as before. On the other hand it may seem best to go over the muscle-groups of the limbs briefly, perhaps in a single period. It is possible to abbreviate the method of progressive relaxation at this stage by omitting the recognition of tensions. In this event, as previously said, a fine control will not be expected.

If the individual has acquired familiarity with the sensation of tenseness, he may contract the biceps-brachial or any other muscle-group without the physician adding passive resistance in order to increase the sensation. Indeed, provided that he readily recognizes sensations from muscular contraction, it is best practice at this point to make the contractions very slight (Fig. 24). For instance, he flexes the forearm, moving the hand half an inch or less, and notes the sensation of tenseness. When this is successful, still slighter contractions are tried, including such as are invisible to the naked eye. That such incipient contractions are actually made by the subject when so instructed has been clearly shown in recent unpublished galvanometric tests.

At this stage, signs of inadequate relaxation can generally be noted. In this event the individual sits with his head only partly bent over, the eyelids wink as in thinking, or the limbs appear somewhat stiff. Some neurotic individuals seem to require the support of a pillow at first, but this is to be omitted as soon as possible. The instructions for relaxation largely resemble those formerly used; those for the limbs and some

Fig. 24.—Flexion of the arm is performed to excite the experience of the muscle sense and permit its recognition. The patient will subsequently let this part relax, usually with eyelids closed, but remaining seated, as the first step toward learning to relax differentially.

other parts need not here be repeated. After the subject has slightly straightened the back (Fig. 25) and has noted the increased tension in the posterior muscles, he lets these muscles go so far as he can without actually falling over or going uncomfortably far forward or backward. Upon coming to the

FIG. 25.—Upon straightening the back in the sitting posture, the sensations from the erector muscles can generally be noted quite readily. Most of the exercises in noting tensions for purposes of differential relaxation resemble those for general relaxation and will not be further illustrated.

muscles of the neck, he should note not alone the tension in moving and inclining the head in any direction but also the slight static tension present almost all around the neck when

[87]

the head is held up in ordinary posture. When the head is allowed to fall over during a prolonged period, the subject generally complains at first of pain from strained ligaments upon the extended side. Adaptation usually sets in after a week or more of practice, and the pain diminishes or disappears. It may be helpful to inform him to this effect in advance when he complains. The average subject does not learn to relax the head adequately until he practices sitting; for in this position rigidity of the neck makes itself apparent by failure of the head to fall in perfect limpness.

Patients of nervous type prior to treatment very often complain of a dull ache in the back of the neck or in the suboccipital region. Examination for foci of infection and other organic sources of distress may prove negative. It is then proper to test if there is chronic contraction of any particular muscle group or groups and if increase of such contraction occasions or adds to the distress. Palpation may prove of service. The patient may be required to report what happens when he bends the head in each direction in the routine way, and also when he turns the chin to the right and to the left, as well as rotates it up and right, and up and left. Leading questions should be avoided. Increased resistance to movements in certain directions, found by the physician upon moving the part, is evidence of persistent contraction. Not infrequently, the patient, after learning to recognize the experience of muscular contraction in the aching region, volunteers that the pain evidently arises from continued muscular contraction there. In the presence of such aches extraordinarily long practice is likely to be required before the particular muscle-groups become habitually relaxed.

Aches or pressing distress in the head due to chronic muscular contraction are called "tension headaches" and sometimes appear to be located at the vertex. As the patient learns to relax these localities, the pains may disappear, without suggestion on the part of the physician that they will disappear and without his indicating what has been their possible cause. Some neurologists state that the peculiar pressure-distress at the vertex which is so often characteristic of "neurasthenia" has never been explained. It seems likely, as here suggested, that local chronic muscular contraction with consequent fascial tension is the cause.

Review of relaxation of the visual and speech organs is particularly important. The subject reports diminution of mental and emotional activity during differential as in general relaxation. In the former the degree of decrease is likely to be less.

Static tensions may require special practice to relax. Even an experienced individual may continue to maintain a moderate degree of rigidity in some part when he believes it thoroughly relaxed. To let go such a tension step by step, each time feeling that one can go no farther yet soon succeeding in doing so, is an accomplishment which, when first performed, marks a step forward in learning to relax.

When the subject has become expert, his posture is typical (Fig. 26). The legs are more or less sprawled out and are flaccid to movement by an observer. The arms and head droop limply, while the trunk may be bent in any direction. Breathing is regular and quiet. There is no trace of restless movement, even of a finger. A certain toneless appearance is characteristic of the eyelids, which do not wink during a prolonged period. This must be sharply distinguished from an earlier stage where the eyelids are held motionless for a time, followed by vigorous winking of the closed lids. Close observation should reveal no motion of the eyeball. The knee-jerk is diminished or absent in the more successful patients. A considerable number of patients unfortunately fall short of this degree of skill.

FIG. 26.—The subject is well relaxed in the sitting posture

As stated before, the presence of the physician often works as an interference with relaxation. In spite of this obstacle his presence is needed in order to observe tensions which the subject fails to note for himself, to call attention to them, and to show the patient how to do away with them.

If success up to this stage has been attained, the patient now is to learn partial ocular relaxation. He has previously

Fig. 27.—Relaxation with open eyelids gives the patient a particularly vacuous appearance. In contrast with this is his appearance if he ceases to relax and begins to think.

learned to open the eyelids and to let the eyes go so that they are not looking in any direction. This gives a vacuous expression and apparently involves a greater degree of relaxation than the oculist demands in distant vision during an examination. Extreme relaxation of the eyes with open lids is soon terminated by a burning sensation due to absence of

Fig. 28.—Relaxation with open eyelids should be illustrated by the physician in order to provide the patient with an example for imitation.

winking and lacrimation (Figs. 27, 28, and 29). Therefore, as a new exercise, he is directed to permit his eyes to wander about to a slight degree and not to relax them extremely. When this is done, a moderate amount of winking ensues, avoiding discomfort and securing relative rest. A warning may be added, if necessary, that this represents no attempt

FIG. 29.—It is possible to have the muscles of the eyes and face extremely relaxed while those of the neck remain contracted, holding up the head, an example of differential relaxation.

to help "to throw away his glasses," in order to avoid such impressions as are given by a certain book on this subject which not alone makes unwarranted claims but does not even advance correct methods to produce relaxation.

[93]

What manner of relaxation is learned in the physician's office is to be repeated for hours daily at home or elsewhere until it becomes a habit. Obviously a measure of practice while sitting can be carried out in the street car or automobile.

The patient now reads, while relaxing the lower limbs; the back so far as sitting posture permits; the chest, so far as inner speech while reading permits; and the arms, so far as is possible while they hold the book or magazine. With the forehead and eyes extremely relaxed the words will of course not be read. This should be carried out in order to make the patient familiar with an extreme form of differential relaxation. A little more tension is then introduced: the words are to be read but the eyes and other parts are to be kept as far relaxed as possible at the same time. Possibly now the subject reports that he follows the words but fails to get the meaning. This still represents too great a degree of relaxation. He is then requested to contract just enough to get the meaning clearly and no more. Close observation enables the physician to detect whether the patient appears to be following the instruction and to stop him and correct him if he is not (Fig. 30).

During the early weeks of practice, fatigue often plays a predominant rôle and the patient may even wonder whether relaxation does not make him indifferent in the performance of his daily duties. We recall, furthermore, that after over-exertion of any kind, such as a prolonged run, the first effects of lying down may be even painful. At such a moment one may prefer to be up and about in order to avoid distress. In the same way a highly nervous individual during the first weeks of practice may resent relaxation and desire to cease. If he continues, the next stage is one of increased rest, during which exertion and therefore initiative is not too much encouraged. Later, when general physical signs disclose that adequate rest has been secured, there may be a return of vigor and initiative, possibly in greater degree than during the state of fatigue or irritability.

Fig. 30.—As the patient appears when reading in a relaxed manner (differential relaxation).

The effect of such practice, as tested by the knee-jerk (chap. ix), is to tend to bring about a quieter state of body during reading, writing and other sedentary occupations. It is possibly true, as trained patients report, that restless move-

[95]

ments and static tensions, through the proprioceptive sensations they arouse, interfere with attention and memory. Sometimes after relaxation has been learned, the patient asserts that he is able to work under noisy or otherwise irritating conditions which formerly discomfited him. He may report less fatigue than formerly after working, and perhaps also a generally increased efficiency. I have not yet tested these reports by objective methods. My clinical impressions, which have little value as evidence but which serve to lead toward further investigations, are that in the course of weeks or months the individual's demeanor as well as countenance shows a change; movements lose their quick, jerky habit; the voice becomes quieter and speech slower; lines of fear become less marked as the anxious or worried appearance gives way to a more placid and restful expression.

In handling particular neuroses, the physician or investigator can devise various details of method which need not be described here. For example, if in a case of speech disturbance, symptoms seem to be characterized by the presence of hypertension in the muscles of the tongue, lips or throat or again in those of the thorax and abdomen (perhaps spasms and inco-ordination as found by Travis [1927]), we are dealing with differential hypertension. Practice at relaxation will then be devoted particularly to those parts in which excessive tension is marked. It may prove of service to have the patient while lying down repeat some sentence, each time with a lesser degree of tenseness in certain parts, until the speech becomes slurred and indistinct. Such exercises in excessive relaxation while speaking serve to overcorrect the habitual hyperkinesis which characterizes the stammerer and to provide him with standards to imitate upon repetition. Evidently the method of diminishing tensions has many applications to neurosis which admit of graphic study. The patient should be encouraged to use his initiative to relax in the absence of the physician.

THEORETICAL DISCUSSION

In theory, differential relaxation permits the presence of the activities of the organism needed for attainment of a purpose, but there is efficient control. It means economy of neuromuscular energy. As the patient reads or writes or engages in other work under observation, certain of his activities are apparently necessary for his occupation. These are called *primary* activities. Included among these are the contractions of those muscles needed for posture or movements of the hand to hold the book or pen, movements of the eyes to follow the words, and in some persons movements of the tongue and lips to repeat the words in inner speech. While all these primary activities continue synergically in the performance of a task, certain others may be observed in the average individual which apparently do not contribute but rather detract therefrom. These may be called *secondary* activities, since they include those not needed for the task in hand. Innumerable examples come to mind: While reading, a noise in the other room may be followed by looking up and turning the head in that direction. Any distracting sensory stimulus may be followed by such secondary activity. Very often the average individual while reading or otherwise engaged has an undercurrent of distracting thought-processes in the form of worries, reflections, irrelevant recollections, intentions to do this or that thing, and very often even songs or strains of music are silently but almost incessantly repeated. Many if not most persons read in this way, so that perfect attention to a book or occupation, even for so brief a time as a few minutes, is probably found only among those select individuals who have attained or are attaining eminent skill in their field.

PROGRESSIVE RELAXATION

Relaxation during activity has two directions. Primary activities may be unnecessarily intense for their purpose. For instance, a person may sing too loudly, pound his fist too vigorously as he converses, peer too intently as he looks about, overexert himself as he tries to study. In such instances a better effect will be produced by not trying so hard, by relaxation of primary activities. This relaxation should be carried only to the point where maximum efficiency continues; beyond this it would interfere with the purpose in hand. However, the aim is to carry relaxation of secondary activities to the extreme point, since these activities are generally useless.

If differential relaxation involves economy of neuromuscular energy, does this mean that the sole purpose is to reduce the caloric expenditure of the organism per diem? Obviously that is not necessarily even the chief purpose. Complete rest with general relaxation would be the method of choice for weakened or exhausted patients for whom muscular contraction and therefore caloric requirements were to be reduced to a minimum. However, the patient who is being treated at the stage of differential relaxation ordinarily is directed to exercise in the sunlight and air; for proper exercise is conducive to subsequent relaxation, and vice versa. But such exercise costs many calories; and if saving them were the only purpose, it would be folly to teach the individual to relax the small muscles of the eyes and forehead, thereby saving perhaps 1 or 2 calories, and then to send him out to swing a golf club at the cost of some 500 or more. Langworthy and Barrott (1920) found with a respiration calorimeter that the individual used 60.7 calories per hour while at rest in a chair. Knitting used an extra 10.1 calories, crocheting 8.3 calories, the average types of sewing 9.4 calories, while sweeping used 40.1 calories. If the total acts of sewing or knitting use only 10 or less calories per hour, it is obvious how insignificant is the caloric expenditure in small movements of the eyes and fore-head. Yet just these parts play a major rôle in the method of relaxation. How explain this difficulty? Obviously because these same parts play a major rôle in most voluntary activities. The latter as a rule continually involve reactions to things seen. Evidently the mere local saving of calories effected by the application of relaxation to ocular unrest is as insignificant as when applied to habit spasm of facial muscu-lature. In both cases the result of relaxation, if successful, is to do away with disorderly tensions and to secure better muscular co-ordination during voluntary activities. The par-ticular importance of relaxation as opposed to unrest of the eye-muscles evidently hinges upon the well-known impor-tance of the ocular nerve centers and tracts for the initiation and carrying-out of voluntary activities. Clinical observa-tions suggest the view that inefficiency of conduct in the aver-age individual, as well as disorder of conduct in many neuroses, sometimes involves unnecessary activities of the eyes and other organs which have important reflex connections with the voluntary activities of the organism. The importance of these activities in the bodily economy evidently depends upon such connections and is not to be measured merely in terms of the expenditure of calories involved by the action of the sense organs or of the associated nerve tracts and centers.

In general, then, contraction or relaxation of a part does not depend just upon calories for its importance. It is the location or distribution which counts, and there lies the sig-nificance of differential relaxation. Another way to put the matter is that an excess of slight or incipient tensions or movements, some co-ordinated and with well-marked func-tion and some not, involving in many instances small but in others great caloric expenditure, seems from one standpoint to constitute the very essence of what is commonly called nervous disorder. From this standpoint, the effect of differential relaxation is to eliminate such elements of motor disorder.

CHAPTER VII
EARLY HISTORY OF THE PRESENT
METHOD OF RELAXATION

U PON applying the method of relaxation to clinical problems, realizing that favorable reports from the patient could not in themselves lead to scientific conclusions, I was repeatedly led to ask whether I was really securing a greater degree of relaxation than what the doctor effects when he simply orders his patient to rest without showing him how. This question led to laboratory investigations to which the method largely owes its growth.

In 1908 certain experiments at Harvard University unexpectedly disclosed the significance of neuromuscular tensions and began this development of interest in relaxation. Few studies had been made, either in physiology or in psychology, of the excessive response which takes place after a sudden unexpected stimulus. Yet the involuntary start is familiar to everyone. It usually occurs after a sudden noise or other strong excitation while the individual is otherwise engaged. The movement, when extensive, generally includes a straightening of the trunk due to the *erector spinae;* but jerks of the limbs or head are also common, and any external muscle-group may be involved. Often the disturbance is so slight that the only external evidence is a rapid contraction of the eyelids; although the individual may report organic sensations of "nervous shock." This is stated to be a sudden gripping or tingling distress in the thorax and abdomen, with palpitation and quickened respiration. The disagreeable experience generally passes after a few seconds, but on rare occasions it is followed by headache. What is the explanation of this neuromuscular excitation that follows a strong unexpected stimulus? It is evident that some of the movements are sensory adjustments to the new stimulus, e.g., turning toward the sound with wide open eyes. But the jerks of the trunk or limbs are obviously not sensory adjustments, and they stand in need of further explanation. Lehmann (1899) had made a study of the effects on the pulse, respiration and volume of the arm produced by a loud unexpected organ note. Féré (1900) had registered contractions by means of a tambour placed on the muscles of the forearm or wrist, while at other times he used a hollow rubber ball held in the hand. He concluded that the contractions were reflexes, which varied in amplitude with the intensity of the excitation, the excitability of the subject, and the unexpectedness of the stimulus. His studies were qualitative and obviously incomplete.

Since experimentation calls for frequent repetition of a phenomenon under like conditions, it was clear that no great progress could be made if during a period of tests with a given subject, we should be unable to produce the involuntary start more than once or twice. This difficulty seemed probable if as previously believed, the start depended primarily upon unexpectedness. It was found, however, that under the conditions of the experiment, the stimulus could be repeated as often as ten to fifteen times during a period and the start would still occur, with some diminution, to be sure, but not so much as to interfere with comparisons.[1] The stimulus was a sudden loud noise, made by raising and releasing one end of a strip of resilient wood (of dimensions somewhat greater than those of a meter-stick) clamped at the other end to a wooden table. A fairly constant intensity of sound was produced by raising the free end to a determined height.

[1] An account of the following investigations has previously been published (Jacobson, 1926).

The experimental study of the involuntary start was begun in order to test a hypothesis which I later had to abandon. It was easy to observe that the start generally occurs when the individual is said, in lay terms, to be "lost in thought." Accordingly it seemed plausible to suppose that, upon sudden interruption of the psychophysical thought-processes, the energy from the latter might be diverted to efferent nerves and their muscles. Preliminary tests lent a certain weight to this assumption. The subject was directed to attend to the details of a coin or to some simple but interesting picture. When the explosive sound was made, a general start was evident. At other times he was directed to be inattentive, that is, not to try to pay attention to the picture or coin or to anything else, if possible. During other tests we sought to distinguish a state intermediate between complete attention and inattention. Apparatus was arranged to record upward and backward movements of the head. Without seeking to generalize, it was found, at least at times, that during complete attention the start was marked (III in Figs. 31, 32, and 33); that during the condition of lesser attention, the start was of lesser extent (II in same figures); and that with mental passivity, the start was much diminished or absent altogether (O in same figures).

Upon closely watching the subjects when they were directed to be attentive, it was noted that the brows were contracted, the eyes were fixated, the limbs were slightly stiffened and the trunk was tensely held. On the other hand, when directed to be inattentive, tension seemed to give way to relaxation. These observations gradually led me away from the hypothesis mentioned above. A simple explanation of the start seemed to be at hand. The muscles contracted with a jerk when the strong stimulus came because they were previously contracting, although in less degree, during attention. If there was no such previous contraction, as during inattention, no start took place. Evidently this simple explanation was more acceptable than my earlier complex hypothesis.

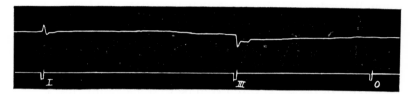

FIG. 31

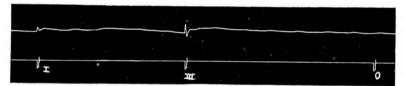

FIG. 32

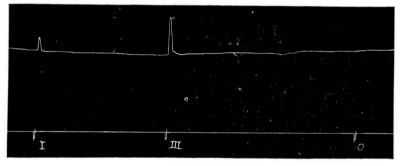

FIG. 33

FIGS. 31, 32, 33.—Movements of the head and trunk are indicated by rises (backward and upward movements) or falls (downward and forward movements) in the upper tracing in each figure. The moment of occurrence of the sudden sound is indicated by breaks in the lower line. At *III* the subject was fully attentive; at *O* he was inattentive, so far as possible; while at *I* there was an intermediate degree of attention. The figures show the most marked start where attention was greatest, and no start during inattention.

The truth of this explanation could be tested by producing in some other way a heightened peripheral contraction or tonus, whereupon we could test whether the sudden noise would occasion an extent of start varying in amplitude with the tonus. Accordingly the subject was directed to stiffen the muscles of the arms, legs, head and trunk; that is, to sit still while partially contracting his muscles. In other tests he was directed to relax his muscles as completely as possible. An intermediate condition was also directed, in which the muscles were held only slightly tense. These grades are respectively indicated by the numbers 3, 0, 1 in Figures 34, 35, and 36. In all instances either his eyes were closed or he looked away from the source of sound. Under these conditions, when the sound was made the subject gave the same involuntary start as before. Indeed, the contraction seemed in these preliminary tests very much greater with extreme muscular tenseness than with mild; while with relaxation, there was no start and no shock, and the sound seemed to lose its irritating character (Figs. 34, 35, and 36). In this way it was suggested that in addition to the factors mentioned by other writers, the involuntary start depends upon the previous neuromuscular state; and that, if a start characteristically occurs when there is interruption of thought-processes or of attention, this is because the mental activity is accompanied by skeletal muscle contractions.

The observations begun at Harvard were continued on several visits to the Willard State Hospital of New York. They seemed too few in number, however, to permit more than a provisional general conclusion, particularly since there were marked exceptions to the rule. Therefore it seemed necessary, in order to get the large number of precise observations required for scientific generalizations, to restrict the study to the movements of some particular part of the organism.

When this was done, as will later be seen, the results proved to be in harmony with the foregoing observations on the nervous start.

In more recent years various investigators have contributed to our knowledge concerning the nervous start, although

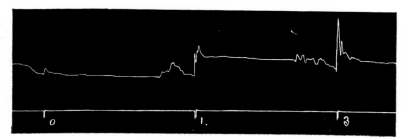

FIG. 34

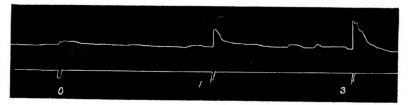

FIG. 35

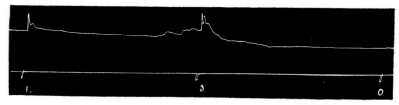

FIG. 36

FIGS. 34, 35, 36.—Tracings made as in preceding figures. At 3 the subject was directed to contract generally the muscles of the limbs, trunk, and head; at O he was relaxed; while at 1 there was an intermediate degree of general contraction. The most marked start is here shown where the preceding neuromuscular tonus is highest, while with relaxation there is no start. These subjects had no training in relaxation, but the experimenter observed that they seemed to follow the directions.

not called by that name. During studies on the nature of the emotions, they have frequently used a sudden unexpected noise or other strong stimulus to arouse surprise or fear. Effects on the pulse and respiration have been noted by Skaggs (1926), on the systolic blood-pressure by Landis and Gullette (1925), and on the cerebrospinal fluid pressure by Dumas and Lavastine (1913). That effects occur on the bladder musculature would appear entirely probable from the observations of Mosso and Pellacani (1882), who conclude that "every psychic experience and every mental work is always accompanied by a contraction of the bladder." The colon (Brunswick, 1924) and the esophagus (chap. xvi) have been found responsive to stimuli far less intense than loud startling sounds, so that we can safely state that these organs also are affected during the nervous start. A slight shove of the front feet downward followed by a retraction of the forelegs marks the general start reflex to sudden noise in the guinea pig (Dodge and Louttit, 1926). We recall the well-known influence of auditory stimuli on the knee-jerk (Bowditch and Warren). In the light of the widespread character of the start reflex, as revealed in the foregoing and other investigations, we may infer that the sensations reported by the subjects as appearing to come from the viscera probably do come from this source. It remains for future investigation to record the effect of extreme relaxation on each visceral organ during a sudden unexpected noise. But so far as can be judged from subjective indications, the start phenomena are diminished or absent in viscera if there is extreme relaxation, just as we found them objectively diminished or absent in skeletal muscles under the same conditions. If this is true, it would seem that the generalized character of the start is paralleled by the generalized character of the effects of extreme skeletal-muscle relaxation.

[107]

Individuals who start naïvely tend to explain their reaction as due to the strong character of the stimulus. However, according to our tests, a violent reaction along with
subjective disturbance seems to depend upon the preceding
state of general muscular tonus. Some individuals do not
visibly start at all as a rule. This was true of Subject 2 until
he was requested to throw off any restraint that he might
intentionally or unintentionally be exercising. In a few tests,

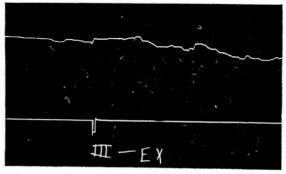

Fig. 37.—In contrast with the foregoing, where the sound occurred unexpectedly, the subject was here forewarned by a special signal that the sound was to take
place. As a rule with normal individuals practically no start is recorded.

in place of the sound occurring unexpectedly, the subject was
forewarned by a special signal: I found that even tense
individuals, thus prepared, might restrain the start (Fig.
37). Possibly this preparation permits the subject to fore-
stall sudden movement by simultaneous contraction of agon-
ists and antagonists. At any rate, if the subject is directed
to hold his muscles firm in spite of the noise, no start may
occur (Fig. 38). Backward extension of the trunk, for instance, can be checked by appropriate voluntary contraction
of the abdominal muscle-group that draws the thorax downward. It is simplest to conceive such restraint as the same
in type as the familiar example of failure of the knee-jerk to

appear upon tapping the patellar tendon when this is due to the subject contracting the flexor muscles of the leg. On the other hand, as said, the feeling of "nervous shock" above-described was weak or absent and the start was wanting or slight when the individual was extremely relaxed. These observations suggested the possibility that all subjective irritation or distress might be reduced if the individual were to become sufficiently relaxed; and this hypothesis remains to-day a beacon-light for further experiments and observations.

Several years later, experiments at Cornell University on a totally different topic led to further unexpected observa-

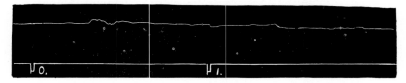

Fig. 38.—The subject is here requested to hold his muscles firm when the noise takes place (at 1 on the signal line). Practically no start is here observed.

tions on relaxation. The intensity of smell in normal human subjects was being studied as affected by simultaneous sounds (Jacobson, 1912). Previous experiments had led me to expect that the sound sensations would diminish the odor, but this did not happen. Perhaps this was due to special effort to attend to the odor during the distracting noises. To do away with such effort, they were now directed "not to continue to exercise effort of any sort at any time during the entire process; to abandon all attitude of effort and to be quite as passive as possible."

This instruction had effects which interest us here, but other results of the experiments may be omitted from the discussion. Considerable variation, I found, can be noted in the behavior of each subject as he changes from voluntary attention to passivity. His brow may be wrinkled and his mouth compressed, or both may be smooth and expressionless; the muscles of his trunk and limbs may be alert and contracted, or they may be flaccid and inactive; he may breathe in the odors jerkily and utter judgment abruptly and vigorously, or may maintain the quietness of sleep and speak in a soft whisper; may talk in lively fashion, or may say practically nothing. In short, he may be alert and tense in this or that part of the body or may be inactive and relaxed.

At first the records of behavior were taken sporadically; but later, as a rule, with every set of eight judgments the investigator sought to estimate the degree of facial relaxation, bodily relaxation (exclusive of the muscles concerned in maintaining the sitting posture), and some other signs of relative inactivity. Sudden palpation was sometimes used to gauge bodily relaxation. The foregoing particulars were each time recorded as of degree 0, 1, 2, 3, 4, 5, the last number representing the maximum. These crude and inaccurate observations of tensions had little more than a suggestive value except that they were forerunners of clinical observations of he signs of nervous hypertension.

It was found that with increase of effortlessness, as reported by the subject, there was an increase of relaxation as externally manifested. Marked deviations were noted, but this correlation seemed to hold as a very rough, but fairly serviceable generalization. There seems to be a good reason why it should hold, for the observers all reported that in order to carry out the instruction they had to "relax"; and it is obvious that relaxation is manifested in behavior as well as in introspection. That is, the investigations suggested that what is commonly called *effort* in our subjective experience consists in part of readily observable contractions of skeletal muscles. During so-called active attention or concentration, outward evidences of effort were clearly discerned by a trained observer.

As muscular relaxation advanced, it seemed evident on the whole that the attention of the subject diminished. In other words, I found that *the cerebral activity of attention apparently diminished in the presence of advancing relaxation.*

A certain personal experience in 1908 gave additional opportunity for observation. At that time, like many other students who carry their work always with them, I had a nightly insomnia which persisted for hours, while mental activity continued regardless of need of rest. Upon seeking to discover what it was physiologically that seemed to keep me awake, I believed that I could always identify processes of muscular tension somewhere in the body, and when these were eliminated by relaxation, sleep took place. Their elimination was not always a rapid process, particularly at first. But as I studied the muscle tensions further, I noted that subjectively I seemed responsible for them and therefore apparently could undo them. This evidently depended upon my carrying the relaxation sufficiently far. These personal experiences, while having in no sense the character of scientific evidence, have proved of service in the selection of experiments and in the training of subjects and patients.

Many studies have appeared (1937) including responses to sudden stimuli in so many organs that it seems safe to conclude that, (depending upon the stimulus) any muscular organ or region may take part. With sufficiently frequent repetition, the response typically weakens and may ultimately disappear. Startle patterns have been recorded in photographs and in moving pictures at high frequencies (Hunt and Landis, 1936; Hunt, 1936). A reflex startle leg response to sound has been studied in the rat by Prosser and Hunter (1936). They attribute the extinction to shifts in excitability of some part of the reflex arc between the sensory neurones and the final motoneurones. Evidently this is only a step toward an explanation.

CHAPTER VIII

THE INFLUENCE OF RELAXATION UPON THE REFLEX REACTION TO SUDDEN PAIN (FLEXION REFLEX)

THE early observations described in chapter vii brought to light a degree of relaxation apparently greater than what is generally called by that name. My interest was increased because graduate students, who were used as subjects, as well as patients in whom the condition was introduced spontaneously reported highly restful effects of even brief periods of this practice. To some of them it seemed, at least from a subjective standpoint, that extreme relaxation was more effective than an equal period of light sleep. We are reminded of the refreshing effects of restful sleep in individuals previously fatigued or exhausted: the improvement in subjective feeling, the better color, changed appearance of the eye, reduction of lines about the face, livelier gait and other signs of recuperation. Such effects were likewise reported and noted after periods of relaxation. On the other hand, patients described the peculiar distress of nervous irritability and excitement as difficult to grasp by anyone who has not frequently felt it: a drive in many directions; a general uneasiness and impulsiveness which may produce anger, tears or hasty action; a consequent feeling of unreliability; a sensation of stiffness or changing tenseness throughout the body, which is sometimes said to be anguish. In a corresponding way, outward effects of such emotional disturbance have been witnessed by every physician: the rapid pulse and irregular respiration, the blanching or flushing skin, the shifting eyes and frequently wrinkling forehead and brow, the stuttering, stammering and general uncertainty of speech, the incessant movements of limbs and fingers and the general restlessness. To secure graphic records of so many and various phenomena of nervous hypertension as influenced by progressive relaxation is a prolonged task which as yet is far from completed: but so far as might be judged by gross observation, aided by records of pulse and respiration, there occurred a marked general change as the patient's appearance passed from excitement to calm.

It was of course highly important to avoid reliance upon the subjective reports of the patient or subject. It was equally important to avoid influencing the belief of the subject as to what the results would be. Therefore throughout the following investigations the subject was so far as possible kept in the dark as to the topic of our inquiry. Indeed, in the experiments to be described, my associates and I could not be confident in advance as to what would occur under the conditions of our tests. Familiarity with the methods of clinical suggestion, as will later be shown, makes it relatively simple to avoid mistakes in that direction.

In 1921, in the hope of a large series of observations on the involuntary start, which might secure quantitative results, experiments were begun by Miss E. Loder and myself in the Psychological Laboratory of the University of Chicago, and later were continued by M. Miller, who used a pistol to set off the start. The inadequate character of our records made it necessary to turn to some simpler form of movement for study. This was found in the sudden withdrawal of the hand from an unexpected painful stimulus. Given a painful stimulus, to be kept constant for each individual, the response obtained during ordinary rest could be compared with that obtained during extreme general relaxation. It was evident that a quantitative inquiry concerning this reflex could be made in terms of speed and extent of movement. If the flexion reflex elicited by a sudden strong stimulus should be found to be diminished by the presence of general relaxation, this would agree with the foregoing observations and conclusions concerning the nervous start (chap. vii). This experiment was accordingly carried on, under my direction, by Dr. Miller, from whose article (1926) the following account is adapted. The shock was given without warning and the subject was instructed not to attempt any voluntary control of his reaction. Graphic records showed the extent of movement and the reaction-time.

An extreme degree of relaxation was required for these experiments. Seven subjects were trained during a practice period of about 3 months with periods of $1\frac{1}{2}$ hours three times a week. One person (Subject G) failed to learn much from the instructions to relax. As a rule, whenever he lay upon the couch he quickly fell asleep. He had always done this in the past, and apparently learned little from the training. It was with some hesitation, therefore, that he was used in subsequent experiments.

Methods of Testing the Flexion Reflex

For the experiments the subjects lay on a couch with two finger-tips of the right hand resting in 2 cups of normal salt solution. A painful, stimulating current could be passed through this solution, unexpected by the subject, causing a quick withdrawal of his hand. Owing to the arrangement of the apparatus, the only movement that the arm could make was flexion at the elbow. This was recorded on the smoked paper of a kymograph. A signal-magnet registered the beginning of the stimulus and a tuning-fork traced a time-line. The interrupted current used to stimulate lasted for a fraction of a second and could be controlled as desired in strength, frequency and duration. For each individual a particular strength and duration of current was selected and kept constant throughout the experiment, with exceptions to be noted later. The frequency did not vary for the several subjects, showed no measurable variation over short periods (3 or 4 weeks) and varied less than 2 per cent from beginning to end of the experiment.

A preliminary series was taken with each subject in order to determine the strength and duration of the stimulus to be used. It seemed best to secure a reaction of considerable extent because the response might conceivably diminish with adaptation. With some individuals this was readily done, for the arm movement increased when the current increased in strength or in duration. With others, the added response took the form of a general stiffening of all the muscles, and in this case the experimenter had to be content with a relatively restricted movement. We found too that a weak stimulus might call out a marked reaction the first three or four times and little or none after that. In each instance the series was continued until a stimulus had been selected which seemed likely to evoke an adequate response over a sufficiently long period of time; but in spite of this precaution it proved advisable with certain subjects to increase the stimulus, either in strength or in duration, at some point in the course of the final experiment. Such changes do not affect the validity of the results presented, for in all cases conclusions are drawn only from the comparison of groups of reactions taken from a single subject under constant conditions of stimulation. Table I shows the duration of the stimulus used for different subjects and gives the data for the factors which determined its intensity. A low reading for the scale of the induction-coil indicates a high degree of current strength.

PROGRESSIVE RELAXATION

The series constituting the final experiment consisted of approximately 200 reactions for each subject. No more than 10 reactions were taken on any day during the hour of test. Half of these were under the condition of general relaxation, and half with the subject in his ordinary or "control" resting state. These two conditions were employed alternately. In some hours, 5 reactions under one condition were followed by

TABLE I

Strength and Duration of Stimulus

Subject	Series	No. of Reactions	Primary Current (Amperes)	Scale Reading of Inductorium	Duration of Stimulus in Seconds
A..........	I*a*	94	0.2	10.2	.400
	I*b*	100	0.2	9.0	.400
B..........	I	200	0.35	7.0	.050
C..........	I*a*	74	0.2	6.0	.050
	I*b*	90	0.2	6.0	.400
D..........	I	200	0.35	7.0	.095
E..........	I*a*	117	0.2	7.0	.400
	I*b*	80	0.2	6.5	.400
F..........	I*a*	79	0.2	6.0	.050
	I*b*	100	0.2	6.5	.400
G..........	I	184	0.35	9.5	.050

This table indicates for each subject the number of reactions, the relative strength of induction shock and its duration. (Miller, 1926.)

5 taken under the other condition. When the relaxed condition came first on one day, the control was given first on the next, and so on. In other cases an hour entirely under one condition was followed on the next day by an hour entirely under the other. This method, which was followed with Subjects A, E and F, obviated the difficulty which they experienced in trying quickly to get out of the relaxed condition, for they found that this tended to hold over for the rest of the period.

The instructions were as follows: "All that you are asked to do in this experiment is to maintain such conditions of relaxation or contraction as may be directed. Several times during the hour you will receive a slight electric shock in your fingers. Please do not prepare to react to this shock in any particular way. If the natural reaction is a movement, let it come freely and spontaneously. Do not try to hasten it, or to inhibit it, or to influence it in any way. Try not to think about the coming stimulus."

For relaxation: "Please relax as completely as you can; do not stop until you are told to. If the stimulus disturbs you, begin relaxing again as soon as it is over. You will be given plenty of time to get well relaxed before any stimulus is given."

For the control condition: "Do not relax this time. You are not asked to do anything in particular except to be sure that you are not relaxing. Please keep this in mind, and if necessary move a little occasionally to be sure that you are not relaxing unconsciously. You may talk as much as you like."

The stimuli were given without warning from 3 to 5 minutes apart; never during the relaxed condition until the subject seemed to be well relaxed; and never during the control condition while the subject was moving or speaking. Whatever bodily response occurred in addition to the arm movement was noted on the record, as well as any observation the subject had to offer at the end of the hour. Each was questioned as to his success in carrying out instructions and after relaxation was asked whether he had been asleep or not. If doubt arose on this point, in either subject or experimenter, the subject was questioned further as to the number of stimuli he could remember and on any other points that might throw light on the matter.

PROGRESSIVE RELAXATION

THE FLEXION REFLEX DURING RELAXATION

The results of the experiments are summarized in Tables II, III and V. They are alike for all subjects except G and the latter will be discussed separately.

Extent of Movement

The movement was markedly less during extreme relaxation than during ordinary rest (Tables II and III). In many

TABLE II

PERCENTAGE OF CASES WITHOUT MOVEMENT

SUBJECT	SERIES	ORDINARY REST		RELAXED	
		No. of Cases	Percentage without Movement	No. of Cases	Percentage without Movement
A........	Ia	44	0	50	100
	Ib	50	0	50	92
B........	I	100	1	100	36
C.......	Ia	37	0	37	19
	Ib	45	0	45	49
D........	I	100	0	100	75
E........	Ia	60	0	57	11
	Ib	40	0	40	25
F........	Ia	39	0	40	48
	Ib	50	0	50	66
G........	I	92	0	92	28

The figures indicate that the flexion reflex almost invariably followed the induction shock when the subject lay on the couch at rest in an ordinary manner (control), but the reflex failed to appear in a considerable percentage of cases during extreme relaxation. (Miller, 1926.)

instances no movement occurred at all during extreme relaxation, although the painful stimulus was present unaltered (Table I). Subject A heads the list with an average of 96 per cent of cases without movement during extreme relaxation. This agrees very well with the excellence she had shown over the others in the speed and skill of learning to relax during the period of preliminary training. Absence of movement during extreme relaxation is seen for all six subjects in a considerable proportion of the cases (Table II); on the other hand (with a single exception) arm movement always followed the stimulus when given during the ordinary resting state.

TABLE III

AVERAGE EXTENT OF MOVEMENT

SUBJECT	SERIES	ORDINARY REST		RELAXED	
		No. of Cases	Movement (cm.)*	No. of Cases	Movement (cm.)*
A..........	Ia	44	5.6	0	....
	Ib	50	6.1	4	0.7
B..........	I	99	8.1	64	1.2
C..........	Ia	37	1.3	30	0.9
	Ib	45	6.1	23	0.8
D...	I	100	2.0	25	0.8
E..........	Ia	60	2.7	51	0.8
	Ib	40	1.7	30	0.4
F..........	Ia	39	0.7	19	0.4
	Ib	50	0.0	17	0.3
G..........	I	92	8.9	67	............†

The figures indicate that the flexion reflex as a rule was considerably less in extent during extreme relaxation than during ordinary rest. (Miller, 1926.)

* Centimeters measured on the drum; the actual movement was about twice as great.

† Average could not be computed because in many cases the extent of movement exceeded the recording capacity of the apparatus.

When arm movement appeared at all during the extreme relaxation, its average extent was less for all six subjects than during ordinary rest (Table III).

It will be noticed that no measure of variation is included in Tables II and III. The movements varied greatly and such a measure would probably not have been useful. However, a test of the significance of the figures was made in another way. Table IV shows this calculation. Here are presented averages of movements during extreme relaxation compared with averages of movements during ordinary rest. Where both conditions were given during a single hour, the two averages for this period were compared with each other; but where relaxation was given in one period and ordinary rest in the next or vice versa, comparison was made between

TABLE IV

NUMBER OF CASES SHOWING MOVEMENT LESS FOR
RELAXED THAN FOR ORDINARY RESTING
CONDITION (CONTROL)

Subject	Series	Total No. of Cases	Movement Less with Relaxation
A...............	I*a*	5	5
	I*b*	5	5
B...............	I	20	20
C..	I*a*	8	6
	I*b*	9	9
D...............	I	21	21
E...............	I*a*	6	6
	I*b*	4	4
F...............	I*a*	4	4
	I*b*	5	5
G...............	I	19	2

In this table may be compared the number of cases in which the flexion reflex was less on the average during advanced relaxation than during the ordinary resting or control condition. As will be seen, excepting one subject, the reflex on the average was diminished in almost all cases during relaxation. (Miller, 1926.)

two successive periods. In drawing up these averages zero cases were excluded; excepting that if all the reactions during relaxation for a given period showed zero movement, the period was included in the table. Table IV clearly shows that the average movement calculated in this way was much less during extreme relaxation than during ordinary rest.

Reaction-Time

In general, the reaction-time was longer during extreme relaxation than during ordinary rest. This was true on the average for four subjects (A, B, C and D in Table V). That this difference is not due to chance but is significant appears from Table VI. Calculations in the latter are made in the same way as above described for Table IV.

The rule that the average reaction-time is prolonged during extreme relaxation met with two exceptions (Table V,

TABLE V

AVERAGE REACTION-TIME

SUBJECT	SERIES	ORDINARY REST		RELAXED	
		No. of Cases	Time (sigma)	No. of Cases	Time (seconds)
A..........	Ia	34	169	0	....
	Ib	38	163	4	.222
B..........	I	94	125	61	.166
C..........	Ia	32	120	25	.142
	Ib	42	130	18	.156
D..........	I	79	188	20	.174
E..........	Ia	53	117	43	.145
	Ib	29	128	22	.150
F..........	Ia	37	120	16	.138
	Ib	36	188	17	.142
G..........	I	82	190	53	.182

The reaction-time of the flexion reflex is generally prolonged during advanced relaxation as compared with ordinary rest. The number of cases represented is less than in Table III, partly because there are no reaction-times corresponding to zero movement and partly because in a few cases defects in the graphs made it impossible to calculate the time. (Miller, 1926.)

Subject D in Series I and Subject F in Series Ib). These were found to be due to a relatively small number of instances in which the reaction seemed peculiarly prolonged during the control resting state. This peculiar prolongation seemed traceable to a tendency of a few subjects to react to the stimulus with a widespread stiffening of the muscles of the body without any very marked arm movement, a reaction which in chapter vii was called the "nervous start." Observation suggested that this consisted of movements initiated by nu-

merous muscles but checked almost immediately by the contraction of their antagonists. Analysis of the curves suggested
that the reaction-time during the control resting condition
in these two subjects falsely appeared prolonged owing to
some instances where contraction of extensor muscles concealed the true time of onset of arm flexion. If this explana-

TABLE VI

NUMBER OF CASES SHOWING REACTION-TIME LONGER FOR
RELAXED THAN FOR NORMAL CONDITION

Subject	Series	Total No. of Cases	Relaxed Exceeds Ordinary Rest Time
A................	I*a*	0	....
	I*b*	4	4
B................	I	18	18
C................	I*a*	8	7
	I*b*	8	8
D................	I	9	5
E................	I*a*	6	6
	I*b*	4	4
F................	I*a*	4	4
	I*b*	4	1
G................	I	16	5

Excepting one subject, the average of reaction-times for
each period is longer, as a rule, during advanced relaxation than
during the "control" rest. (Miller, 1926.)

tion is correct, the two exceptions do not disprove the rule
that the average reaction-time was prolonged during extreme relaxation.

The results for Subject G did not agree with the others.
During extreme relaxation his reaction either was practically
zero or else far greater than during the resting condition.
When he tried to relax without sleeping, he found it almost
impossible. This might be explained as follows: He did not
succeed, as said above, in learning to control his relaxation,
but simply went to sleep during the experiments as he might

and doubtless would have slept if he had never received any training.

Cultivated relaxation does not differ from natural relaxation except that the individual to a certain extent learns to do it at will and to control the degree. Falling asleep may lead the individual into deeper relaxation or he may become less relaxed, particularly if untrained. The results for this individual therefore represent a relative failure on his part to learn to relax, along with a natural tendency to sleep.

General Bodily Response

It has been mentioned that the arm movement called forth by the stimulus was usually accompanied by some other bodily response, and the question arises whether a decrease in the extent of arm movement represented a generally diminished reaction. So far as could be determined by gross observation, this was uniformly the case.

Subjective Effect

A further difference between the results obtained with the two conditions appears in the reports of the subjects. Without being questioned on this point, each subject (Subject G still excepted) reported the stimulus as less painful, less disagreeable, or apparently weaker with relaxation than without. Each individual remarked upon this effect more than once, and they were all questioned about it at several later sittings as well as at the end of the series. There were two occasions, occurring with the same subject, when the effect was not observed; otherwise it was reported as the invariable accompaniment of relaxation.

Effect of Sleep

During the experiment proper, as well as in the practice period, relaxation tended to induce sleep; and it may be asked whether the apparent effects of the relaxed condition may not have been due to sleep rather than to relaxation as such. In order to answer this question, the results were tabulated separately for sleep and waking. It was difficult to determine in every case whether the subject had been asleep or not; but the benefit of the doubt was always given to sleep. If the subject was believed to have been asleep for any part of a period, all the reactions of that period were put under the "sleep" heading. The reactions so listed included all those (1) where the subject said he had been asleep, (2) where he was not sure whether he had slept or not, and (3) where, although he was not aware of having slept, the period seemed unusually short to him or he was confused about the number of stimuli received. Of a total of 472 "relaxed" reactions for five subjects, 140 were thus classified as "sleeping." Subject E did not sleep at any time.

Exclusion of these sleeping-reactions did not alter in any respect the conclusions drawn from the original comparison between the ordinary resting state and the relaxed condition. The differences were unchanged in direction and remained significant. So far as could be judged, however, from the limited number of cases available, the effect of sleep was slightly to exaggerate the differences found with relaxation alone. But this effect was not invariable with all subjects and all criteria. More specifically, it was found that with two subjects sleep increased the percentage of zero reactions, decreased extent of movement, and lengthened reaction-time. With one subject the opposite effect was found; with one, sleep increased zero cases and lengthened reaction-time but also increased extent of movement; and with another it decreased extent of movement but also decreased zero cases and shortened reaction-time.

The attempt to isolate the effect of sleep was based on the

assumption that sleep and relaxation might be independent factors. There remains, however, the possibility that the effect of sleep may at times be due to a heightened degree cf relaxation prevailing with that condition.

DEPENDENCE OF THE NERVOUS START AND THE FLEXION REFLEX UPON THE FOREGOING STATE OF NEUROMUSCULAR TONUS

The results tend to confirm the earlier ones on the nervous start. In both, the extent of muscular contraction (arm flexion or general reflex start) following a sudden stimulus evidently depends upon the general state of neuromuscular tonus. They indicate that greatly diminished general tonus tends to bring diminution or absence of response and that heightened general tonus tends to bring increased response. Likewise, the sudden stimulus is found subjectively shocking and disagreeable when the general tonus is high, but does not prove disturbing when the general tonus is very low or zero.

These matters lead to interesting considerations. The involuntary start is particularly marked in so-called "irritable" or "excitable" or "nervous" persons. It tends to be marked not alone in neuroses but also, as is well known, after prolonged illness, operations, or other disorders. It can, of course, readily be masked by contraction of antagonistic groups. Nervous individuals are readily disturbed by noises which tend to interrupt their cerebral processes; but as is well known, they are prone to be distracted by many other types of stimuli (in addition to those that are sudden and unexpected) which do not affect phlegmatic individuals to the same degree. Events of slight importance and pains with slight objective pathology are particularly likely to arouse distress and complaints from such individuals to the point of interrupting their useful pursuits. In fact, their subjective symptoms of distress appear to mount according as they become increasingly irritated and excited. It seems probable that the underlying physiology is a heightened neuromuscular tension (increased tonus and contraction of striated and smooth muscles, with corresponding increased impulses in their afferent and efferent nervous supply) which would readily account for many characteristics of individuals in a "nervous" state.

The "Suffering" and "Pain" Reactions

It has become traditional to believe that "pain" and "suffering" which are universally met in clinical medicine are merely sensory experiences. The findings in the present investigations afford a new view. As illustrated above, the subjective disturbance created by a strong, unpleasant stimulus is greatly diminished upon extreme relaxation. Freeman has confirmed some of these results (1933). When a patient with wire electrodes under the skin shows complete local muscular relaxation, pain is reported absent, but commonly sets in with the appearance even of slight action-potentials. We conclude that "pain" and "suffering" are sensory-motor reactions commonly diminishing upon relaxation, although this, of course, does not remove the stimulus.

CHAPTER IX
THE INFLUENCE OF RELAXATION
UPON THE KNEE-JERK
GENERAL RELAXATION

IF THE extremely limp state produced by the method of progressive relaxation really differs in degree from what usually is called "relaxation," it would seem reasonable to expect a corresponding difference in the knee-jerk. In 1924 Professor A. J. Carlson kindly joined me in this fundamental investigation.[1]

So far as we knew, the effect of extreme relaxation on the knee-jerk had not previously been studied. In 1875, Westphal and Erb described the use of the knee-jerk in diagnosis. In 1886, Mitchell and Lewis concluded that "the responsive jerk brought about by striking a stretched tendon is the most refined measure we possess of deciding as to the tone of muscle." Since that time the knee-jerk has been widely used as a test of functional as well as of organic condition of the nervous system. It is generally accepted that any reflex that can be studied with quantitative accuracy can be used as an index of functional conditions of the central nervous system. Accordingly it should be possible to measure progressive relaxation by its influence on the knee-jerk.

Lombard had noted in 1887 that the knee-jerk was reduced on the average about one-half during quiet or, as he termed it, cerebral inactivity and sleep. Absence of the jerk during sleep has been reported also by Bowditch and Warren (1890), Noyes (1892) and Piéron (1913). More recently, Lee

---

[1] The account which follows in the first section of this chapter has previously appeared in the *American Journal of Physiology*, Vol. LXXIII, No. 2 (July, 1925).

and Kleitman (1923) observed that no jerk was evoked during sleep; but Tuttle (1924) found the jerk absent during
deep sleep, while during light sleep low kicks were registered.

It is well known that in "relaxation" as commonly produced in clinical examinations the knee-jerk is not diminished. Indeed, it is usually increased. Simply requesting the
untrained individual to relax usually increases the knee-jerk;
this, as commonly explained, is because he usually holds
his limbs tense before the request to relax, thus mechanically
decreasing the jerk, as can be verified upon palpation of the
flexor muscles. When requested to relax, the jerk becomes
more brisk in character because the contraction of the quadriceps is unimpeded by their antagonists (see chap. xiv).
Sometimes it is necessary to talk to the patient or to have
him perform some simple act such as counting or looking
out of the window in order to distract his attention from
the limb so that he does not hold it too tensely. When patients with an exaggerated knee-jerk are told to relax in this
manner, the kick usually becomes very marked.

Method of Testing the Knee-Jerk

The subject lay in a semi-reclining chair or on a canvas
cot, with the right leg, the movements of which were to be
recorded, at an angle of about 70°. Support was provided for
the left leg so as to make the general position comfortable.
At desired moments the patellar tendon was struck with a
spring-hammer. This hammer was attached to a metal stage
which moved vertically on a spiral thread, so that the height
of the blow could be conveniently arranged for each individual (Lee and Kleitman). The spiral thread rested on a heavy
tripod stand which could be moved on the floor in any direction for purposes of adjustment. The spring-hammer had an
excursion of about 5 centimeters and was set off, mechanical-

ly and noiselessly, by pulling a string. Seven subjects were used. Two were the writers, and five were former patients who had received prolonged training at progressive relaxation.

The procedure was as follows: Control tests were taken with the subject resting quietly and relaxed in the ordinary sense. He generally chatted or engaged in mental arithmetic in order to avoid extreme relaxation. From the beginning of the experiment to the end, no shifting or change of position of the trunk or limbs took place, for this would interfere with the accuracy of results. After the control records were obtained, the subject was instructed to relax extremely, including the muscles of the trunk, limbs, neck, as well as the small muscles of the head, such as those of the eyes and speech. Generally a pause of about 15 minutes was permitted for the subject to get extremely relaxed, but sometimes 5 minutes sufficed. On the other hand, disturbing noises from the outside sometimes led us to lengthen the period. Likewise on occasions when the subject appeared more tense than usual at the outset, a relatively longer time was given him to become relaxed. We assumed that the subject was relaxed when he appeared generally flaccid and in complete repose while the eyeballs were motionless and the closed lids did not flutter or wink. In the completely relaxed state, the muscles of the tested limb were entirely flaccid to palpation and unresisting to a slight push from the operator, which would easily flex it or extend it because of the limp state. The stroke of the hammer came noiselessly and unexpectedly at intervals of irregular length. After such a stroke during the relaxed period, it was necessary each time to allow several minutes for the subject to relax again. Sometimes apparent restlessness made it necessary to increase the length of this interval. Needless to say, the subject did not, as a rule, interrupt the period of relaxation by speaking, and the operator tried to act quite noiselessly. Following the period of relaxation, a control series was again taken. Here it was generally necessary to enliven the subject with conversation, for the relaxation tended to hold over.

The Knee-Jerk during General Relaxation

The records show a decrease of the jerk parallel to the degree of relaxation (Figs. 39, 40, 41, 42). In three of the

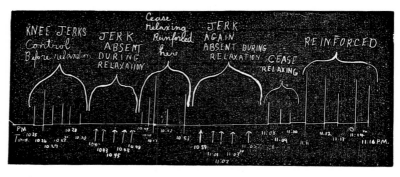

Fig. 39.—(E. J.) Absence of knee-jerk is shown during advanced general relaxation. Arrows indicate when no jerk occurred, although the same stimulus was given as previously. Numbers below the base-line indicate the moment when the tendon was struck. The subject is experienced, and the conditions have been rendered favorable to relaxation.

subjects when relaxed to the greatest possible extent, the knee-jerk was usually absent. Our observation that the jerk, at least in some subjects, is absent in extreme relaxation and present but of diminished extent in relaxation not quite so extreme evidently parallels Tuttle's observations that the jerk is absent in deep sleep but is present though diminished in light sleep. It should be emphasized here that our subjects, even when extremely relaxed, were not actually asleep, although in some cases they may be said to have been on the very border of sleep. Our results indicate that *by vol-*

*untary relaxation while still awake, one may attain a degree of
neuromuscular tonus lower than that of light sleep.*

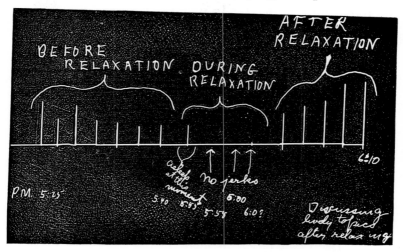

Fig. 40.—(L. D.) Absence of knee-jerk is shown during advanced general relaxation. As will be seen, the reflex was present during light sleep, and the subject
(experienced) was awakened by the stimulus, starting slightly.

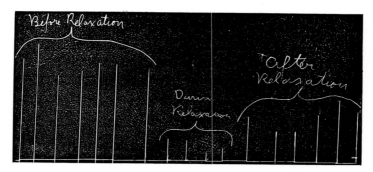

Fig. 41.—(F. S.) Decrease of the knee-jerk occurs during incomplete general
relaxation. Subject experienced.

During advanced relaxation there was at least in some
of the subjects a diminution of the respiratory and heart-
rate.

[131]

Study of individual records discloses various points of interest. A persistent noise, like a crackling steam-radiator in the room, may be relatively unnoticed by the subject and perhaps may nevertheless interfere with complete relaxation. The jerk may be diminished during the noise, as occurs in partial relaxation, and disappear when the radiator is turned off so that the room becomes quiet. In general, noises like whistling or shouts of passing students, or the starting of an automobile outside the window, tended to interfere with re-

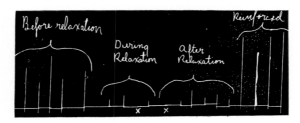

Fig. 42.—(A. J. C., untrained.) Showing decrease of knee-jerk to complete absence (X, X) by relaxation. At X, X the subject "was on the border of sleep but not actually asleep."

laxation. They tend to arouse reflexes which the subject has to relax away, if he is to be successful.

Conclusions from These and Other Investigations

It is important to note that the relaxation used or aimed at in these tests is more than that secured in the usual studies of the knee-jerk. It is not sufficient to have the subject sit down or recline. Under such conditions the knee-jerk will of course be present, as recently reported by Lee and Kleitman (1923).

As judged by the knee-jerk, there is great variation in the ability of individual subjects to sustain a relaxed state even after it has been secured. But objective as well as subjective criteria support the conclusion that the knee-jerk decreases *pari passu* with increasing relaxation. The diminution of the knee-jerk to zero thus furnishes a simple objective criterion of the relative completeness of the relaxation.

Our results may explain the absence of the knee-jerk in a small percentage of healthy persons when no reinforcement is used (1.567 per cent according to Berger, 1879). Some of these individuals at the moment of examination may be too phlegmatic or relaxed for the knee-jerk to appear without reinforcement.

After cutting the efferent nerves to the vasticrureus muscle in the spinal cat, Sherrington (1909) found that no knee-jerk could be elicited. With a similar preparation, Viets (1920) described a method "to decrease its tonus until finally a point was reached where no reflex was obtained. . . . . It was observed that it was not possible to obtain a knee-jerk in an absolutely toneless muscle." These observations are in harmony with our own.

In the preceding experiments, then, the effect of extreme relaxation (without sleep) was tested with seven individuals who, excepting Professor Carlson, had received prolonged training in progressive relaxation, with the following results: (1) The knee-jerk decreases parallel with advanced relaxation. (2) In some subjects a degree of relaxation was attained in which the knee-jerk was abolished without the individual going to sleep. (3) There is usually a "hang-over" of the effects of extreme relaxation, since the jerk tends to remain diminished after the subject is instructed to cease relaxing, and only reinforcement or moving the leg brings the response back to normal. As will be recalled, during the tests on the flexion reflex (chap. viii) a similar hang-over effect of extreme relaxation was reported by the subjects.

The finding that the knee-jerk diminishes with advanced relaxation has been confirmed, according to Tuttle and William (1925), in their investigations where they produced general relaxation by excessive heat due to electric current of high frequency. Since 1925 the writer has had frequent occasion to confirm that the knee-jerk is diminished or absent in extreme general relaxation. Furthermore this has proved to be a useful clinical test as an aid to determine whether a patient during practice at relaxation has reached an advanced stage. Clinicians will readily understand that the absence of the knee-jerk in advanced relaxation is only temporary: in normal individuals, as will be discussed later, it reappears if the patient pulls vigorously and repeatedly, after interlocking the fingers of both hands after the usual manner of Jendrássik; although milder forms of reinforcement may fail. Reappearance of the jerk under such conditions enables us to distinguish between functional absence and organic absence, as in tabes dorsalis, where no degree of reinforcement proves effective.

DIFFERENTIAL RELAXATION DURING READING, WRITING, AND OTHER ACTIVITIES

The purpose of the following studies was to inquire whether activities like reading or writing may go on even if the individual is at the same time so relaxed that tests reveal a diminished knee-jerk, that is, whether moderate activity in some muscle groups nevertheless may permit a certain relaxation in others.[1] Obviously, general relaxation is not here in question, for with completely relaxed arms and eyes and other parts such activities would cease; accordingly, if any degree of relaxation were to be possible under the conditions, it must be "differential," a term defined in chapter vi. Here, then, we need to test whether we can find any foundation for

[1] A description of these investigations has recently been published (Jacobson. 1928).

the assumptions and practices in that chapter concerning differential relaxation.

Methods of Testing Knee-Jerks during Differential Relaxation

These tests were performed with the aid of forty subjects, of whom twenty-one were healthy medical students not trained to relax, five were individuals in a markedly nervous condition likewise not trained to relax and fourteen were former patients who had been trained to relax according to methods described in chapters v and vi.

During the tests on all subjects, the attempt was made to render conditions favorable to relaxation: Walls of Celotex building-board were installed so that sounds in the laboratory room might be diminished. Noises were so far as possible avoided by the experimenter excepting when he gave instructions. Since it was not unusual to find an increase of the knee-jerk appear upon approaching the subject or moving before him, the experimenter also tried to keep out of sight much of the time.

Apparatus served three purposes:

1. *To strike the patellar tendon*, a spring-hammer with a 5-centimeter stroke was used in the tests first made, namely Series 3. The stimulus could be set off at any moment by pulling a cord attached to a lever. In all the other series (1, 2, 4 and 5) a more convenient device was employed—an electromagnetic hammer which delivered a stroke automatically every 30 seconds (Johnson, 1927). To prevent slight slipping, due to its own forces upon impact, the hammer was secured with special clamps. In all of the tests, an attempt was made at the outset so to adjust the hammer and so to gauge the strength of blow that a marked jerk would be produced. I was not certain that the initial jerk was maxi-

mal with any particular subject, but probably in the majority of instances it was nearly maximal.

2. *To record the magnitude of jerk*, records were made on a slowly revolving kymograph, securing linear graphs whose

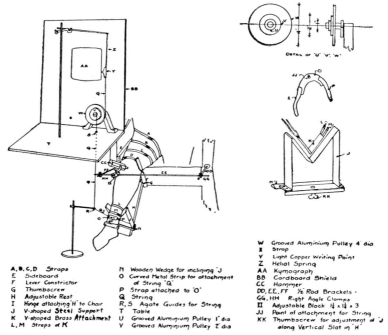

A,B,C,D	Straps	N	Wooden Wedge for inclining 'J'
E	Sideboard	O	Curved Metal Strip for attachment
F	Lever Constrictor		of String 'Q'
G	Thumbscrew	P	Strap attached to 'O'
H	Adjustable Rest	Q	String
I	Hinge attaching 'H' to Chair	R,S	Agate Guides for String
J	V-shaped Steel Support	T	Table
K	V-shaped Brass Attachment	U	Grooved Aluminium Pulley 1" dia
L,M	Straps of K	V	Grooved Aluminium Pulley 2" dia

W	Grooved Aluminium Pulley 4" dia
X	Strap
Y	Light Copper Writing Point
Z	Heliol Spring
AA	Kymograph
BB	Cardboard Shield
CC	Hammer
DD, EE, FF	½" Rod Brackets ·
GG, HH	Right Angle Clamps
II	Adjustable Block 1¼ × 1¼ × 3
JJ	Point of attachment for String
KK	Thumbscrew for adjustment of 'J' along Vertical Slot in 'H'

Fig. 43.—Diagram of apparatus designed for the quantitative registration of the knee-jerk. See text. A detailed description will be found elsewhere (Jacobson, 1928).

lengths in centimeters are considered in the present chapter. In Series 3 (which, as previously said, was the first to be performed) a system of rods and levers was attached to the right leg (Johnson), but in the other series the string-and-pulley system recently described (Jacobson, 1928) was used. The string (Fig. 43) was attached to a curved metal strip firmly strapped a few centimeters above the ankle and led through agate guides to an aluminum triple-wheel pulley. According to the size of the two wheels used for any test, a magnification could be secured of 4 or 2 or 1. From the pulley wheel selected, the string passed vertically to a metal writing-point placed in contact with the revolving drum. A spiral spring attached to a rod above tended to draw the writing-point upward at the moment of knee-jerk. As the leg returned to its former position, the string drew the writing-point back to the corresponding point. Thus a vertical line was secured with each knee-jerk.

Much effort was expended to prevent possible shift of position of the knee during a period of tests, since the hammer striking at a different spot might thereby effect an alteration of extent of jerk: it was feared that the thigh might slip as the individual relaxed. Various measures failed to restrain this part until we finally strapped it firmly to a sideboard attached to the right arm of the morris chair occupied by the subject. In addition, the knee was pressed firmly against the side-board by an adjustable metal lath placed on the left. After each jerk, the leg was brought back to its former resting position by its own weight. These matters have been given in detail in the previous reference.

3. *To record the position of the knee,* testing whether in spite of all precautions there was any slipping, required a second kymograph. Forward deviation of the knee moved the end of the rear arm of a long lever, causing the end of the forearm with its writing-point to draw a vertical line. The dimensions of this lever were such as to yield a magnification somewhat more than sixfold. A second lever was attached at the end of its rear arm to the inner aspect of the thigh a few centimeters behind the patella. This served to record lateral deviation or rotation of the thigh. Its dimensions were such as to yield a magnification somewhat more than

twofold. Accordingly the writing-point of each of these levers made a vertical tracing with each knee-jerk; but thereafter, provided that the knee did not shift, each lever returned to its former position of rest. If no shift took place during the entire period, each lever (apart from the vertical tracings) drew a fairly horizontal line. As standards to test how far, if at all, these two lines deviated from the horizontal, base-lines were drawn with each lever at the beginning of the period, revolving the drum by hand.

The Knee-Jerk during Differential Relaxation

Untrained subjects: Series 1, normal students.—Twenty-one normal students of both sexes were tested during periods of 40–70 minutes. Of 27 test periods, the subject continuously copied from a magazine in 17, wrote numbers to the beats of a metronome in 5, made an ergograph tracing with the right hand in 2, read aloud in 1, conversed actively in another, while in 1 instance no occupation was furnished. During writing, the instruction was given to proceed at a rate which seemed possible to continue for an hour. A more satisfactory occupation was writing numbers from 1 to 99, inclusive, then repeatedly, to the beats of a metronome. This was found to hold the attention better than the other occupations (at least when the rate was made rapid) and permitted us to secure a constant rate of mental work throughout the period, if desired, or to vary the speed at will. We found the speed of writing numbers during a 2-minute period before and after each session with each subject and from this determined the average rate of speed per minute. It was then possible to set the metronome at a rate well below his speed, if desired, so that uniform writing could be secured throughout a session.

No request to relax was made; in fact, as a rule the word

"relaxation" was not even mentioned and the purpose of the investigation remained unknown to the subjects. Taps to the patellar tendon were delivered every 30 seconds throughout the entire period of test without intermission, excepting to eight subjects. For the latter, periods of 5–15 minutes occurred during which they continued at their task but with no stimuli from the hammer. In almost all instances—24 out of 27—the jerk tended to decrease progressively following

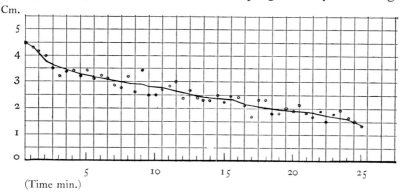

FIG. 44.—Graph shows decline of knee-jerk averaged for thirteen medical students during a 25-minute period reading aloud or writing. Stimuli at 30-second intervals. The curve has been smoothed, but original points are indicated.

the initial strokes. This result was not as a rule influenced by the presence of 5–15-minute intermissions when no stimuli were delivered to the patellar tendon. The only effect after such a period was a still further drop in the extent of the knee-jerk.

The curve of relaxation.—It is of interest to plot the averages for the first 50 jerks of thirteen subjects engaged in reading or writing, as shown in Figure 44. The curve has been twice smoothed by the method of averages, using 5 points at a time, but the original points are indicated. We have what closely resembles an exponential curve, showing the diminution of neuromuscular tension as time proceeds. If the reciprocal of the foregoing 50 values is multiplied by 10 and plotted (Fig. 45), a curve is secured which shows the extent of relaxation. Changes in the character of the curve near $X=0$ and $X=25$ can be neglected, since they are due to inadequacy of the method of smoothing near these points. The

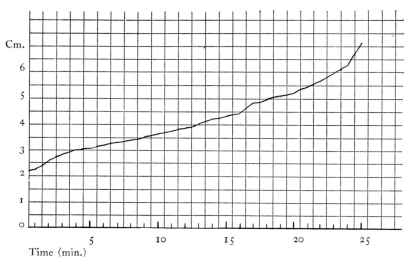

FIG. 45.—The curve of relaxation for the same students. This has been plotted from the reciprocal of the values shown in Fig. 44 multiplied by 10.

curve as a whole may be compared with that of the learning process as noted by Bryan and Harter (1897; cf. also Ruger, 1910; Thorndyke, 1923). Just as we can in a very general way speak of a curve of fatigue (Kronecker, 1871; Mosso, 1888) and of a curve of memory (Ebbinghaus, 1885, 1897), so we can speak of a curve of relaxation. Evidently the characteristics will vary greatly with individuals, their particular state of nervous hypertension or hypotension, the nature of the occupation and of the distractions present, the length of time seated before the first stimulus occurs, the strength of blow and various other factors.

We conclude that normal individuals engaged while seated in physical and mental tasks involving no direct activity of the leg muscles and no emotional excitement tend to become relaxed during the course of an hour as measured by the knee-jerk. To use a phrase common in biological literature, it is a process of "adaptation," whereby the organism, at first using neuromuscular energies not essential to a task, soon omits them, so that more approximately only those energies which are required come to be in use.

Light is thrown on the foregoing rule by the exceptions (3 out of 25). One subject (R. S.) when copying from a magazine or writing numbers to the metronome at a rate well below her greatest speed had shown characteristic diminution of the jerk as discussed above, at times almost to the point of disappearance. At such times she reported that she felt "relaxed" or "drowsy"; one could with care observe a difference in her facial expression and attitude in the direction of greater relaxation, most clearly in her forehead, which was markedly wrinkled at the outset but became smooth later. Accordingly, in order to test whether the apparent relaxation was brought on by the relative ease and unexciting character of the task, during the next period a companion of hers was permitted to be present during the entire period so that she might converse spontaneously and interestedly. When this was done, practically no diminution of the jerk took place. Another test was made while writing numbers with the metronome rate well below her maximum speed, which was attended by diminished knee-jerks: a change to a high metronome rate, stimulating her to speedy writing, produced a change in the direction of heightened knee-jerk. At the new rate, however, after a series of relatively high jerks, the height tended again to decline, although not as a rule to the low level previously attained with the slower writing. A second exception occurred with subject I. L. J., while using the ergograph. Here the diminution of the jerk was but slight. It is readily understood that lifting a weight repeatedly will tend to have a reinforcing effect; accordingly we should anticipate that at least with some individuals little or no diminution of the jerk would take place under such conditions. The third exception occurred with a male medical student engaged in writing. No fall of the jerk took place; indeed, the tracing is distinguished by high jerks that are unusually uniform to the end. Such uniformity is not, in my experience, characteristic of neurotics (see below); and the subject affirmed that he was not of nervous temperament. The explanation doubtless would have to be sought in further study of the habits and reactions of this individual.

Untrained subjects: Series 2; neurotic patients.—Neurotic patients with no training to relax were tested under the same conditions. They, like the foregoing subjects, were not requested to relax, with one exception noted below. No psychosis was present in any of these and no organic pathology, so far as revealed by clinical investigation. Three out of five failed to exhibit a diminished jerk (Fig. 46). This result contrasts with that obtained with most of the normal untrained subjects. Observation of the behavior of these three suggested that they failed to relax, as indeed the knee-jerk test indicated.

Two of the neurotic subjects appeared to relax during the activities. Of one the first 10 jerks averaged 4.41 centimeters on the graph; after a 10-minute interval free from stimulation, the second 10 jerks averaged 3.21 centimeters. After another 10-minute interval free from stimulation, he was requested to relax so far as he knew how. He later reported

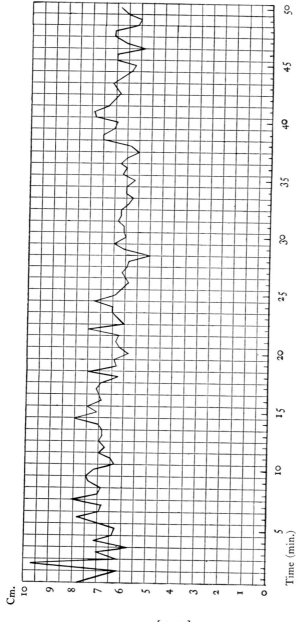

Cm.

Time (min.)

Fig. 46.—Subject J. K., untrained neurotic, reading aloud, July 21, 1927. Stimuli at 30-second intervals. Failure is noted of knee-jerk to diminish, probably indicating absence of relaxation (see text).

[143]

that after a period of experience of hypertension ("I felt as if I would blow up"; "my right leg felt tense"), he tried to let go naturally, became interested in what he was reading and seemed to relax. Fourteen jerks of this period averaged 1.61 centimeters, considerably less than those of his first series. Such diminution was not due to displacement of the hammer at the patellar tendon, for the position of stroke was finally changed four times without affecting the low type of jerk.

The other neurotic individual who appeared to relax complained of marked fears when he tried to recite in his classes and had shown a marked knee-jerk at the time of a general clinical examination; but during the present tests he appeared undisturbed, with a relatively low jerk from the outset, while subsequently there was a diminution.

These studies indicate that even neurotic individuals may relax at certain times, with or without request, although not previously trained to relax. This finding does not conflict with the thesis that as a rule the neurotic disposition is characterized by habitual hypertension or by frequent hypertensive responses not exhibited by normal individuals. The present two neurotic individuals who were able to relax to the extent indicated by the tracings, nevertheless apparently in their daily lives failed to use and apply such ability, as their clinical histories abundantly disclosed. Evidently they could and did relax upon some favorable occasions at least; but they were not habitually relaxed.

Trained subjects: Series 3, 4, 5.—Of fourteen trained subjects, three (C. H., H. S. and E. J.) were normal in their previous nervous history, while eleven had been patients with medical or nervous hypertensive disorders or both; but these patients had been previously discharged from treatment excepting two: M. F., epileptic, still in the earlier stages of treatment; H. A., with previously spastic esophagus and nervous hypertension, recovered except for occasional moderate relapses. All had received training to relax lying down (general relaxation), and all but three (M. F., H. S. and C. H.) had received training to relax in the sitting posture (differential relaxation), but there had been considerable variation in the periods of training and in the apparent skill attained. Excepting two (D. M. and E. J.), no special training to relax while reading had been given.

Throughout the entire period the subject read aloud or copied from a magazine, being instructed to write as quickly as seemed reasonably possible for a prolonged time. The order of record was generally as follows: (1) a set of jerks "before relaxation," occurring irregularly from about $\frac{1}{2}$ to 2 or 3 minutes apart during a period usually about 10 minutes long (rarely as brief as 3 minutes or as long as 28 minutes); (2) an interval (usually 10 or 15 minutes, rarely as brief as 3 or as long as 17 minutes) to relax, during which the occupation was continued but the tendon was not struck; (3) a set of jerks "during relaxation" occurring irregularly from about $\frac{1}{2}$ to 3 or 4 minutes apart (occasionally longer, up to 9 minutes) during a period on the average about 20 minutes, with extremes of 5 to 36; (4) usually but not always an interval of 3 to 5 minutes to cease relaxing, during which the occupation was continued but the tendon not struck; (5) a set of jerks "after relaxation" with conditions similar to those of the preceding period of relaxation. Occasionally the foregoing was followed by another interval of 5 minutes to relax, a second set of jerks "during relaxation," a briefer interval to cease relaxing, and a final set of jerks "after relaxation."

After I had taken most of the records in this series (which was performed before Series 1 and 2), I found in the last nine that the tendon could be tapped regularly every 30 seconds during a test period without making it impossible to relax. It is true that such relatively frequent blows tend to disturb relaxation and for this reason are disadvantageous, tending to destroy the very condition whose presence it is desired to register. But this difficulty is inherent in the knee-jerk as a test of relaxation and therefore I have sought to

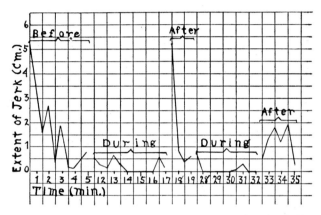

FIG. 47.—Subject H., experienced at relaxation, reading aloud, December 4, 1926. Stimuli at 30-second intervals, with pauses as indicated on ordinates. Graph shows diminution of knee-jerk (often to zero value) during differential relaxation.

develop certain galvanometric tests and have found them a better measure. However, I adopted and used during the remainder of the present work the 30-second interval because of the advantages of uniformity, relative frequency of test, continuity of curve, and the possibility of using an automatic hammer.

Series 4 was performed with five of the above trained subjects, but the occupation was writing numbers to the beats of a metronome as previously described. In this series the taps on the tendon were always 30 seconds apart, and generally no intermission of the taps took place during or preceding the relaxing period. Levers as above described were used in this series, but the graphs so secured will be discussed later along with similar ones with untrained subjects.

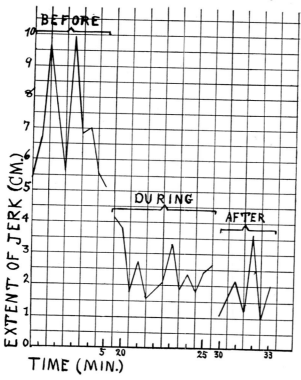

FIG. 48.—Subject I. B., experienced at relaxation, reading aloud, November 21, 1926. Conditions as in Fig. 46. Graph likewise shows diminution of knee-jerk during relaxation.

Typical curves of the knee-jerks before, during, and after relaxation periods for various trained individuals are shown in Figures 47 and 48. As will be seen, the average jerk generally is decreased during relaxation as compared with the preceding period and often also as compared with the succeeding period, but relations are more clearly brought out in

Table VII. Here we are to compare arithmetic mean values
(A.M.) for groups of knee-jerks before, during, and after relaxation. But the successive knee-jerks of groups for any individual often differ greatly from each other in extent, while
the A.M. of two groups compared with each other may differ
much or little. Therefore we need some criterion to determine what difference between A.M. values should be considered significant rather than due to chance. Accordingly the
standard deviations and probable errors are calculated in the
usual manner:

$$\left(\sigma=\sqrt{\frac{\Sigma D^2}{n}};\ \text{P. E.}=\frac{0.67\sigma}{\sqrt{n}}\right).$$

The ratio of the difference between the A.M. values of adjacent groups for each individual (for example, before and
during relaxation, or during and after relaxation) to the sum
of the probable errors of the same groups is shown in the
columns headed "R." Where $R \gtreqless 1$, we can doubtless safely
say that the difference is *significant*. It will be convenient
to use this term in the following paragraph.

We first compare the A.M. value during relaxation with
that of the preceding period where the subject was requested
not to relax: The jerk decreases during the first period of
relaxation in all but 1 of 36 instances; it decreases during the
second period of relaxation in 12 out of 12 instances; likewise
it decreases during the one instance we have of a third period
of relaxation. In sum, a significant decrease of jerk is found
in 48 out of 49 instances of relaxation as compared with
the preceding period where the subject is requested not
to relax.

We now compare the A.M. value *after* each period of relaxation with that of the preceding period of relaxation: In
the first after-period an increase occurs in 19 out of 36 in-

TABLE VII

Arithmetic Mean Values of Knee-Jerk of Trained Subjects During Relaxation Compared with Values Before and After Relaxation

NAME	DATE	OCC.	BEFORE			R	DURING			R	AFTER			R	DURING			R	AFTER		
			A.M.	σ	P.E.		A.M.	σ	P.E.		A.M.	σ	P.E.		A.M.	σ	P.E.		A.M.	σ	P.E.
I. B.	Nov. 21, 1926	r	3.20	1.50	2.80	7.10	0.32	0.45	0.08	0.13	0.34	0.40	0.07								
H. G.	Nov. 16, 1926	r	3.69	0.37	0.09	7.40	1.75	0.94	0.16	0.48	1.63	0.37	0.09								
H. G.	Nov. 20, 1926	r	6.43	0.61	0.09	2.05	3.76	0.14	0.04	0.22	3.78	0.22	0.05								
L. G.	Nov. 29, 1926	r	0.16	0.10	0.02	8.00	0.00	0.00	0.00	6.00	0.06	0.04	0.01								
G. H.	Dec. 4, 1926	r	1.72	1.54	0.32	4.14	0.23	0.22	0.04	2.20	1.77	1.98	0.66	2.21	0.22	0.24	0.06	6.40	1.18	0.59	0.11
G. H.	Dec. 11, 1926	w	1.33	0.92	0.19	6.04	0.36	0.39	0.09	0.63	0.55	0.88	0.21	0.00	0.55	0.58	0.14	1.16	0.26	0.52	0.11
M. F.	Nov. 29, 1926	r	2.25	1.50	0.26	4.84	0.75	0.37	0.50	3.40	1.70	0.92	0.23								
M. F.	Dec. 13, 1926	w	5.25	1.62	0.31	4.35	2.51	1.54	0.32	0.72	2.12	1.24	0.22								
M. F.	Dec. 7, 1926	r	2.32	1.94	0.43	2.67	0.53	1.02	0.24	1.82	1.44	0.97	0.26	0.77	1.06	0.99	0.23	0.58	0.88	0.31	0.08
C. B.	Nov. 17, 1926	r	0.85	0.24	0.06	2.83	0.51	0.31	0.06	1.28	0.74	0.51	0.12								
E. J.	Nov. 13, 1926	r	2.77	0.62	0.17	7.08	1.07	0.43	0.07	0.71	1.27	0.93	0.21								
E. J.	Nov. 14, 1926	r	2.32	0.75	0.20	6.14	0.53	0.35	0.09	1.69	0.31	0.14	0.04								
E. J.	Nov. 14, 1926	r	3.51	0.88	0.22	7.23	1.05	0.77	0.12	1.90	0.67	0.30	0.08								
E. J.	Nov. 16, 1926	r	2.90	1.34	0.37	4.00	1.18	0.39	0.06	0.23	1.21	0.33	0.07								
E. J.	Nov. 19, 1926	r	1.58	1.03	0.24	2.96	0.75	0.40	0.04	3.62	0.46	0.14	0.04								
H. K.	Nov. 26, 1926	r	1.24	0.28	0.07	5.06	0.43	0.51	0.09	0.12	0.41	0.31	0.07								
H. K.	Nov. 20, 1926	r	2.56	0.38	0.03	1.53	1.49	0.41	0.04	0.13	1.43	0.41	0.04								
D. M.	Nov. 27, 1926	r	7.93	1.21	0.38	5.05	5.45	0.96	0.21	3.17	4.12	0.94	0.21								
D. M.	Dec. 3, 1926	r	0.93	0.75	0.13	5.18	0.05	0.19	0.04	1.25	0.00	0.00	0.00								
E. R.	Dec. 8, 1926	w	5.64	1.52	0.32	14.63	0.52	0.14	0.03	2.52	1.20	1.16	0.24	3.70	0.20	0.10	0.02	2.16	0.33	0.20	0.04
E. R.	July 16, 1926	r	3.32	0.24	0.07	8.64	2.11	0.33	0.07	0.50	2.04	0.24	0.07								
S. W.	Dec. 8, 1926	w	10.84	1.07	0.22	9.19	7.90	0.48	0.10	5.30	8.33	0.81	0.17	3.61	7.50	0.28	0.06	5.92	8.27	0.31	0.07

TABLE VII—Continued

Name	Date	Occ.	Before A.M.	σ	P.E.	R	During A.M.	σ	P.E.	R	After A.M.	σ	P.E.	R	During A.M.	σ	P.E.	R	After A.M.	σ	P.E.
S.W.	Dec. 15, 1926	w	13.55	0.46	0.13	11.00	10.58	0.50	0.14	1.08	10.32	0.36	0.10								
B.R.	Nov. 30, 1926	r	3.84	1.65	0.37	3.33	1.91	0.96	0.21	1.50	1.34	0.73	0.17								
B.R.	Dec. 27, 1926	w	5.17	1.82	0.38	5.20	2.36	0.79	0.16	1.93	2.96	0.74	0.15								
B.R.	Dec. 13, 1926	w	5.26	1.42	0.30	2.20	4.07	0.99	0.24	0.70	3.69	1.49	0.30								
B.R.	Dec. 7, 1926	r	2.09	1.43	0.26	2.05	1.25	0.86	0.15	6.07	3.68	0.75	0.25								
B.R.	Jan. 14, 1927	w	3.80	1.05	0.22	1.07	2.69	1.22	0.25	0.58	2.44	0.93	0.20								
M.F.	Jan. 3, 1927	w	3.04	0.87	0.19	2.19	2.23	0.85	0.18	0.31	2.32	0.52	0.11								
H.S.	Jan. 15, 1927	w	1.15	0.65	0.14	8.21	0.00	0.00	0.00	....	0.00	0.00	0.00	0.56	2.78	0.81	0.17	0.07	2.81	1.19	0.25
J.C.	Jan. 15, 1927	w	0.99	0.73	0.15	1.10	1.33	1.15	0.24	3.70	3.29	1.36	0.29	2.15	2.66	0.87	0.18	1.02	2.11	1.20	0.36
B.R.	Jan. 23, 1928	m	2.75	1.58	0.33	2.66	1.39	0.89	0.18	5.04	2.80	0.45	0.10	2.46	2.72	0.63	0.14	1.82	4.36	1.01	0.27
E.J.	Aug. 5, 1927	m	2.92	1.73	0.38	5.29	0.38	0.54	0.10	2.72	1.63	1.76	0.36	2.50	1.82	0.42	0.09	1.67	2.19	0.62	0.13
I.B.	Nov. 1, 1927	m	3.23	1.50	0.28	8.08	0.32	0.45	0.08	0.13	0.34	0.40	0.07	1.21	2.78	0.63	0.13	5.55	5.40	1.08	0.26
B.R.	Oct. 30, 1927	m	1.23	2.14	0.31	1.58	0.52	0.77	0.14	2.56	0.11	0.14	0.02	10.86	0.52	0.52	0.11	1.00	0.74	0.52	0.11
M.F.	Nov. 2, 1927	m	3.96	1.77	0.32	1.88	2.89	1.91	0.25	1.87	2.00	1.15	0.22	1.19	1.39	1.45	0.30	1.59	3.00	3.17	0.71

Abbreviations: Occ. = occupation of subject during tests; r = reading; w = writing copy from a magazine; m = writing numbers regularly to beats of a metronome; A.M. = Arithmetic Mean value; σ = Standard Deviation; P.E. = Probable Error; R = ratio of the difference between the adjacent A.M. values to the sum of the corresponding probable errors. Where $R \geqq 1$, it is presumed to be significant. So calculated, the A.M. jerk during relaxation is diminished as compared with the foregoing period in about 98 per cent of the instances; while after relaxation the jerk increases over that during relaxation in about 42 per cent of the instances.

stances, but only 12 of these are significant; while in the second after-period an increase occurs in 9 out of 12 instances, 8 of which are found to be significant. Accordingly the A.M. jerk after relaxation significantly exceeds that during relaxation in 20 out of 48 instances.

Study of the individual records of the foregoing two series reveals the following items of interest: (1) In the case of J. C., where, contrary to all other instances, the jerk increases during the first period of relaxation as compared with the preceding period, marked discomfort and consequent failure to relax owing to distress from unusually tight strapping was reported. However, during the second period of relaxation, the A.M. jerk is less than that of the preceding or succeeding periods where the subject was requested not to relax. We can readily understand that the individual may fail at times to relax in the presence of distress. My aim was to avoid drawing the straps so tightly as to cause distress, because I did not wish to have this obstacle to relaxation; but in this one instance, I failed. (2) A subjective experience of relaxation (i.e., the absence of sensations from muscular contraction) was often reported by the subjects, agreeing in time with the period of diminished jerks. More precise tests on this point are planned. (3) Comparing records of the knee-jerk secured in different months without any voluntary attempt to relax, there are indications that as relaxation is practiced and becomes habitual during the course of months, the knee-jerk tends to undergo a gradual diminution and becomes more uniform. The records of subject M. F., epileptic, before training were marked by extreme inequality in the extent of jerks, low and very high ones occurring irregularly; that this was due to her emotional state seemed probable from her subjective reports, her occasional sighs, and the frequently contracting and relaxing muscles of her forehead and brow. A year later, when she no longer appeared or re-ported being emotionally disturbed, her tracings show great-er uniformity and some diminution. The jerk of another sub-ject, L. G., although not high in 1925, had become so low in 1927 that it could scarcely be elicited without augmentation. D. M. appeared to give a lower jerk in 1927 than in 1925, and in E. J. very marked diminution was noted in 1927. However, the characteristics of knee-jerk curves of neurotics, we find, vary considerably from day to day. Accordingly, these observations are incomplete and too few in number to permit more than a tentative conclusion. (4) Three subjects (H. S., partially trained; E. F. and E. J., well trained) ordi-narily gave low jerks. One (E. F.) expressed doubt in ad-vance that she would be able to carry out the instruction to read without relaxing, for she stated that for more than 6 months previously she had practiced at relaxing while read-ing and she believed that the combination had become ha-bitual. Another (H. S.) volunteered that for years she had habitually relaxed while reading but also had "gone to sleep that way each night," and she too believed that she had formed an association. To the third (E. J.) it seemed evident that as a rule his leg muscles were highly relaxed. Assuming that these individuals are correct in their autosensory obser-vations, we can readily understand their low jerks as due to the habitual relaxation they report. However, it would be hasty and unwarranted to assume that a low knee-jerk is a necessary and sufficient mark of a habitually relaxed indi-vidual, always indicating the absence of nervous hypertensive symptoms. For in the writer's clinical experience, although high jerks predominate in nervous hypertensive conditions, low jerks are not infrequent, even in the absence of all or-ganic nervous derangement.

Series 5 was performed with five trained subjects for purposes of control. They were requested not to relax through-out the period, an instruction which was occasionally repeat-ed. Nevertheless all reported a tendency to relax at times. One (M. F.) tried to correct this by special attention to the occupation. Another (D. M.) was observed to yawn. The occupation was writing numbers to the beats of the metro-nome, except with one subject (M. F.) who reported that she experienced a relaxing tendency from the repeated beats and preferred to write copy from a magazine. In the cases where the metronome was used, the rate was set well below the speed of each individual, except in one (E. J.). Here we secured almost no jerk with a low rate from the outset and in order to produce a jerk it was necessary to set the metro-nome rate well above his speed. Even then, the subject re-ported a very strong tendency to relax. The subjects did not seek to overcome this tendency by voluntary stiffening of muscles. Likewise, the general conditions maintained were the same as those in the foregoing four series: we sought to avoid noises, conversation and disturbances on the part of the experimenter that might stimulate or excite the subject. In each case the subject was instructed not to relax; but if, nevertheless, he tended to relax, he was permitted to do so. If we except the one record (E. J.) where failure to keep from relaxing was reported, the records are in obvious contrast with those of Series 3: Where the subject was requested not to relax and reported relative success in following the instruc-tion (4 instances out of 5), the knee-jerk was not markedly decreased (Fig. 49). These records of trained subjects re-quested not to relax bear comparison with those of untrained neurotic subjects who in Series 2 failed to relax: they are alike in that the apparent absence of relaxation was attended by failure of the knee-jerk to show the diminution character-istic of periods of relaxation. The results of Series 4, 5 and

6 therefore tend to confirm our conclusions from Series 1 and 2 that differential relaxation produces a diminished jerk.

Graphic tests of position of knee.—In about half of the records with trained subjects and in most of those with untrained subjects (excluding the neurotic, and under quiet

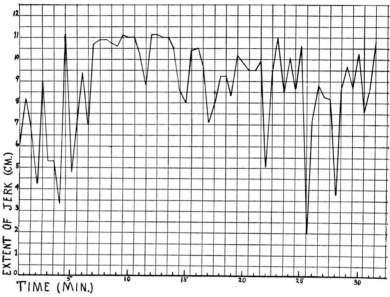

Fig. 49.—Subject same as in Fig. 48, writing numbers, November 28, 1927, but requested *not* to relax. Graph indicates failure of knee-jerk to diminish.

conditions), a general decline in the magnitude of jerk had been observed during the course of the period. This aroused our suspicion that, unknown to the subject, slight shifting of the knee might have occurred, diminishing the knee-jerk when a less sensitive area of the patellar tendon was stimulated. We sought to minimize the effects of any possible forward or backward shifting of the knee by making the unimpeded stroke of the hammer relatively long, namely 66 millimeters. Apparatus to record movement of the knee,

[154]

either forward or lateral or lateral rotation, is described above under "Methods." With four trained and three untrained subjects, records of knee-jerks were taken on one kymograph, while the position of the knee was recorded on another.

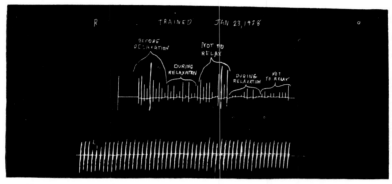

FIG. 50.—Upper tracing: The heights of knee-jerks at 30-second intervals in a trained subject before, during, and after relaxation are indicated by the vertical rises. As will be seen, the jerk is on the average diminished during the periods of relaxation.

Lower tracing: The horizontal straight line (base-line) is drawn before the record is taken in order to have a standard for measurements. Forward shifting of the knee is shown by rises above this base-line, backward shifting by falls below this base-line. (Magnification 6×, photographic reduction 5×). Accordingly, as the series of knee-jerks is taken, the series of vertical lines is made, corresponding in order with the lines in the upper tracing. As the knee and the lever resting against it return to the position of rest between stimuli, it will be seen that what appears to be a second but slightly irregular horizontal line is drawn near the base-line and almost coincides with it. It is evident that the knee each time returns practically to the same position of rest as before, indicating that no noteworthy shifting has taken place.

We wish to test whether the knee-jerk can diminish during a period of relaxation with a trained subject, then increase in height after relaxation, while all the while the resting position of the knee remains practically constant. An instance in point is shown in Figure 50, where, as will be seen, virtually no shifting took place. With untrained subjects, the absence of shifting is illustrated in Figure 51. In all instances no shifting occurred laterally or as lateral rotation;

[155]

and in all instances, with several exceptions, the forward deviation of the knee was zero or approximately less than 1.5 millimeters, which evidently was negligible in comparison with 66 millimeters of possible thrust of the hammer. In the exceptional cases, where, in spite of all strapping, the resting position of the knee shifted forward or back a short distance, such as 2.5 millimeters, examination of the record disclosed the absence of correlation between decline of the

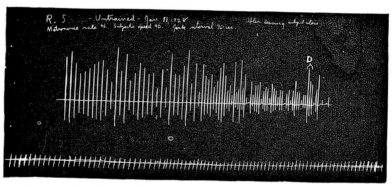

FIG. 51.—Untrained subject relaxing under favorable conditions but with no request to relax. Tracings same as in Fig. 49. Here again evidently little or no shift occurs. At "D" there was a disturbing noise.

jerk and change of position of the knee; that is, if the knee deviated 2.5 millimeters forward for a time, the jerks during this time were not characteristically diminished but were often equal or in some instances increased. Accordingly, we secured a double line of evidence that the diminution of the jerk during our periods of relaxation could not be attributed to shifts of the knee.

Results with reinforcement.—As has been well known since the time of Jendrássik (1883) and other early writers (Mitchell and Lewis, 1886; Lombard, 1887), vigorous voluntary movements tend, if appropriately timed (Bowditch and Warren, 1890), to increase the knee-jerk. Their evidence makes it seem safe to say that the marked contraction of almost any skeletal muscle-group tends to have this effect. But the present investigation discloses that this does not always hold true for the extremely relaxed state. After extreme relaxation has once been induced, not only does it tend to hold over in the subsequent period as we previously found, but moderate contraction of almost any muscle-group (not of the leg where tracings are being taken) may fail to restore the jerk to the extent it had before relaxation: Even Jendrássik's method of pulling apart the two hands with fingers interlocked, when moderately performed, occasionally fails to produce a high kick, although this effect generally occurs if the pulling is vigorously repeated. This method of Jendrássik was used in nine untrained subjects of Series 1. In the majority of these (five) such reinforcement failed to restore the jerk to the original height; in two instances it became greater than the mean of that set. One subject, who was tested on two occasions in this manner, showed a difference of effect of such reinforcement, for on one occasion the jerk considerably failed to be restored to its previous extent, while in the other the opposite condition took place. Likewise other types of reinforcement (clenching the left fist, rotating the left hand, several minutes of general muscular exercise, wrinkling the forehead, deep breathing, conversing in addition to writing) were used, but these as a rule failed to restore the jerk to the original, being less effective than the Jendrássik type of reinforcement; except when reinforcement took the form of voluntary stiffening of the legs or thigh, particularly the right, which seemed a very effective form. In Series 3 and 4 with trained subjects, Jendrássik's reinforcement, which was performed toward the end of the period in 19 instances, restored the jerk to more than the extent present before relaxation in 13 instances, to about equal in 3 instances, while in 3 others the reinforced jerk was considerably less than the original extent. The greater restorative effect of such reinforcement in Series 1 as compared with Series 3 and 4 is perhaps to be explained because, in the latter two series, reinforcement took place not during the relaxed period but at a time when the subject had been directed to cease relaxing. However, since we cannot be certain that this is the correct explanation, the experiment should be repeated with reinforcement occurring during the relaxing period with trained subjects, and the conditions should be made precise so that the muscular work performed in the act of reinforcement could be calculated in ergs.

In harmony with the foregoing findings on the effects of reinforcement during relaxation are some observations of Bowditch and Warren (1890) under conditions where their subjects became drowsy. During this state the jerk of one subject, which without reinforcement normally averaged 83 millimeters, decreased during 18 tests to an average of 17 millimeters. The reinforced jerk during the drowsy state as compared with the wide-awake state was greater than the non-reinforced jerk, but 60–100 per cent smaller than the reinforced jerk.

Evidence That the Present Diminution of Knee-Jerk Is Due to Relaxation

That the diminution of the knee-jerk in trained and in untrained subjects under our conditions was due to muscular relaxation and not to other factors is evidenced by the following considerations:

1. As has previously been shown, the knee-jerk tends to diminish with advanced general relaxation, furnishing us with a possible test for relaxation.

2. Subjects experienced at relaxation reported a diminished sense of tenseness during the periods devoted to relaxation, when the corresponding tracing revealed a diminished knee-jerk. Certain of the untrained subjects likewise spontaneously reported during such periods that they had felt relaxed or drowsy or "had nearly gone to sleep."

3. Both the trained and untrained subjects seemed to the investigator to manifest external signs of relaxation, as shown by a less alert manner, more drowsy appearance, lessened contraction of visible muscles of the forehead, brow and face, and lessened stiffness of posture. In some instances there was a manifest appearance of "not working so hard" as they wrote, or a certain lassitude of voice when reading aloud during differentially relaxed moments.

4. When trained subjects were requested not to relax, and apparently did not relax, the knee-jerk did not show marked diminution, although other conditions were maintained as in previous series where diminution of the jerk had occurred.

5. Diminution of the jerk could be prevented by whatever apparently prevented relaxation, as exciting occupation, increasing the metronome rate so that the subject wrote more rapidly, disturbance by the investigator, and muscular reinforcement.

6. Three neurotic subjects, who evidently did not relax, correspondingly did not show a diminished jerk.

7. The observed diminution of the knee-jerk evidently was not due to slipping of the knee or slipping of the hammer from its original position as shown by various tests:

a) Tests made with levers as above described disclosed no movement or practically no movement under our conditions with the thigh strapped in place.

b) When the jerk had become diminished apparently

through relaxation, change of position of the hammer failed to restore the jerk to its initial height. This was true for both trained and untrained subjects. In some experiments the hammer was thus adjusted from four to ten times in vain.

c) The hammer was prevented from slipping by clamps and adjustments of a particularly firm nature. Movable joints, where not absolutely required, were soldered.

8. Fatigue was not the source, or at any rate not the principal source, of the observed diminution of the jerk:

a) The subjects did not as a rule report fatigue either generally or at the site of stimulation.

b) General fatigue probably did not occur, since reading, which is a lighter occupation than writing, surely could not as a rule cause such a condition within 15–25 minutes. Furthermore, no evidence is found in past investigations which would lead us to expect fatigue arising in consequence of brief periods of mental work to result in diminution of the jerk. We owe to Lombard (1887) our current conception that fatigue can have such a result, but he emphasized that he meant "the amount of fatigue such as would ordinarily occur during the course of the day" (p. 48). He is careful to say, "In our experiments we have not found that short periods of mental work have any effect on the knee-jerk" (p. 45).

c) Diminution of the knee-jerk could be made to occur relatively early in the curve if the request to relax was given at such a time—evidently too early for fatigue to have set in as a rule, either general or local.

d) Fatigue of the sensory end-organ (muscular, according to Sherrington, 1915) certainly was not generally the cause of the observed diminution of jerk in untrained subjects, since in those instances where an interval of 5–15 minutes took place free from stimulation by the hammer, the drop in the height of jerk following this interval was fully as marked as when the hammer continued to strike without interruption.

e) Likewise, fatigue at the site of stimulation evidently was not the responsible factor, since, when the jerk was diminished during relaxation, changes of the position of the hammer failed to restore the jerk, while

f) Reinforcement due to voluntary movements of the subject or the occurrence of noises (exceptions noted above) brought back a strong jerk.

g) In trained subjects who were requested not to relax and who reported success (Series 5), diminution of the jerk did not set in or was not equal to the foregoing.

h) In dogs, after a preliminary diminution, Johnson and Luckhardt (personal communication) have secured knee-jerk curves for periods as long as 17 hours without marked further diminution. Evidently the dog's end-organs do not readily fatigue. While this result may not hold equally for man, it seems unlikely from this that end-organ fatigue in man should set in within a fraction of an hour.

i) With trained subjects, as recounted above, there were 21 out of 49 instances where, subsequent to relaxation, the jerk increased. This would not be expected if fatigue, rather than relaxation, were to be regarded as the cause of the diminution of the knee-jerk.

9. Interference with circulation or other effects of constraint due to strapping were not responsible factors; for when the jerks had become diminished they remained so if the straps were removed, and if the straps were not used at all during the entire period the diminution of the jerk occurred nevertheless.

Luckhardt and Johnson (personal communication) have observed in spinal animals a fall in the jerk owing to causes as yet unanalyzed. It is likely that such falls can be produced by causes other than relaxation, such as (1) diminishing vitality or approaching death of the animal, (2) local injury from blows of the hammer, (3) shifts of position of the hammer or of the patellar tendon, (4) fatigue at some stage, (5) the action of anesthetics or other toxins and (6) possibly other unknown causes. Since manifestly there are numerous possible causes of fall of the knee-jerk, we cannot under any particular set of conditions state the cause unless there is evidence that enables us to discriminate. Under the present conditions with human subjects, evidence was given above that the responsible factor was relaxation.

Lombard recorded his own jerks 2,321 times during 14 successive days. Although he does not mention the fact, his figures show a marked diminution of the average height of jerk per day as the days pass. Thus the average height on the first day was 63 millimeters, while on the fourteenth day it was 35 millimeters. It is difficult to account for this decrease except on the basis of progressive differential relaxation.

We have found that differential relaxation, as tested by the knee-jerk, is produced in trained subjects and exists under favorable conditions in the untrained normal individual. Our observations with untrained neurotic subjects indicate that some of these fail to give what we have regarded as a normal curve of relaxation under our conditions, suggesting that this is a possible test for nervous hypertension or neurosis. We know, however, that sometimes the neurotic individual succeeds in relaxing, giving a normal curve, and that therefore the test, while promising to be useful, is not infallible, but must be considered along with other data.

Conclusions on Differential Relaxation as Tested by the Knee-Jerk

1. In experiments on twenty-one subjects without preliminary training to relax and with no request to relax, it was

found that continuous reading or writing numbers or copying generally permitted sufficient relaxation of the mechanism of the jerk to register marked decrease in its amplitude during the course of the period.

2. Decrease of the jerk tended to be prevented by whatever seemed to have a stimulating effect, including exciting conversation, noises, disturbance by the investigator and vigorous voluntary movement.

3. Decrease of the jerk, even under favorable conditions, fails to occur in some untreated neurotics.

4. Fourteen subjects trained to relax engaged in the same occupations during a control period, a subsequent period of relaxation, and a period after relaxation. During relaxation the jerk almost invariably diminished as compared with the preceding period, and was frequently less (42 per cent of instances) than during the subsequent period.

5. The effects of relaxation generally tended to hold over into the after-period.

6. It is generally possible to prevent the diminution of the knee-jerk in trained subjects during a prolonged test by instructions not to relax.

7. Evidence is presented that the foregoing decrease in the knee-jerk is due to differential relaxation.

CHAPTER X

THE INFLUENCE OF RELAXATION
UPON MENTAL ACTIVITIES

IF PHYSIOLOGIC relaxation were without effect upon the higher centers, it evidently could not reasonably be expected to quiet the nervous system. Yet does muscular relaxation unaided by "suggestion" bear any relation to thinking, emotion, and other so-called mental activities? This is the problem of the present chapter.

In 1922 and 1923 experiments were performed in the Psychological Laboratory of the University of Chicago on the quieting of psychological activities.[1] Clinical records on this subject also had been made with great detail and care and the interests of investigation rather than merely of therapeutics had been paramount. About fifteen qualified patients received prolonged training in psychological observation; their reports were supplemented in 1922 and 1923 with the aid of thirteen subjects in the psychological laboratory and two in the physiological laboratory.

The problem of diminishing cerebral activities may prove of interest to every psychological school of thought. Students might differ in describing the purpose of the present investigations. Clinicians might see it as a search for a sedative for overactivity of the cerebrospinal and autonomic nervous system or for a means to diminish the psychic processes of an individual at a particular moment or over a period of time, including what James might have called the state of

<hr>

[1] The following account in part has previously appeared in the *American Journal of Psychology*, January 31, 1925. Privileges of the Laboratory were very kindly extended by Professor H. A. Carr.

attention and perhaps also the processes of mental imagery and kinesthesis, what Marbe and Ach and Watt might call the thought-processes, what Watson and behaviorists might term subvocal speech, and what everyone knows as the emotions. In all of these various expressions for the whole or some part of experience, there is one point of general agreement: each of these writers conceives the occurrence he names as having essential physiological accompaniments. It might therefore be possible to find among such various physiological accompaniments some universal component which might be diminished, or momentarily eliminated, in order to produce a corresponding psychological effect. Doubtless a study of the data later to be presented will make the problem more clear and open to interpretation in terms of the reader's choice and experience.

Interest may perhaps be added to the present discussion if we first recall some views of various authors that harmonize with our results. Long ago Hughlings Jackson said (quoted by Ribot, 1879), "In the anatomical substratum in which the feeble discharge corresponds to that which we call thinking of an object, there is a motor as well as a sensory element." The present writer is not seeking to prove what psychologists call a "motor" or any other theory of consciousness, nor was Külpe (1895) when he said,

It is important to emphasize . . . . that reproduced sensations are by no means the only aid to recollection. . . . . Very intensive impressions are usually sensed not only by way of the organ to which they are adequate, but by others as well. *Movements are everywhere important.* It is perhaps not too much to say that voluntary recollection never takes place without their assistance. When we think of intense cold, our body is thrown into tremulous movement as in shivering; when we imagine an extent of space, our eyes move as they would in surveying it; when we recall a rhyme, we mark its rise and fall with hand or foot. Most important of all, however, are the movements of speech, which stands in unequivocal relation to the perception of every department.

[165]

Külpe likewise indicates that reflection never takes place without the assistance of movements, for later he adds, "Reflection, psychologically considered, is nothing else than a more or less complicated series of reproductions, associatively originated, and possibly abbreviated by the exclusion of intermediary terms." Subsequent experiments in Külpe's laboratory disclosed to him that reflection is made up not alone of reproductions but also of certain unanalyzed mental elements. Nevertheless Külpe evidently believed that reflection cannot go on without reproduction; and voluntary reproduction, he observed, never takes place without the assistance of movement.

It is but a step from Külpe's views if we should find that the intensive relaxation of movement brings with it a subsidence of voluntary recollection and reflection.

Wundt (1907) recalls that Fechner observed during attention to an outer stimulus in the sense organ—for example, in the ears during hearing, or about the eyes during vision— a slight tenseness (*eine leise Spannung*). Assuming that Fechner was right, it would seem logical, if a means might be discovered to relax the tension, that thereupon the attention to the outer stimulus might be diminished or done away with altogether. Such an assumption would harmonize with the results presently to be described.

It seems unnecessary for the present purposes to discuss the views of Wundt, beyond recalling in a general way that he maintained that feelings of strain and relaxation are always connected with the processes of attention.

A familiar quotation from Binet (1886) also comes to mind at this point:

It is enough to remember that all our perceptions, and in particular the important ones, those of sight and touch, contain as integral elements the movements of our eyes and limbs; and that, if movement is ever an essential factor in our really seeing an object, it must be an equally essential factor when we see the same object in imagination [Ribot].

The observations of Lange and James are well known (see chap. xii). James, for instance, believed that, if in fancy we abstract all bodily symptoms from a strong emotion, there is nothing left behind. First among such symptoms he named "rigidity or relaxation of this or that muscle." However, he lacked an experimental method to test his beliefs. Had it occurred to him to try out the effect of intensive relaxation on the emotions, he would have come upon the method herein described.

In the same vein Washburn writes in her book on *Movement and Mental Imagery*. In advancing her motor theory, she discusses associations such as occur with a series of nonsense syllables or when a man is about to write a check, and she makes what she calls "a crucial assumption for her whole hypothesis,—that there probably are, going on in the muscles, slight actual contractions." This theory is similar to the views of Watson and others that thinking is essentially a play of muscles.

Such assumptions as those of Washburn and Watson are theoretical. In the work to be described, the question is investigated whether thinking may take place where there is extreme muscular relaxation.

INITIAL INSTRUCTIONS

The general method of chapter v was used in these investigations, but it is necessary to specify what directions were given. Each individual was at first merely informed that he was to learn to relax to an extreme degree. Therefore it was useful first to show him what it is to be tense. Lying with eyes closed, he flexed the right or left forearm, while the experimenter held it back in order to make the contraction more marked. The individual then reported whether he noted the sensation of tenseness and where. When successful, it was agreed that this experience was to be called "tenseness." Whenever this same sensation or experience appeared in this or other parts of the body, it was to be called "tenseness." As he contracted the part, it was said to him, "This is you doing! What we wish is the reverse of this—simply not doing." Each subject was frequently reminded that he was to make no effort to relax, for, as he found, making an effort is being tense and therefore is not to relax. As previously said, this is one of the most delicate points of the entire technic. It is quite similar to learning any other physical art, such as voice culture or billiards. In these, also, effort spoils the results, which must be secured in a differentially relaxed fashion.

Since the stimulation of discussion was to be avoided, few words passed between the experimenter and the subject. From time to time, when the investigator returned to the room, instructions were given as tersely as possible. Never telling the subject suggestively what is happening or what will happen, the investigator avoids the use of suggestion. For instance, it is not said, "Now your arm is becoming limp!" or "Your arm will become limp!" Our experience shows that the presence of the investigator often proves disquieting, and that it is necessary to tell the subject not to bother about the other's coming and going, but just to keep on relaxing or to relax all the more when disturbed in this or any other manner.

The subject is not requested to distinguish between the sensation and the image of muscular contraction. Either is to be called "tenseness" or "the experience of muscular contraction." In the present work the well-known "stimulus error" is carefully avoided (chap. xi). Each subject is trained

[168]

to report whether or not he notes the presence of a certain kind of sensation. At no time is he asked, "Is your muscle contracting?" but only "Do you perceive the sensation?" After each successful observation of sensations from contraction of a particular muscle-group, the subject is given a period of 15–45 minutes of practice at relaxation of this group along with other groups that have gone before. This should be kept in mind while reading the following account.

SAMPLE RECORDS

Only two records will be summarized, since they are typical of practically all of the others who learned to relax well:

Subject P. W.

There were sixteen periods of about $1\frac{1}{4}$ hours each, beginning January 8 and ending February 16, 1923. The subject had some knowledge of anatomy. On the first day following the initial instruction as outlined above, practice was devoted to the left biceps alone. At first the subject failed to note tenseness there even when instructed, but finally succeeded, $-2+1$I. He became fairly quiet for a beginning, but the eyelids winked frequently. During the second period, all the muscle-groups of the arms were covered. Tenseness in four muscle-groups of the left arm and in the right biceps was recognized $+1$I, but toward the end of the hour he succeeded upon the first attempt spontaneously with other groups of the right arm. On one occasion when the experimenter entered the room, it became obvious that the subject did not really know what was desired, for he lay upon the couch with eyes open. Thereupon the directions were given *first to make the whole arm stiff, but without flexion or extension; next to let go a little, then on and on, past the point where it seemed to him fully relaxed.* During the third period the muscle-groups of the legs were covered, and in some cases he succeeded in recognizing tenseness +1S. He lay fairly quiet and did not commit the error of opening his eyes. Upon arising at the end of the period, he appeared somewhat drowsy. Period 4 was devoted to the trunk and shoulders. He noted tenseness +1S upon drawing in the muscle-groups of the abdomen; similarly in the intercostal muscles upon inspiration, and in the *erectores spinae* upon bending the trunk backward. When the arms are drawn forward, the pectoral muscles come into play. This was noted −1+1S. Drawing the shoulders backward, he noted tension posteriorly −2+1S. Shrugging the shoulders furnishes the instructor with a test of the subject's ability to localize tenseness, for the average person does not know that the muscles for this act are partly in the neck. Upon shrugging, the present subject noted tenseness −1S+1I. Period 5 was devoted to the neck, for he usually lay with his head held stiffly. He was instructed, "Let all parts go as limply as a rag! When they become so, then whatever you have done up to that point, continue on and on!" Later, when he still held his neck stiffly, it was added, "Let the head go just like a rubber ball!" At the sixth period when he was directed to wrinkle and frown, he noted tenseness +1S. For the first time he now began to appear relaxed. When the investigator raised the arms, they fell limply.

The next three periods were devoted chiefly to the eyes. Since this stage is important, it will be described at length, although repeating some points that have gone before. While the subject lies with eyes closed, he is instructed to close the lids still more tightly. Afterward the usual question is asked, "What took place?" As a rule the correct reply is made that tenseness is felt in the eyelids. Next, with eyes still closed, he looks to the right and then reports. If he mentions a feeling of "tenseness or muscular movement here," indicating the eyeball, it is recorded +1S. This is repeated looking to the left, then up, then down. Later the instruction is given, "Just let the eyes go: do not bother to look in any direction!" The investigator notes the degree of success by the flaccidity and absence of winking of the closed eyelids. After practice at relaxing the eyes in this way has continued for perhaps 30 minutes, it is interrupted and the eyes are opened for the next task. Standing at the foot of the couch about 6 feet away from the eyes, the investigator holds his index fingers on a level but about 3 feet apart. After looking from finger to finger horizontally from side to side, the subject is desired to report what happened. It must be emphasized that care is taken to avoid influencing the subject's observations. He is to give a general report of what took place at a specified moment. Not alone do his accounts often mention elements other than tenseness, but in the advanced work he is not even directed to look for tenseness. For this reason, the subject may fail upon the first exercise of a part to mention tenseness either as present or absent. Precaution is of particular importance in the periods on imagery to be described in the following pages. Throughout the work leading questions are to be shunned, but spontaneous observation is to be cultivated. If he reports muscular sensation of eye movement, the fingers are neared to about one-half of the former distance, when the subject repeats the movement and the observation. This is later done with the fingers held about half an inch apart. Next a single finger is held up while the subject looks at it for a moment or longer. The report is recorded +1S if, for instance, he reports that he saw the finger and notes tenseness in the eyeballs as in convergence. Thereafter the fingers are held one above the other and the subject's eyes move vertically from finger to finger through the three distances.

Finally the lone finger is again held up. After each test the subject reports. At this point it is convenient to give the subject an example of relaxing the eyes. Sitting with eyes open where he can readily be seen, the instructor lets the eyes relax so that they appear not to be looking. Both forehead and brows must also be relaxed, along with the rest of the body, so far as possible, in order to give a clear illustration. The subject may describe the appearance of the director as empty and expressionless (Figs. 27, 28, and 29). After he had witnessed the example, P. W.'s relaxation of his open eyes was visibly improved.

With the seventh period, work was begun with imagery. The subject is not told what he is to look for, but lies with eyes closed and is directed to see in imagination the fingers held horizontally the widest distance apart. As a rule, when this is done, the eyeballs can be seen to move a little beneath the closed lids from side to side. Furthermore, observations by many experienced patients and subjects have concurred that there is a feeling of tenseness in the muscles of the eyeballs, quite similar to that when the movement is made with open eyes, save that it is less intense. Accordingly, if the subject reports a feeling of tenseness in muscular movement, the record is +1S. This took place in the present instance. Leading questions at this stage are particularly to be avoided. The same results were had after imagining the fingers at the middle distance and at the smallest distance apart. From our experience it appears that the tenseness which is momentarily present when the stationery finger is imaged is more difficult to note than that when looking in imagination from finger to finger. At first P. W. was doubtful about it, but finally he reported it present. Following such practice with ocular tensions, the subject who is progressing with the method will lie generally limp, and the eyelids will be strikingly relaxed, without winking. As a rule, we find that those who have been most apt in recognizing slight tensions are the earliest to reach this stage.

The next step requires greater delicacy of observation than any of the preceding. While the subject still lies with eyes closed, the instruction is tersely given: "Imagine a motor car passing." In the present instance he simply reported that he had no definite experience. Comment is avoided by the investigator so as to offer no leading questions. The subject was again requested to imagine a motor car passing, and thereafter to say what happened. Now he reported a very hazy visual image of a motor car moving, along with a feeling of tenseness in the muscles of the eyeballs. Instructions to imagine "a man passing" and later "a flower fluttering in the wind" also brought reports of very hazy visual images with slight tenseness in the eyeballs. Following this was a period of about 15 minutes for ocular and general relaxation. After trying to "imagine a bird flying in the sky," P. W. reported "No experience." Upon repetition he reported a vague visual image with slight tenseness in the eyeballs. "Eiffel Tower in Paris" brought the report of a more distant visual image, along with slight ocular tenseness. Imagining "a tree in the distance" gave similar results.

Period 9 began with review of imagining a "motor car passing." P. W. reported that he did not see the car, that he had no experience. After imagining "a man passing," he stated that there had been a visual image of a man, with slight tenseness of eyes moving to the left. Instructions to think of "a blade of grass" and "a bird in the sky" elicited similar accounts. Here he was asked, "What happens when you relax your eyes?" He replied, "There is no tension, no strain, but a tendency to go to sleep." "What else?" He was silent, and the question was postponed, to be repeated later.

[173]

Various contractions were now performed, each followed by a report of experience and then by a brief interval of relaxation: smiling, pouting, protruding the tongue, retracting the tongue, closing the jaw tightly. A correct report, following protruding of the tongue, is one of the tests whether a subject is really succeeding in localizing tenseness, for otherwise the subject, provided that he does not know anatomy, is not likely to report tenseness both in the tongue and in the floor of the mouth.

During the tenth period, P. W. counted aloud from 1 to 10 with eyes closed to facilitate autosensory observation. This produced an experience of tenseness of the muscles of the tongue, lips, jaws and larynx. Instructed to whisper from 1 to 10, he made the same observations, except that the tensions were less marked. Upon whispering half as loudly as before, there were analogous results. He whispered so that the investigator could just detect slight movement of the tongue and lips. This brought similar reports, except that the tenseness was much diminished. The next request was to repeat the act, but so faintly that it could not possibly be seen. A feeling of tension in the muscles of the tongue, lips and larynx was again reported, although now very faint. Another repetition was requested, but very much more feebly still. He stated that he was able to note this tension only in the larynx and tongue. Following this, the instruction to relax was given, with special mention of the muscles which are engaged during speech.

The next exercises were with auditory imagery. Entering with no previous announcement, the investigator tapped repeatedly but lightly with a pen on the table to the right and then asked if the subject could report what happened. Reply was made that the eyes moved toward the sound. Tapping in front and to the left of the subject brought the same report. In some of these instances it was possible for the investigator to observe the eyes move under the closed eyelids. Following this, the subject was asked to imagine the sound of tapping. He reported that he had a similar but less intense experience—auditory images with tenseness as from ocular movement. Instructed to imagine the sound of a motor car on the Midway, he later reported that he had seen the car, with no auditory image, but with slight ocular tension. He was instructed to imagine the sound of a horn in the distance, but replied that he failed to have an image. Following the direction to imagine a running faucet, he reported "the image of a groaning sound with a feeling of tension in the ear-drum, as well as a visual image of a faucet with a faint tension in the eyeball." Requested to imagine the voice of a certain university instructor, he replied that he heard the peculiar sound of the voice and saw the classroom, but reported no tension. After again imagining the sound of that instructor's voice, he corrected his previous report, saying that there had been slight tension in the eye-muscles.

At the eleventh period, during attempted relaxation, the subject still held his head a little stiffly. To make him aware of this, his head was pushed to the side, and he was urged to let it go a little more. Following an interval of successful relaxation, he was requested in turn to imagine the following things: a motor car passing, a man passing, 13 times 14. In each instance he reported visual images with tension in the muscles of the eyeballs. Next he was requested to think of "the relation of interest to knowledge." He reported "visual images of the terms and feelings of tenseness in the eyes, but no thought of the relationship." The questions were asked, "What is mean deviation?" "What is the relation of idealism to monism?" "What is probability?" Reports on these complex experiences were evidently beyond the capability of this subject. After considerable hesitation, he evidently could not report further than to say, "Tension in the vocal apparatus." Again he was requested to relax completely, including the vocal apparatus.

At the twelfth period his neck still appeared to be held stiffly when the investigator gently pushed the head to and fro. P. W. was instructed to practice at letting it go completely. Asked whether his eyes seemed to him to relax completely, he replied that they did so occasionally. He was then instructed to "let go until there is not the faintest trace of tension."

At the thirteenth period the subject's arms appeared extremely limp. Without any leading question he was asked, "What happens when you relax your eyes completely?" He reported that relaxation seems general and complete for a brief period during which no thought at all is present; that "thoughts occasionally slip by" during the intervals when he is not so relaxed; that with complete relaxation of the vocal apparatus there is no inner speech; while with complete relaxation of the eyes, along with general relaxation, there is no attention to sound. In short, with the approach of general complete relaxation, all imagery ceases.

The remaining periods were devoted to permitting the subject unaided to perfect his technic, letting go further and further, and increasing the period of complete relaxation.

At the fourteenth period or later he reported that the times during which no imagery was present were increasing in length. He failed to go to sleep during the periods. During the final meeting, the investigator noted that he appeared well relaxed but not to the utmost. The eyelids did not have the perfectly toneless appearance that distinguishes complete relaxation in the most successful subjects.

Subject B

Subject B., graduate in psychology, was present at about twenty 1¼-hour periods, ending March 7, 1922. From the outset tenseness usually was recorded +1S. Likewise at each stage her progress at learning to recognize tensions and at relaxation was excellent. She practiced considerably at home but was somewhat late in learning to distinguish between strain and tenseness. Only portions of her record will be quoted. When the arms were extended forward, contracting the pectoral group, she stated that she felt tenseness behind and strain in front. The method of correcting this error is to have the subject relax, while the director pulls the limp arm forward, thus arousing strain in the interscapular area. In the present instance, the subject then spontaneously noted that this sensation was not the same as tenseness. Apparently her localizing sense was developing, for, after shrugging the shoulders, she localized the tenseness +1S, and similarly after movements of the eyes, open and closed.

Her observations on visual imagination were clear-cut: When staring at the finger in imagination, there was "a slight feeling of contraction, not so much as when open, along with a visual image of the finger." Requested to imagine "a man moving from right to left," she had a visual image, along with a sensation from ocular movement left to right. Upon imagining "a motor car passing," "a boat passing on the lake," "a street car moving," "a man standing still," she had visual images, with sensations as from contractions of the ocular muscles. Her first account of imagining the Eiffel Tower in Paris was of a visual image, "but no contraction, for I can see it so far off." After visualizing the figure of a man nearby and then far off, she said of the latter experience, "I hardly can describe it. There is not quite the same contraction as before; so slight that I can scarcely think of it as

[177]

contracting." Repetitions with Eiffel Tower made her feel certain that she did not note contraction in these instances.

When a subject fails to note tenseness, which the investigator suspects to be present, the latter must carefully avoid prejudicing the subject's judgment, for observation should be spontaneous. After such failure, it is generally better to have a period of general relaxation, and then to return to the exercises. Accordingly the subject relaxed at this point and reported that she almost slept. At first there were many visual images, but they gradually disappeared as she became quieter. She would feel a contraction and would then try to get rid of it. Contraction of the eyelids often left together with the frown. On the previous occasion she had found that visual imagery interfered with relaxation, for "the eyes shot around toward the things that were thought of." Imagining "a ball falling" was followed by the report, "downward motion of the eyes, with visual image of the ball." Imagining "a blade of grass" brought a similar account. However, upon imagining "a green ball in the distance," she noted a visual image with practically no contraction.

While the subject was an excellent observer, she still remained doubtful whether tenseness is present in imaging *distant* visual objects. Accordingly she was requested to imagine the green ball again in the distance, but off to the left. A visual image was noted, but this time along with a sensation of slight contraction; similarly when off to the right. Upon imagining it in the center of the field, she observed the image of the ball with a very faint contraction. Images of Eiffel Tower off to the right and then to the left seemed to be accompanied with slight contraction, as in the eyes. She said that the contraction seemed less than in the case of the ball, for the tower was farther off. When the tower was imaged in the center of the field, she found it difficult to say whether tenseness was present. At the next period she reported that during successful relaxation visual imagery disappeared. Her experience was that she seemed more likely to go to sleep when the eyes relaxed. During the period she slept. Review was given of looking from finger to finger with eyes open, and later again in imagination. Further practice was devoted to imagining a rubber ball about 10 feet away, and a building in the distance. In each instance she noted a visual image, with an experience as from a very slight muscular contraction. Later she was requested to imagine London Bridge and reported a visual image, but practically no experience of contraction, or at least so little that she remained in doubt. At this stage she was not certain but that there was some contraction, which she believed she ought to call the faintest contraction that she could distinguish. She was now requested to imagine London Bridge, but at the same time to relax completely and to see if "the visual image" remained or disappeared. Under these conditions, with advancing relaxation, she found that the visual image disappeared. Likewise an image of Eiffel Tower seemed to behave in the same way. Finally she observed very slight muscular tensions present upon imagining even very distant objects. In distant vision, as is well known, the ocular muscles are relaxed, in the ordinary sense of the word. If this observer is correct in her final report that slight tenseness was present during visual imagination of distant objects, this may perhaps be due to slight shifting of focus during that act, or to the tendency to visualize the objects out of the central line of vision.

During the period of February 15, the behavior of involuntary visual images was studied. The instruction was, "Do not seek any visual images, but if any should come, report!" The following are her observations: "First an image of the director, accompanied by a slight contraction of the ocular muscles. Second, a vague visual image of dropping water, with very slight contraction of the eye muscles. Third, an image of a scene out of the window, along with slight contraction. Fourth, the lake at Jackson Park, with almost no contraction. Fifth, a certain motor car, with very slight contraction." At the time the investigator entered the room, she said, there were no visual images, and her eyes seemed perfectly relaxed. Periods from February 15 to 21 were given largely to the face, lips, tongue and speech, with success usually +1S. On February 21 the question was asked, "What takes place when you relax fully?" She replied, "I lose visual and vocal imagery." "What do you have left?" "Nothing. As a rule I go to sleep."

The period of March 7 was devoted to observation with eyes open. She was instructed to relax her eyes and to imagine a man walking by. She reported that the eyes did not remain relaxed! They tended to move in the direction the man was walking. Instructed to imagine a motionless motor car, she described a very slight downward tendency of eye movements. "The motor car was seen out there" (indicating). After imagining a tree in the distance, she recounted, "There was no perceptible motion. The eyes seemed, however, as if held still, with a slight tenseness." Another task was to report when anything occurred spontaneously to the visual imagination. Her statement was that when her attention turned to any visualized object, the eyes seemed to tend to focus there. As a rule, when her eyes are open, she does not spontaneously have visual images. Requested to imagine a certain graduate student, she reported a slight tendency toward convergence along with a visual image of the individual in front of her. When she imagined a person walking upstairs, she noted a slight tendency to move the eyes up. She visualized a man walking by, whistling, and reported with it a tendency to move her eyes toward the side on which he was going. To summarize her observations, she said that when visual images appeared, there was a slight tendency away from relaxation and toward convergence. This was more noticeable when she was instructed to have a visual image than when the image came spontaneously. She found the tenseness also more noticeable if the visual image was projected than if it was not.

To recapitulate: This subject learned to relax excellently. She found that imagery, visual and vocal, both voluntary and involuntary, was accompanied by sensations of tenseness. This was true of visual imagery with the eyes both closed and open. Most difficult of all to note was the faint tenseness present upon visually imagining an object in the distance. Visual and other images disappeared upon progressive relaxation of the corresponding tension.

SUMMARY OF EVIDENCE THAT MENTAL ACTIVITIES
DIMINISH AS MUSCULAR RELAXATION
PROGRESSES

The foregoing records are, for the interests of the present discussion, typical of all the other subjects or patients who learned to relax very well. Four subjects are not considered further, since one failed to attend, another did not seem earnest in following instructions, the third was evidently in too excited a condition to make the necessary progress in the required time, while the fourth failed to attain control of muscular relaxation and the finer skill in observation that was desired. This leaves eleven laboratory subjects and ten highly experienced patients who attained the requisite skill in their observations in connection with imagery. Seven of the laboratory subjects seemed to have clear visual imagery and were of the visual-motor type. Four, including P. W., were less clear in their visual images, inclining more to the motor type. One of these at first seemed lacking in visual imagery but later reported some very faint forms.

The earliest observations of each subject, of course, may not be counted. *It is repetition that makes the sensation of muscular contraction easy to observe.* At first the individual who contracts his biceps may scarcely be confident of the sensation he gets from this large muscle, but with experience it becomes easy. A more advanced task is to recognize the muscular sensation from a slight movement or from slight convergence of the eyes. Here again practice improves, until the experienced subject reports the tenseness every time. When the stage of visual imagery is reached, the subject may here also at first fail to report the experience of muscular tenseness. It must be remembered that the director, according to the methods of the experiment, does not ask the direct question, "Did you note tenseness?" but only requires the subject to report what happened. There seem to be four sources of omission with the inexperienced subject: (1) Omission through lack of training. The subject admits that he does not know how to recognize the sensation of tenseness. (2) The visual image is observed, but the attendant fainter tenseness is ignored. (3) The subject wrongly identifies tenseness with movement, and fails to report static tensions in convergence and other acts. (4) Tenseness is observed but not reported; error of the incomplete report. However, with repetition of the imagery, every subject finally reported the tenseness in a spontaneous way.

Our conclusions must be subject to certain reservations: (1) The agreement of our observers that the experience of muscular contraction accompanies imagery does not permit us to infer that physiological contraction takes place, unless there is additional evidence by objective methods (*vide infra*).

(2) There is a possibility of error on the part of our observers. For they may conceivably have developed a prejudice or habit of reporting tenseness. However, it does not seem likely that there was such error: (*a*) Our observers were mostly of a critical type of mind. (*b*) The investigator took precautions to avoid leading questions or hints, and the observer did not know what matter was under investigation. (*c*) The subjects who have shown the greatest skill in relaxation and in their reports on the locality of contractions do not require the same repetition as the others. Upon the first occasion, they generally report tenseness during imagination. (*d*) Subjects whose technic is still a little lacking in relaxation of slight eye movements, as externally manifested, do not report that imagery has vanished. (*e*) Every skilled subject was requested to relax the muscles of the eyes, forehead and brow extremely, but at the same time to have visual images. Some reported failure with chagrin, and tried it over again. Everyone found the combination impossible. (*f*) I attempted to rule out the error in question by using a relatively large number of observers, independently of one another, covering a period of years. Their observations agreed.

There seem to be additional grounds for believing that the experience of contraction accompanies imagery. The profoundly relaxed individual has a characteristic appearance (chap. v); does not look or act like an imagining or thinking person; shows no movements of the eyeballs, however slight, if the lids are closed; later reports that he was not mentally active; gives no evidence of the results of thought-processes; and often goes to sleep. Preliminary tests of action currents during experiments now under way lend further objective support to the foregoing conclusion.

Relaxation of the residual tension of the small muscles of the sense organs is a delicate matter that requires practice.

All observers agreed that, when this was done, mental imagery dwindles or ceases in a corresponding way. The face of the individual who is so relaxed makes a lasting impression on the beholder; his eyes, if open, are vacant in appearance and practically motionless; they seem to be "not looking." His countenance is expressionless. One student, watching the investigator illustrate such relaxation, stated that it looked "as if the individual could not possibly be thinking." The thesis that progressive relaxation brings with it absence of thinking is apparent, literally, on the face of it.

This thesis also harmonizes with the experience of all the subjects and patients who considered that it was impossible to be relaxed extremely and to have images at the same time. With the advent of the one condition, the other invariably ceased.

If it is true that the experience of contraction conditions imagery, attention and thought-process, we might expect to find some confirmatory evidences in the work of previous investigators. This expectation is justified. Peters (1905) studied the process of attention when stimuli acted simultaneously on two different sense organs. If a sound was given with a light, the former was perceived later. Such alteration of time perception failed to take place if attention was inactive. Peters concluded from his investigations that there are muscular contractions, like ocular accommodation, which are necessary in order to produce clear sensory effects; to omit such contractions reduces the attention. Moore (1903) observed that "sensations from eye-movements play a predominant rôle in the control of the memory image." Meakin (1903) concluded that "the stability of an image, or internal sensation, depends on the activity of its motor accompaniments or conditions." Stricker (1880, 1882) comments on the importance of feelings in the eye muscles for recollection: "If I wish to represent to myself the flight of clouds, I must connect with the image of the clouds the feeling as if the eyes would follow them." Slaughter (1902) concludes, in part, "(*a*) that the factors which keep visual images in clear consciousness are their own internal organization combined closely with motor elements; (*b*) that auditory images appear only in connection with an organized associate situation, in which motor elements usually play a predominant part." Kuhlmann (1906) noted eye movements in connection with the development of visual images but did not commit himself to generalization. Murray (1906) attacks Meakin on points which do not interest us here, but her findings permit her to agree "that this localization of images is correlated with the presence of motor elements, actual or ideated, has abundant evidence." She particularly emphasizes the importance of visual fixation for reproduction; in our terms this would be the experience of "static contraction."

Perky (1910), by an ingenious test with lights, found that in his experiments memory images generally involved actual eye movements, while imagination generally involved none. This was true of auditory as well as visual experiences. He believed that imagination also involves an element of contraction, however, for he states that there is steady fixation, but lack of general kinesthesis.

Certain other investigations mentioned in chapter vii (Jacobson, 1912) bear in the same general direction. Sound and odor sensations were produced simultaneously; and the subjects all reported independently that during strong voluntary attention to the one or the other sensation, there was a pronounced motor element.

During an investigation of another matter, the nature of perception (Jacobson, 1911), a motor element in this process also was brought to light: Printed letters or words or sentences were exposed for 1–2 seconds to the vision of trained subjects. They were directed to stare at the object and later to report in full what had taken place in temporal order. Each subject reported that as he stared the sight of the print persisted but the perception or meaning of what was seen seemed to come and go. At the instant when he perceived the print, getting the meaning of it, something more took place than the mere visual experience; for instance, he then repeated the letter or word to himself or had a visual image of it, or in imagination pointed to it, perhaps with a faint but apparently actual contraction of a finger. Some such process, occurring in addition to what was seen, was apparently requisite for perceiving or getting the meaning; when the one disappeared, the other went with it, leaving but the blank print. My experiments were not sufficiently numerous to warrant the generalization that all perception depends upon such motor elements, supervening upon what is seen or otherwise sensed, but this seems likely. These investigations were in part repeated by Wheeler in 1923 and by Shimberg in 1924, and the results were confirmed.

Thorson (1925) has sought to record movements of the tongue during internal speech with Lashley's modification of Sommer's movement-analyzer. Under these conditions she observed that the tongue did not hold still, so that a continuous record of slight movement was obtained. Because she found no constant patterns during successive repetitions of the same word, she concluded that movements of the tongue are not an essential element in internal speech. However, the insufficient delicacy of the apparatus and the indirect character of her reasoning leaves the issue unsettled.

With the preceding background of various investigations, we find, therefore, the experience of muscular tenseness a *sine qua non* of imagery, attention and thought-process. This is not so strange as at first it seems; for everyone will admit that, when a sense-organ—for instance, the eyes—are active, the muscles that control them also are active, and the sensation from the controlling muscles evidently plays a useful rôle in vision. Accordingly it seems a matter of course that when the original visual experience is repeated, taking the form of imagery, the muscular experience that goes with it also is repeated. In this way the individual doubtless at least in some measure *controls* his images and thought-processes. The cultivation of this factor has been used in clinical work for the control of undesirable mental activity or over-activity.

Résumé

In the foregoing experiments on the relaxation of mental activity:

1. The subject was trained to continue the process of relaxation of skeletal muscles, or *negative* of contraction, to an extreme degree.

2. The muscles of the ocular regions and of the speech apparatus received special prolonged practice.

3. The director judged the relaxation of the subject by certain objective signs.

4. The subject judged the relaxation by the absence of the sense of muscular tenseness. Particular training was given with the chief muscle-groups of the entire body in order. Observation of this sense was gradually cultivated, being most difficult to observe in the small muscles of the sense organs, especially in their slight and fleeting variations.

5. The subjects made their observations independently and unknown to each other, in order to avoid prejudice.

6. For the same reason, the director carefully avoided suggesting to the subject what he was to look for. The subject did not know in advance that we were investigating

whether imagery diminishes with advancing relaxation of muscles. The only instruction given was to relax certain muscles and to report.

7. All the subjects and patients who attained high skill in progressive relaxation spontaneously arrived at, and agreed in, their conclusions regarding psychological activities. With visual imagery there is a sense as from tenseness in the muscles of the ocular region. Without such faint tenseness, the image fails to appear. With complete ocular relaxation, the image disappears. This may be done by individuals of greatest skill and experience, not alone lying down but also sitting up with eyes open.

Motor or kinesthetic imagery likewise may be relaxed away. "Inner speech," for instance, ceases with progressive relaxation of the muscles of the lips, tongue, larynx and throat.

Auditory imagery also is attended by a sense of tenseness, sometimes perhaps felt in the auditory apparatus, but characteristically in the ocular muscles. The individual tends to look toward the imaged source of sound. With the relaxation of such looking or other tension, the auditory image is absent.

Progressive relaxation is not, as a rule, perfect or complete save perhaps for brief periods of time. It is during such brief periods that imagery seems altogether absent. However, when the relaxation of the muscles of the sense organs seems to approach completeness, there takes place the diminution of image-processes. It appears that natural sleep ensues after the imageless state is maintained for a relatively prolonged time.

With progressive muscular relaxation—not alone imagery, but also attention—recollection, thought-processes and emotion gradually diminish.

RELAXATION AND MENTAL ACTIVITIES

From 1929, when the foregoing chapter was published, until 1937, reports have been secured more or less independently from about one hundred additional patients. The conclusions stated above have to this extent been repeatedly confirmed. However, for many years I had hoped that it might some day become possible to test the validity of the reports by objective methods. If neuromuscular tensions are indeed present, as I seemed to experience as far back as 1908 and as described subsequently by many subjects, records of action-potentials should be secured, provided that an instrument could be constructed sufficiently sensitive and stable to detect them. The aim is no less than the electrophysiological measurement of conscious activities. Possibly this might provide one basis for an experimental psychology.

First attempts as early as 1922 proved promising but not in themselves successful. In 1927 suitable apparatus was finally developed The results will be recounted in chapter xvii.

CHAPTER XI
PSYCHOLOGICAL FACTORS RELATED
TO RELAXATION

IN APPLYING the method of relaxation to medical practice, there are times when the physician needs to employ something more than the type of "psychology" common in present-day medical journals. He should know something about the nature of thinking and other mental processes so far as investigated to date and he should know something about the laboratory method of approach. It is perhaps fitting here to recall some topics in the history of experimental psychology and some methods which lend background to our present discussions. We may touch upon what have been called "association" and "behaviorism," for the method of relaxation, as seen in the preceding chapter, throws some light upon these phenomena.

The day of the old-time neurologist who had no interest in laboratory psychology is beginning to pass. Preparation for work on relaxation will be aided by courses in experimental psychology, including the thought-processes; while courses in muscular anatomy and on nervous and sense physiology are prerequisite.

HISTORICAL

Psychologists as well as physiologists have studied the sensations of sight, hearing, taste and smell and have distinguished between these and kinesthetic (muscle and joint) sensations (see p. 36). It has been previously mentioned that muscle sensations generally are relatively faint and difficult for the beginner to note and distinguish. To aid him, the eyes are generally closed and a quiet place is selected. The sensation of touch, as from contact of clothes, must likewise be discriminated. If contraction is maintained, a local sensation of fatigue sets in, which also differs from the sensation of tenseness. Sensations from tickle, temperature, pain, vibration, the gastrointestinal tract and various vague experiences from other viscera, as well as the glands and blood-vessels, scarcely need more than mention in this connection. The patient will be aided by very brief practice in distinguishing these various types of sensation, but this is not indispensable and is often omitted.

The importance of sensations in mental activity was emphasized by Locke (1690) and Hume (1740). They sought to explain all mental phenomena by laws governing the succession of ideas, which they called "association." Hume concluded that the principles of such connection are resemblance, contiguity in time or place, and causality. This conception was further developed and modified by Hartley (1749), Mill (1829), Bain (1855) and others. Hume distinguished between "sensations" or "first impressions" and "faint images" which subsequently arise in thinking and reasoning. To the present day we use this distinction: we call it "sensation" when there is a stimulus and a sensory experience with afferent nervous discharge, but "image" when that experience recurs, usually in a fainter and modified form, in the absence of a similar stimulus.

The views of these early psychologists were severely criticized by their successors in two respects: (1) It was pointed out that logical connections between ideas, such as their mutual resemblance, cannot explain their occurrence as thought-processes (Wundt, 1894) and that sensations are not disparate things which can be compounded like chemical elements (James, 1890). (2) Experiments and subsequent reasoning made clear that the principles of association fail to account for all the phenomena (Wundt, 1894; Watt, 1905; Ach, 1905 and many others). In particular, they provide no adequate explanation of original reasoning and of voluntary activity.

A new development of method for the examination of mental processes was begun by Marbe in 1901 and Binet in 1903. Observers under set conditions were requested to report upon their experiences in connection with intellectual processes such as passing judgment and remembering. Further developments of method were made by Watt (1905), Ach (1905) and other German investigators, and in this country by Woodworth, various Cornell investigators and others. They and their followers have made detailed descriptive studies of attention, pleasure and the disagreeable, conscious attitudes, the so-called *Aufgabe* and other experiences.

It is not so much the discoveries and conclusions of these investigators as their methods which interest us here. We desire to have our subjects or patients learn to observe certain features of their thought-processes, reflections, worries and emotions during their occurrence. Most psychologists agree that the only way to learn the inner phase of mental activities is from subjective reports made subsequent to those experiences. As is well known in sciences on man, it is difficult to attain through subjective reports knowledge of scientific worth. In order to overcome this difficulty, various psychological laboratories have sought to train observers, much after the manner that histologists learn to use microscopes. When this has been properly carried out, each individual has learned under conditions made precise with appropriate apparatus to give as accurate an account as possible of his sensory and other psychological experiences in their temporal order. The beginner generally fails at this at first: he relates what he has been thinking of in lay fashion, pauses, hesitates, omits and speculates in place of giving the terse descriptions of processes which are desired. Where the reports of trained observers working independently during an experiment have agreed, the experimenter has felt justified in assigning to such observations a certain measure of scientific probability.

I shall use the term *autosensory examination* where an individual is accomplished in the method of observing his sensory and imaginal experiences. Such training of a student ordinarily requires months or years. He should make himself familiar with the observation and description of practically all his sensations and their recall in image form. The instructor needs to know anatomy, in order to lead the student through the various muscular experiences.

Psychologists of certain laboratories have in past years examined thought-processes and other intellectual activities and have reported that these always involve experiences of sensation and images, including sensations from muscles Such experiences are commonly ignored by the layman, who can tell you, if you ask him, what he was just now thinking of, but cannot describe just how he did this thinking. It is convenient to use the term *meaning* to designate the significance which a mental act bears for the subject, and *process* to apply to the manner of its occurrence as he can observe it. That laymen and even psychologists tend to overlook their psychological mechanisms (processes) was strikingly illustrated by the discovery of Galton in 1883. He found that the scientists of his university, although accustomed daily to state the meaning of their thoughts to one another in their respective vocations, were nevertheless unprepared to tell him in what types of mental imagery these same thoughts were usually couched—whether visual, auditory or motor. When observers are highly trained in autosensory examination, it has been found (Jacobson, 1911; and others) that the wealth of detail of even very brief experiences beggars verbal description. This is not strange by comparison with any visual scene: a complete description requires a photograph and is impossible in words. However, the principal points can be stated. Such accounts may describe the processes and state in parentheses the corresponding meanings. A subject who writes a full account for even so brief a period as 2 seconds may use several sheets of paper. Complete descriptions, which were advocated by Külpe and Baird, are now seldom attempted.

THEORY OF THE ACTIVITY OF THINKING

If thought-mechanisms are so readily overlooked, it is not strange that some psychologists failed to identify them and therefore asserted that thinking can take place without imagery; or as we should say, they asserted that meanings occur unaccompanied by conscious processes. It seems that these psychologists lacked the prolonged and detailed training and perhaps also the innate skill needed for the detection of very faint and brief experiences. About this point controversy waged for a decade, but few adherents of the foregoing view now remain. In the experience of the writer and his subjects, there has been no such thing as an "imageless thought-element." Meanings when present have been carried by sensations, images, attitudes, postures and situations, according to their patterns and combinations; somewhat as the meanings of a book are carried by the printed words or the meanings of conversations are carried by the articulated sounds. My results on these matters, as previously said, have been confirmed by subsequent workers.

The activity of thinking, then, to the best of our present knowledge consists physiologically (chiefly or entirely) of a series of sensations, images and neuromuscular (including striated and smooth) tensions. These processes may be described by the subject who experiences them in terms of sensations, images, and perhaps feelings; for the tensions become known to him through proprioceptive impulses. However, in place of this, the subject may state the meaning of his thought. This is what we are accustomed to do in everyday life, when we are not acting as psychologists or physiologists. Analogously, we read a book for its meanings; but the typesetter and publisher may consider each sentence only in regard to the printed characters and their relations. Thinking may be aptly described as a series of acts, quite similar to the organism's overt reactions to the environment but differing in that images more or less replace the direct sensory experiences, while the muscular tensions consist of relatively feeble contractions, with little or no effect upon the environment. From this point of view, the reactions in thinking are abbreviated, more-or-less-adequate reproductions of past experience or combinations of reactions to imaged new experience. Accordingly a function can be found for any thought, which might be expected since physiologists find a function for every reflex (cf. Sherrington, 1915).

The student of relaxation will not go into the philosophic nature of thought as discussed, for instance, by Dewey (1910). Likewise he need not concern himself with psychologic studies such as that of Heidbreder ("An Experimental Study of Thinking," 1924), who defines thinking as "denoting any activity which by some means other than overt motor reaction produces responses which are adequate, novel and invented." We are not now interested in such distinctions, and the responses produced may be inadequate, old and imitative, yet fall under our sense of the term "thinking." Any reader who would in this or other ways restrict his sense of the term would perhaps do better to substitute the word "reflection," in the sense of Külpe, in the foregoing discussion. For, as will be evident, our needs in the present methods of investigation and treatment are largely satisfied by gross physiological descriptions of reflection, so far as known up to date.

THE STIMULUS ERROR

It seems appropriate to remind the reader of what has been called the "stimulus error." Martin and Müller pointed out (1899) that the reports of their subjects engaged in comparing weights differed according as they referred to the weights themselves or to the sensations aroused by these stimuli. Since their time, psychologists have generally trained their subjects to make reports refer to subjective sensations rather than to the objects or stimuli presented. This distinction, often insisted upon by Titchener (1909), has recently aroused discussion (Boring, 1920); and its importance has been experimentally confirmed by Fernberger (1921). It is twice of interest in the practice of relaxation. First, the subject is to confine his report to the sensation of muscular contraction; he is not called upon to say whether the muscle is actually contracting, or to palpate it in order to determine its physiologic state. Such determinations are the business of the physician or the experimenter alone. Second, when describing what happens during the process of thinking, the subject is trained to make his report primarily in terms of what is in this volume called autosensory examination: he should report first the processes experienced and second what he is thinking about. If the latter alone is reported, he fails to distinguish the process of tenseness and therefore does not know what and where to relax.

PSYCHOLOGICAL FACTORS IN RELAXATION

BEHAVIORISM

A few words seem pertinent as to the modern movement led by Watson and known as "behaviorism." Despairing of the reliability of introspective methods, Watson and his followers have sought to confine themselves to the study of the behavior of animals and human beings. Their studies, in so far as they have led to positive observations, are of obvious value and of direct interest to the student of relaxation. Indeed, certain work on relaxation (Jacobson, 1911) may be counted as among early laboratory studies of human behavior. But their condemnation of what is here called auto-sensory observation even where carefully and critically employed as useless and unreliable is not sustained by the results of many preceding and subsequent investigators. It must be admitted that the latter method, even in the hands of the most skilful, brings results that bear woeful comparison with the accurate determinations familiar in chemistry and physics. Nevertheless such results, if carefully secured and critically checked, may have value as scientific evidence; they offer us the only possible route to secure records of sensory experience as examined by the subject; in some instances they suggest the presence of phenomena which can then later be tested and confirmed by objective methods.

SENSATIONS AS PSYCHOLOGICAL ELEMENTS

Much confusion seems to have arisen in past literature through attempting to analyze psychological experience into so-called sensational and other elements. Psychologists often have arranged their experiments on the assumption that the mind is a sort of mosaic of elements which could be dissected in the laboratory. James protested against this view. They seem to have neglected the obvious fact that any sensation is the experience of the whole organism, more

[197]

or less as an integer. The adequate study of sensation, therefore, requires a study of the whole organism at the moment of such experience, in order to find out what physical and psychic factors then come into play. This point of view need not lead us to go the whole way with theories of Kafka and others on so-called *Gestaltqualität*. Members of this school deny the validity of analysis into component sensations, believing that "forms" and "patterns" are of fundamental importance. Washburn (1916) has pointed out that they have too largely neglected movements. In any event it is agreed by most psychologists, regardless of other views, that movements are important for the occurrence of at least some mental activities.

RELAXATION AND THINKING

The practical significance of some of these points is apparent. When, in résumé, the patient relates what he has been thinking about of a disturbing character, it would be futile abruptly and without preparation to request him to relax away such thoughts. He would not know how and what to relax. The layman in his daily affairs is interested in the matter or content of thought, but not in how he thinks. Accordingly the beginner who, while lying down in partial relaxation, is to imagine or think and at once thereafter to report what took place, commonly relates the content of his thought, or as I should say, his meanings, but ignores the images and sensations, including those from muscular contraction, which take place when he thinks as a part of that process. He ignores the sensations from activities of the tongue, lips and laryngeal muscles, which graphic tracings reveal during the thought-processes of many or most individuals, and the sensations from slight eye movements to which I have called attention, as well as many other motor processes which go on when thinking takes place. Yet it is precisely these tensions, hitherto ignored, which he will need to relax if the undesirable thought-content is to be deleted. Doubtless this can and does take place at times automatically during general relaxation in untrained as well as trained individuals: conscious direction is not always necessary. But where this fails and persistent or recurrent mental activity proves disturbing, some knowledge of the distinction between the matter of thought and the experience of tensions will, I believe, prove of importance both to investigator and subject and to physician and patient, if the former is to teach and the latter is to learn control of conscious mechanisms.

CHAPTER XII

THE EMOTIONS AND THE NEURO-
MUSCULAR ELEMENT

THE importance of the emotions for the science and practice of medicine was recently (1927) emphasized in a symposium of the Section on Medicine of the American Medical Association. Writers well known for their work in other departments (Woodyatt, Hunter, Foster, McLester, Neilson and others) discussed emotional and psychic factors in the production of digestive disorders and in the course of arterial hypertension heart disease, diabetes mellitus, exophthalmic goiter, laryngology and ophthalmology. Foster believed that in some cases an unhealthy mental attitude is a direct result of the arterial hypertension or heart disease, while in others it is a consequence of the knowledge that disease is present. McLester was inclined to believe that unhealthful emotions were seldom causative in digestive disorders, and were to be traced largely to faulty constitution. At this and other points there was evident room for debate.

Considerable interest was developed in the foregoing discussions and the present signs of awakening of the medical profession to the recognition of the psychic aspects of disease were generally applauded. But what should be the next step? Evidently it is not sufficient to report clinical experience and views concerning the emotions without definite knowledge of the underlying physiology. Mere exchange of opinion in this field would be as unwarranted as in the departments of electrocardiography or of diabetes. The physician should not be satisfied with a mere popular psychology but rather should insist upon carefully controlled investigations. As previously urged, the application of experimental psychology to the practice of medicine is greatly to be desired (Jacobson, 1921).

Obviously, we cannot hope to understand the emotions in relation to the science of medicine if we ignore the results of the many past years of investigation both in physiological and psychological laboratories. The closer we can come to a definite understanding of the nature of the emotions apart from disease, the better will be our grasp of their attributes in medicine. It is therefore fitting at this point to outline some features of past studies of emotion; the more especially since, as will readily be seen, the manifestations of emotion, in certain bodily localities at least, are the reverse of what we here know as relaxation.

INVESTIGATIONS

Among the first of many investigators on circulation in connection with emotion was Mosso (1881). The cerebral circulation (blood volume), he concluded, shows greater changes in response to emotion than to intellectual activity, however energetic. Influence of the emotions is much less on the blood volume of the forearm than of the brain. In the brain there occurs a dilation, in the forearm a constriction. Changes in the pulse and heart-beat during emotion have been found by a host of other investigators, including Wundt (1903), Lehmann (1899), Menz (1895) and similarly Bickel (1916) and Blatz (1925). White and Gildea (1937) find marked increased pulse-rates upon immersing the hand in ice water during "tension" and anxiety. Transitory rises in spinal-fluid pressure during emotion have been demonstrated by Rechède (1913), who found these greater than during moderate muscular movements. Respiration is rendered irregular (Wundt, 1894; Meumann and Zoneff, 1901; Blatz, 1925 and others). Contractions of the bladder (human and dog) respond very delicately to emotional stimuli (Mosso and Pellacani, 1882).

It is well known that the knee-jerk commonly increases during emotional excitement. Upon emotional disturbance, leukopenia results in rabbits, but leukocytosis in cats (Nice and Katz, 1936). Nice and Fishman (1936) explain the increased specific gravity of the blood which they find in these animals and in pigeons during emotional excitement as due to water loss from the plasma into the tissues (Nice and Katz, 1934), the inpouring of blood cells into the general circulation (Binet, 1927; Barcroft, 1930), and the addition of metabolites into the circulating blood, causing an increase in sugar, urea, uric acid, total and preformed creatinin, cholesterol and hemoglobin.

Increase of oxygen consumption during emotional stimulation (fearful expectation of a shock) was found by Totten (1925) in six out of fourteen subjects, varying from 4.9 to 25 per cent. The investigation left in doubt whether this increase was due to hypothetical dilatation of bronchi and bronchioles during emotion or to the increased contractions of skeletal muscles which were demonstrated with tracings. Increased energy requirements have been found by Schroetter (1925) with one subject during laughter and sobbing as compared with inaction. He notes the importance of the contractions of the voluntary muscles associated with those acts. Landis (1925) observed variable influences of emotion on the basal metabolic rate. Fearful anticipation generally caused a noteworthy increase, but anger had the same or an opposite effect. Ziegler and Levine (1925) obtained more marked changes in psychoneurotic patients during emotion than during rest. In nine out of fifteen cases there was an increase amounting to more than 10 per cent but in two cases there was a decrease amounting to more than 10 per cent. They emphasize the energy expenditure thus observed during emotion and conclude that *"lying apparently still in bed is not to be taken as a criterion of rest."* This evidently confirms what was said in chapters i and iv.

Gastrointestinal changes during emotional states are probably the best known. They include the excitations and inhibitions of salivary and gastric secretion observed by Pavlow and his associates (1902) and of pancreatic juice and bile observed by Oechsler (1914). Kleitman and Crisler (1927) measured the salivary flow during emotion (nausea) in dogs. Cannon's observations of inhibition of gastric and intestinal movements in cats during fright are familiar (1902). The esophagus has been found in two cases to contract readily during emotion (chap. xvi). Clinical observations during mucous colitis have suggested that there is an association of the chronic spasm with a chronic emotional overexcitability (chap. xvi). A systematic study of the effects of emotional stimuli on gastrointestinal tone has been made by Brunswick (1924), who took records with balloons in the stomach, duodenum and rectum. He observed tonic changes in the stomach and rectum. Certain contractions of the stomach occurred particularly with pleasant states, and a decrease in gastric tone was observed in one subject during fear. The rectum often failed to respond during emotion, giving variable results. Brunswick feels unable to conclude whether changes in gastrointestinal tone are felt as emotion.

Since the galvanometer was first adapted for this field (Féré, 1888; Tarchanoff, 1890; Veraguth, 1908), many studies have been made therewith on the emotions (cf. Prideaux, 1920). During the passage of a slight constant current, emotion produces a deflection which doubtless is at least partly due to secretion by local sweat glands (Darrow, 1927). The matter is well reviewed by Wechsler (1925) who recites the reasons for believing that the psychogalvanic reflex is due to a diminution of the counter electromotive force of polarization produced by fresh secretion of sweat (Gildemeister, Aebly, Piéron, Gregor and Loewe). Wechsler concludes that the psychogalvanic reflex primarily indicates affective tone, sometimes "unconscious experience" (Jung, 1919; Smith, 1922). Landis (1929, 1930) finds, from an extensive review, no one-to-one correlation between the reflex and emotion or affective experience. Electrical changes in the skin are of no more psychological significance than changes in blood pressure or in basal metabolism.

That increase of blood sugar, sometimes with glycosuria, may result from pain and emotion was shown by Boehm and Hoffmann (1878), confirmed by Cannon and his associates (see his book, 1915), and by Gildea *et al.* (1935). One study by Derrien and Piéron (1923) showed that the increase of blood sugar was most marked in the three patients who were emotional but who almost completely inhibited the outward manifestations of their emotion. Prideaux finds similarly with the galvanoreflex that inhibition of muscular movements causes a deflection.

The effects of emotion upon the glands of internal secretion need not be thoroughly discussed here. Cannon and his associates during years of investigation have given various (more or less indirect) types of evidence that emotion increases the output of adrenalin, which favors appropriate reaction to emergency by increasing the blood-pressure, quickening the heart, inhibiting the stomach and intestines, favoring the coagulation of blood and increasing the blood-sugar available for energy requirements. Their findings have been confirmed by various investigators, but have been sys-

tematically contested, as is well known, by Stewart and Rogoff (1917–23), whose precise and direct quantitative examinations of the blood for the presence of adrenalin have failed
to find any increase such as would produce the above-mentioned phenomena.

In exophthalmic goiter, which probably involves excessive
secretion from the thyroid, emotionality is commonly increased. The patient may have the appearance as of chronic
fear. Thyroxin fed in excess generally produces hyperemo-
tivity. That the nervous system can excite thyroid secretion seems probable from the work of Asher and Flack (1910);
direct evidence was found by later investigators (Rahe,
Rogers, Fawcett and Beebe, 1914). Nevertheless we still
lack a direct test for thyroid secretion in the blood, and we
do not yet know what rôle, if any, the normal thyroid plays
in the production of emotion. It is suggestive, however, that
in marked hypothyroidism emotivity is characteristically decreased. Signs of restlessness and anxiety commonly follow
the removal of the parathyroids before death sets in.

The sex organs undoubtedly play an important part in the
production of emotion. Erection has normally a strong emotional tone, as have the desire for intercourse (which may
produce general neuromuscular tension) and the after-state.
Removal of the gonads does not necessarily bring with it a
loss of psychic sexual feelings. Restless activity characteristic of female rats has been shown to be due to a stimulus
arising from the internal secretion of the ovaries (Wang,
1923). In ovariectomized rats, which display decreased voluntary activity, injections of ovarian follicular hormone are
followed by increase of voluntary activity (Bugbee and Simond, 1926).

In underdevelopment of the anterior lobe of the pituitary
gland (*dystrophia adiposa genitalis*) there is failure of devel-

opment of the genital glands, with consequent emotional abnormality. In hyperfunction of the posterior lobe leading to acromegaly, there is emotional abnormality.

The tendency in various clinical and pseudophysiological books to make unproved statements about the glands of internal secretion in connection with emotion is to be discouraged. We know relatively little about the subject under normal conditions, and its bearing upon the physiology of relaxation must await future development. Evidently in states of disordered secretion such as Graves' disease a variable is introduced which acts contrary in direction to voluntary relaxation, which also may be regarded as a variable: if both variables are present, the effect will be a joint function.

Photographs of facial expression during emotional reactions were studied by Landis in 1924. By inspection of these he believed that he could determine the presence of action or inaction of particular muscles, e.g., the risorius. As a rule he found, according to this method, no correspondence between the muscle-groups involved and the name applied by the subject to the emotion. For example, there was no confirmation of Darwin's statement that "frowning is the natural expression of some difficulty encountered or of something disagreeable experienced either in thought or in action."

In a later series of experiments (1926) Landis aroused what he calls "severe emotional upset" by having his subjects go without food for 46 hours and without sleep for 36 hours followed by continuous stimulation with an induced current as strong as the subject could bear without struggling Records were made of blood-pressure, thoracic and abdominal respiration, gastric and rectal contraction and basal metabolic tests before, during and after the state of emotional disturbance. He discovered no "pattern" of blood-pressure or of inspiration-expiration ratio, no certain change of metabolism and no type of gastric or rectal contraction which bore a one-to-one correspondence with the subjectively named emotion or with the type of stimulation. He concludes that on the basis of such changes alone, one is unable to differentiate emotions to which the subject gives such names as "anger," "fear," "horror" and others (cf. Strehle, 1935).

THEORIES OF EMOTION

What is an emotion? Early writers on this subject, as Landis remarks, depended upon divers observations, folk lore, tradition, art, literary opinion and common observation (Bell in 1806, Lavater in 1804, Bain in 1859, Darwin in 1890). These writers tended to accept the popular point of view that the manifestations of emotion, like tears, laughter or frowning, are the expressions or results of foregoing psychic states in which the content of a situation was grasped. From this point of view Wundt studied the pulse (1904) as revealing the course of feeling. Changes in facial expression during emotion were commonly noted and explained by a theory of radiation of nervous energy from higher centers (Spencer, Darwin, Wundt, Dumas).

Into this state of things Lange and James (1884–85) entered their theory of emotions, now so well known because subsequent discussions and investigations have centered about it. Lange gave detailed descriptions of the physiology of grief and of other emotions. James stated his view as follows:

Our natural way of thinking about these coarser emotions is that the mental perception of some fact excites the mental affection called the emotion, and that this latter state of mind gives rise to the bodily expression. My theory, on the contrary, is that *the bodily changes follow directly the perception of the exciting fact, and that our feeling of the same changes as they occur is the emotion.* Common-sense says, we lose our fortune, are sorry and weep; we meet a bear, are frightened and run; we are insulted by a rival, are angry and strike. The hypothesis here to be defended says that this order of sequence is incorrect, that the one mental state is not immediately induced by the other, that the bodily manifestations must first be interposed between, and that the more rational statement is that we feel sorry because we cry, angry because we strike, afraid because we tremble, because we are sorry, angry or fearful, as the case may be.

Why has the James-Lange theory attracted such widespread attention? First of all, I believe, because it called attention and added impetus to a view which was already well known among students and which they were advocating: Modern psychology largely arises and develops in opposition to the previously dominant theory, commonly attributed to Descartes (1650), that mind is in most respects the opposite of body and lacks almost all connection with it. To be sure, this view was declining during the last century, but here was another blow, which everybody could understand, against a merely spiritual interpretation of the emotions.

In fact, the theory went a point farther than this: These physiological responses, it said, which people have been inclined to regard merely as the expressions of their inner meanings, really are found to be something more important. Their rôle is to give body to the psychic phenomenon of emotion, and thereby to make it real. This point is well brought out in a passage from James which reveals his capacity for self-examination:

The next thing to be noticed is this, that *every one of the bodily changes, whatsoever it be, is FELT, acutely or obscurely, the moment it occurs*. When worried by any slight trouble, one may find that the focus of one's bodily consciousness is the contraction, often quite inconsiderable, of the eyes and brows. When momentarily embarrassed, it is something in the pharynx that compels either a swallow, a clearing of the throat, or a slight cough; and so on for as many more instances as might be named.

A third feature lay behind the new theory to make it interesting, although the development of this remained latent for years. It contained the seeds of revulsion against another ancient theory, dating from the time of Aristotle and still current even among many scientists, that the mind's action depends solely upon the brain and nervous system. It called attention in a clear way to the study of peripheral phenomena as important, not just as effects, but also as causes of the reactions of the brain and nervous system; in other words, to the fact that peripheral phenomena can be from certain points of view the causes or even the aspects of psychic conditions. Thus it anticipated the modern doctrine of "behaviorism," but it is not necessarily identified with this doctrine. The ancient view that peripheral phenomena are of minor importance is contrary to our findings in connection with the study of relaxation. What we seem to show in chapter xi is that peripheral phenomena can influence the occurrence of emotion, thinking and other psychic processes. Whatever our debt to the James-Lange theory in this particular, however, we are not obliged to adhere to it in general.

A fourth reason why the foregoing theory seems to have attracted widespread attention is its startling assertion that the psychic phase of the emotion *follows* the bodily manifestations—that we "feel sorry because we cry." This point will be discussed later.

Many investigations, including some not mentioned above, have been designed to test the James-Lange theory or some modification. Sherrington as well as Landis and Gullette, for example, inquire whether Sergi's theory that emotion depends upon vasomotor changes holds true and report negative results. Less convincing are the conclusions of Landis based on experiments previously quoted that characteristic physiologic changes cannot be found for various emotions: (1) Landis employs a large variety of stimuli and emotions, but the number of repetitions of any one emotion appears insufficient to prove his point. The conditions of "emotional upset" that he produced involved various factors including fatigue and nausea, but no characteristic state of what we are accustomed to regard as one type of emotion. (2) Detection of slight muscular contractions, concealed by cutaneous tissues, is not dependable by means of the camera. Yet precisely this is what we need to be most sure about. That Darwin's statement is sound would doubtless be almost universally conceded on the basis of general experience. He did not deny that frowning occurs in a bright light or during reading as well as under the conditions he mentions; among various correlations, perhaps, Darwin draws our attention to one. Accordingly the point of view which he represents does not assert, as Landis seems to suggest, that there is a one-to-one correspondence between frowning or other particular action and some more or less constant emotional pattern. Presumably what is meant is that in a considerable percentage of instances of difficult or disagreeable experience some measure of frowning is present and might be detected with sufficiently precise apparatus. Whether this is true requires more delicate measures of test than any employed to date. But the proper formulation of the problem to be investigated requires that we bear in mind that it is not a question of one-to-one correspondence between specific types of muscular contraction or other physiological phenomena and what a subject calls a particular emotion. We recall that a disease entity, such as typhoid fever, presents various clinical "pictures," which nevertheless correspond to one diagnosis. This sort of correspondence evidently is present in our daily experience, since we are frequently able correctly to discern from the facial expression of those about us whether they are, for example, angry or amused or in deep grief; in short, this sort of correspondence evidently exists between emotions and facial or other bodily "patterns" in the general experience of mankind. As investigation proceeds to reveal the character of these patterns underlying various emotions, we must not be confused by their manifold variety, their relative inconstancy with different individuals and with the same individual upon different occasions; through these complicated relationships, certain threads of correlation evidently run, and our task remains to seek these out, along with the limitations and variations. Such variations of physiological patterns have been found to exist for thought-processes, as described by trained subjects, and have been generally admitted by experimental psychologists, who nevertheless are agreed that correlations exist between types of such processes and types of "bodily patterns."

This brings us to the subjective characteristics of emotional experiences as described by trained subjects. Conklin and Dimmick (1925) made a brief study of fear in this manner, and their discussion is to be recommended in order that this aspect be not neglected. As a rule, they indicate, emotions are classified in the literature according to the meaning which the situation has for the individual. If it is danger, we may add, the classification is under "fear." But the emotion of fear differs according to the individual, the time, and the feared object. My fear of a burglar differs from my stage-fright, my fear of giving offense or of suffering financial loss. Just as the conscious patterns differ in these instances, there are also beginnings of evidence, as previously said, that the physiological patterns for fear may differ greatly. In Brunswick's experiments, fear in one subject seemed carried in part by experience from relaxation of the stomach; in my subject by contraction of the esophagus. Many such observations are needed before drawing final conclusions. Eventually we may inquire whether some factor is common to the various types of physiological reactions called fear (possibly the contraction of one or more of the viscera, not necessarily the same).

Sherrington contests the view of James and Lange that the cerebral and psychological processes of emotion are secondary to an immediate reflex reaction of vascular and visceral organs of the body suddenly excited by certain stimuli of peculiar quality. He cut the vagi and cord in dogs (1900a, 1906) so as to do away with the sensation from all the viscera and the circulatory apparatus as well as from the skin and muscles behind the shoulder. Such dogs continued to display undiminished signs of joy, disgust, anger and fear. From this he concludes that it is likely that the visceral expression of emotion is secondary to the cerebral action occurring with the psychical state.

There are several points at which this evidence is inconclusive: (1) The first, which is admitted by Sherrington, is that although the visceral and vascular and much of the muscular mechanism of emotional expression was cut off, a small but notable fraction of the latter, namely, the facial (and also, as Kempf adds, the diaphragm), still remained open to react on the centers with which consciousness is colligate. Indeed, the most important sense organs, including the eyes, which contribute by their movements so largely to conscious processes, were left intact. Did his operated dogs retain the full gamut of their former emotions or was their emotional experience lacking in some respects, corresponding to their crippled voluntary movements and their diminished sensorium? Removal of large portions of one or both kidneys or lungs, as is well known, permits nevertheless, under favorable conditions, practically a normally sufficient urinary secretion or respiration. From this it becomes evident, if important muscles or viscera are not included among those severed from the cerebrum, that we must be cautious in drawing conclusions about the part played in emotion by the totality of intact muscles and viscera. In any event, partial severance in animals fails to rule out the possibility that the emotional experience is present, at least in part. (2) At best, animal experiments fail to yield a complete answer to the present question, since we require human reports in order to learn how the experience of the operated individual compares with the normal. (3) Sherrington assumes that "to the ordinary day's consciousness in the healthy individual the life of the viscera contributes little at all, except under emotion." There is considerable evidence in various of the above-mentioned investigations as well as in our own for regarding this assumption as unsafe. (4) Sherrington seems to regard the striated muscular expression of emotion, like the visceral expression of emotion, as secondary to the cerebral reaction occurring with the psychical state. But the action of voluntary muscles, we seem permitted to infer from sundry experiments cited in preceding chapters, probably is no longer to be regarded as a mere "expression" or "aftereffect" of conscious activities, as was commonly believed during the last portion of the nineteenth century, but through the sensations aroused, forms part of the general play of psychical processes. Since this matter is open to experimental inquiry, an assumption one way or the other is not justified.

Further evidence that emotion is primarily a cerebral reaction is commonly sought in observations where the hemispheres of the brain have been removed. Goltz (1892) observed a dog kept many months in that condition. Save for expressions of anger, it was indifferent and neutral to its surroundings, not even showing sexual emotion. Somewhat contrary evidence is that in anencephalic infants, stimuli which

to the adult human being are unpleasant bring forth characteristic emotional reactions, such as drawing down the angles
of the mouth and the lower lip, puckering of the mouth,
withdrawing the head and whispering or crying (Sternberg
and Latzko, 1903). But if we conclude that at least in man
some grade of emotion is in certain instances possible in the
absence of the cerebrum, it still seems almost certain that
in the emotional experience of the intact human being cerebral activity generally plays an essential part. For in order
that a stimulus may arouse an emotion, it must as a rule
be perceived and discriminated from other objects of indifferent character; this is particularly true when the source of
emotion is something seen or heard. Perception and discrimination undoubtedly depend upon cerebral functions
(without prejudice to the question whether perception involves the activity of striated muscles, particularly those
of the ocular organs). These quite obvious considerations are
occasionally forgotten in discussions of the rôle of the cerebrum. To this extent at least we must therefore modify our
acceptance of the James-Lange theory.

A more serious blow to the latter comes from the observations of Wilson (1924) on pathological laughing and crying.
This may occur in double hemiplegia, pseudobulbar paralysis, disseminated sclerosis and in the presence of tumors,
infective conditions and vascular degenerations in certain
locations. Under these conditions some of the patients show
laughter or tears but protest that they do not feel accordingly. The conclusion, suggests Wilson, is unmistakable that
the bodily reverberation, as James calls it, is not per se the
emotion; the latter is not, so to speak, the mental symptom
of the former.

It is clear, therefore, that the James-Lange hypothesis must be mate-
rially modified if it is to be brought in line with observations such as have

been recorded. . . . . Some of my own and of the reported cases of others indicate the possibility of dissociation between the psychical and the physiological elements in the emotion [Wilson, 1924].

Wilson deals a further blow at the peripheral theory of the emotions when he considers pathological conditions in which free movement of facial musculature is impeded by organic disease. Examples are facial diplegia, facial myopathy, myasthenia gravis, paralysis agitans and postencephalitic Parkinson's disease. "The conclusion in each instance of bilateral facial impairment has been that the patient can readily feel and be acutely conscious of experiencing a particular emotional state such as that associated with hilarity and joy in spite of the minimal expression in the face."

Of similar import are the observations of Marañón (1922) on the emotive action of epinephrin. After injection of this substance in certain dosages, peripheral effects such as tremor of the limbs, internal tremor, chills in the back, coldness of hands and shedding of tears were described by some of his subjects (particularly those with Graves' disease) as arousing true subjective emotion, but by most of them as failing to produce this type of experience.

Evidence concerning the central control of emotional activities has been accumulating. In 1887 Bechterev observed affective responses in decerebrate animals. In 1925 Cannon and Britton noted spontaneous fury in decorticated cats. Even after ablation of the brain anterior to the diencephalon, the "sham rage" continued, according to Bard (1928). He and Rioch removed the neocortex and additional portions of the forebrain in 1937 and found rage, fear and sexual excitement more arousable than normally, unless also the hypothalamus was removed, when rage was no longer elicited. Ranson, also, furnished evidence (1927; Kabat et al., 1935) that the hypothalamus is the center for the integration of the visceral and somatic components of emotional expression (but see Bard and Brooks, 1934).

"The peculiar quality of the emotion is added to simple sensation when the thalamic processes are aroused." This is Cannon's well-known theory (1929). But whatever centers take part, muscular and visceral elements are clearly disclosed in the emotional experience of intact man, when examined by subjects trained in present methods.

Realizing the importance of the observations of Sherrington, Wilson and Marañón, we may ask whether they have quite proved their case. The issue is not crucial from the standpoint of our present interest in relaxation, because as Wilson states, "under normal conditions practically all writers (including Wilson and Sherrington) agree on the reinforcing and intensifying of the emotional cerebral state by the advent of somatic and visceral impulses." Accordingly the effect of relaxation in preventing such advent of somatic and visceral impulses would be to tend to diminish the emotional experience. Nevertheless we may retain at least an academic interest in the question propounded. Wilson's experiments seem open to another interpretation besides the one he offers: The patient who laughs or cries pathologically does not experience a complete emotion, precisely because he has nothing to laugh or cry about; perception of an object appropriately causing laughter or tears is absent, and this may be all that we can safely infer that the patient means when he protests "against the laughter or tears being taken as the index to his actual affective state." An emotion like laughter in an intact individual admittedly is a highly complicated synergic series of physiological reactions: possibly a localized lesion is incapable of stimulating the complete physiological response, just as electrical stimulation of areas of the exposed cortex generally evokes some relatively simple motor response, but fails to evoke the complexity and concatenation of responses of the normal animal under voluntary initiation. Perhaps some of the "visceral reverberation" whose occurrence may be essential to complete emotional experience fails to take place when Wilson's patients, lacking an intact nervous system, perceived nothing appropriate as the cause of their tears or laughter. As for the subjects of Marañón following the injection of epinephrin, their reports might for the most part be equally well interpreted that they *did* feel true subjective emotion, except that the experience was rendered incomplete owing to the absence of any perceived cause or reason for emotion. This may very well be what they mean when they say, "I feel as if I were afraid" or "as if I were expecting a great joy" or "as if I were going to weep without knowing why" or "as if I were in great fear, yet I am calm." In Graves' disease, as is well known, where peripheral effects are stimulated by disordered secretion, the perception of appropriate cause for emotion more readily excites emotional experience than in normal individuals. Accordingly it seems justified in the present state of knowledge to suspend judgment whether peripheral phenomena are the direct inciting cause of subjective emotional experience. Most of the evidence to date is too incomplete and too indirect to permit final conclusions.

A few individuals have investigated the time-relations involved. Golla (1921) measured the latent period of the psychogalvanic reflex in response to emotional stimuli and concluded that it is much too long to lend any substantiation to the James-Lange theory. However, we must take into account that an emotion does not have constant characteristics either physiologically or subjectively from its initiation to its decline and therefore that the psychogalvanic reflex may be and probably is a late response preceded by other integral elements of the emotion.

The Influence of Relaxation on the Emotions

Whatever be the conclusions regarding the theoretical aspects discussed above, the importance of efferent neuromuscular activity for the emotions is emphasized by practically all writers. Examination of the phenomena uncovered by most of the investigations reviewed early in this chapter furnishes the experimental evidence for the importance and generality of this neuromuscular element in emotion. In my own observations, reports have been secured independently from subjects and patients under normal conditions as well as during states of nervous excitement. They agree that the emotions subside as the individual *completely* relaxes the striated muscles, particularly those which he seems to find specifically concerned in the emotion at hand: the esophagus in one instance of fear; the forehead and brow as a rule in worry or anxiety. As a further test during emotion, the instruction has been given to relax completely yet retain the emotion. When this was done without intimation by the experimenter of possible result, all highly trained subjects independently reported it impossible to carry out the double instruction. They found it impossible to be emotional and relaxed at the same time.

Accordingly, present results *indicate* that *an emotional state fails to exist in the presence of complete relaxation of the peripheral parts involved.*

Present Conclusions Concerning Emotions

Some of the general conclusions to be drawn in a provisional manner from a consideration of the investigations in this chapter, as well as my own, are as follows:

THE EMOTIONS

1. Emotions involve hypothalamic and cortical centers but also peripheral localities, including all types of muscle. It seems most likely, according to our present knowledge, that the subjective experience of emotion is largely derived from intensive proprioceptive impulses.

2. The time-relations of the emotional experience have not yet been adequately investigated.

3. Emotions are of many kinds, and what is called the same emotion shows marked physiological differences between individuals and at different times for the same individual and in response to various situations. Because of the many variables involved, the search for a single uniform measure will probably prove futile. In this I agree with Landis (1926). The aim should be to find many quantitative tests for emotions in a particular test situation and to seek to correlate them where possible.

4. Muscular contraction (smooth, cardiac or striated) occurs in most of the phenomena studied: pulse, blood-pressure, respiration, tremor, viscera and in the glands of internal secretion. Theoretically, therefore, muscular elements appear in every emotion, and relaxation would seem an opposite state.

5. Certain tests suggest that if the outward manifestations of emotion are suppressed or concealed by the individual, the affective phenomena are all the more increased. This means that certain regional reactions become all the more responsive. The physiology of suppression or concealment has not yet become established, but these phenomena evidently differ from extreme relaxation, where no attempt is made to conceal or suppress and the subjects report that the emotional experience diminishes, disappears or fails to appear.

[219]

CHAPTER XIII

TONUS AND THE NERVOUS REGULATION
OF MUSCULAR CONTRACTION

GENERAL CHARACTERISTICS OF MUSCULAR ACTIVITY

SINCE muscular relaxation is in effect the negative of
contraction and tonus, it will be profitable to review
recent investigations in this connection. A discussion
of various topics not here included will be found in Fulton's
comprehensive monograph (1926).

As is well known, muscles fall into three groups—smooth,
striated and cardiac. Striated muscles are supplied from the
cerebrospinal nervous system and their contraction is customarily called voluntary. These are the muscles that receive
practice when human beings are trained to relax. So far as
we have been able to judge from subjective observations,
our subjects do not relax smooth (visceral) muscles directly:
when this is accomplished, it seems to be a sequence of the
inactivation of the striated type. The muscles of the larynx
and tongue in our experience have relaxed like other striated
muscles.

Since the diaphragm and other respiratory muscles are in
part under the influence of the autonomic center, they will
not be expected to respond completely to voluntary relaxation; but breathing becomes uniform and quite moderate in
extent. In our experience a muscle may be fully relaxed in
any position that can be maintained with no expenditure of
energy. For example, the biceps may be fully relaxed with
the arm either in flexed or in extended resting position.

The muscular contractions dealt with in the present investigations in man are of the tetanic type. As is well known,

a series of rapidly repeated stimuli to a muscle produces a fusion of single twitches into a continuous contraction. That the uniformity of the contraction conceals an interrupted series has been shown by applying a telephone to the muscle (Wedensky, 1891) as well as by galvanometer records (Piper, 1907). According to Hoffmann (1922), tendon reflexes are either twitches or tetanus depending upon the duration of stimulus: a blow as brief as 0.01 second causes a single twitch. If stimuli are repeated, e.g., 100 per second, the 100 twitches fuse into tetanus. The tendon reflex differs from what this writer calls the *Fremdreflex,* such as the scratch-reflex, in that the latter type always has a specific character not dependent upon the duration of the stimulus.

In contrast with skeletal muscle, visceral muscle can enter a state of prolonged activity which is automatic or independent of the receipt of excitatory impulses from nerve centers. Of late, however, controversy has arisen as to what the muscle tissue itself can do if internal nerve plexuses are stripped away (see Bayliss and Starling, 1899; Carey, 1921; Alvarez, 1922).

Muscle Chemistry

Problems of muscle chemistry have not yet been undertaken in connection with the present work, but a few matters may be profitably recalled. Little increase of metabolism has been found during tonic contraction of smooth muscle as compared with skeletal (Parnas, 1910; Bethe, 1911). Evans (1923) agrees that smooth muscle uses less O_2 when in tonus than when relaxed, but explains this as due to reduction of the surface of the fibers. He adds that after a period of diminution, the O_2 consumption is increased.

It was formerly believed that likewise there is little increase in metabolism in skeletal muscle tonus, for Roaf

(1912) asserted that the CO_2 output and O_2 intake were no greater during decerebrate rigidity than after abolishing such contraction by curare. But Dusser de Barenne and Burger (1924) found the O_2 consumption of muscle during decerebrate rigidity increased 10–25 per cent, as compared with the atonic state. Furthermore, in 1926 Evans reported results contrary to those of Roaf with curare.

Energy for muscular contraction perhaps arises from hydrolysis of phosphocreatine, discovered by Fiske and Subbarow in 1927. The freed acid unites with hexoses derived from enzymic hydrolysis of muscle glycogen, forming labile monoesters. These, upon hydrolysis, set free phosphoric acid, some of which is resynthesized into phosphagen. During the breakdown of reactive muscle-sugar, methylglyoxal appears, and finally stable lactic acid. Evidence for this view rests especially upon Lundsgaard's finding that contraction can occur in muscle, poisoned with monoiodoacetic acid, wherein no lactic acid is formed (1930).

Earlier than lactic acid, ammonia is formed, apparently by enzymic deamination from adenylic acid (Embden and Zimmerman, 1927), affording also inosinic acid (discovered by Liebig in 1847). Whether the hexosephosphoric acid ester in muscle discovered by Embden (1923, 1924, 1925), and named "lactacidogen" is split and accounts for increase of orthophosphate and lactic acid has been questioned (Lohmann, 1933; Eggleton, 1935).

Probably later than the lactic-acid formation occurs esterification and hydrolysis of the pyro fraction of the "nucleotide" which consists of hypoxanthine and phosphoric acid attached to d-ribose (Levene and Jacobs).

During oxidative recovery (see Milroy's review, 1931) there occur: resynthesis of phosphagen; removal of lactic acid; restoration of glycogen (minus the lactic acid equivalent that has been synthesized); reamination of the deaminated nucleotide; and later, resynthesis of adenylic and orthophosphoric acid, if the pyro fraction has been broken off in prolonged activity.

Contrary to the current view, Sacks (1937) believes that glycogen yields the energy for normal muscular activity, primarily through oxidative reactions. He finds that resynthesis of phosphocreatine proceeds at so small a rate as to invalidate the hypothesis of Lundsgaard.

Urban (1937) finds evidence from oscillograms that cytochrome-*c* furnishes energy for the early phases of contraction, while cytochrome-*a* is probably active during the relaxation phase.

In progressive fatigue, muscle shows, among other features, increased content of lactic acid, of iminazol (Massione and Donini, 1937); increased equilibrium concentration of orthophosphate (Stella, 1928); lessened e.c. of creatin (Eggleton, 1930).

Previously fatigue had been variously investigated. Ranke (1865) suggested that the failure of contractile power during fatigue is due to the liberation of metabolites. Extracts of fatigued muscle of one frog when injected into the circulation of another brought on fatigue, while control experiments with unfatigued muscle failed to give this result. Lee (1907) confirmed these observations. Following severe exercise, lactates are found in the blood, evidently having diffused through the capillary walls (Hill, Long and Lupton, 1924; Hines, Katz and Long, 1925; Furusawa, 1925). But in normal muscle tissue with intact circulation, the fatigue products from one muscle fiber do not pass over into the adjacent fibers and this explains why postural reactions can be persistently maintained (Sherrington, 1892; Fulton, 1925).

An excess of creatin was reported for invertebrate muscles in tonus (Pekelharing and van Hoogenhuyze, 1910), and Pekelharing in 1911 made his well-known discovery that creatinin is increased in the urine of men after prolonged voluntary tonic contraction but not after walking. Investigators agree that creatin plays a part in muscular metabolism. Spiegel admits (1923) that there is increase of creatin with muscular work and Tiegs (1925) furnishes evidence that creatin is produced by contracting muscles. Looney (1924) observed an increase of blood-creatin after prolonged cata-tonic contraction and a decrease after relaxation. He sug-gested that the amount of creatin in the blood may be an index of the tonicity of the muscles of the individual. But Cobb (1925) contends that Looney failed to take into ac-count conditions of impeded circulation and concludes in harmony with Spiegel (1923) that as yet no satisfactory evidence has been produced that creatin metabolism is in any way associated with tonic muscular contraction.

The Nature of Tonus

In the active organism, muscle tissue frequently exhibits a condition of moderate sustained contraction which is com-monly called tonus. As is well known, this disappears in skeletal muscle when the nervous elements are stripped off, although the same is not true of the smooth variety. J. Mueller (1838) and others of his early period believed that skeletal muscle tonus depended upon the action of nerve cen-ters, but they were in doubt whether this was automatic or reflex. To answer this question, early experimenters looked for a lengthening of the muscle after severing the motor nerve. This was not found by various investigators, but Brondgeest (1860) severed the afferent roots of one hind limb of the frog; when the animal was held vertically the affected limb showed a loss of tonus, since it was less flexed than the other. This showed that the greater flexion of limb with normal innerva-tion was due to muscular tonus which was reflexly maintained through the afferent nerves. His conclusions have been re-cently tested anew, but it is commonly agreed that an intact arc is required for the maintenance of tone in skeletal muscle. The current view of the reflex arc helps us to understand the mechanism of progressive relaxation, where there occurs a diminution of afferent impulses in connection with a diminution of tonus.

Sherrington (1915) calls attention to the ambiguity of the term tonus. A "slight constant tension" is not always present in healthy muscle. It is absent in the extensor muscles of the hind limbs of the squatting spinal frog, as well as in those muscles of the decerebrate cat which bend the spine downward, droop the neck and tail, flex the head and depress the jaw. In the latter animal, tonus is distributed to just those muscles which maintain erect posture: if you bend the head down, the tone disappears from the extensor muscles of the forelimb and the fore quarters sink, while the hind limbs stiffen so that the attitude is that of looking under a shelf (Magnus and De Kleijn, 1912). Therefore it becomes evident that a slight steady contraction is not universally characteristic of healthy muscle; its presence or absence in a particular group of muscles is simply a part of the posture of that animal.

Present-day views of tonus have been greatly influenced by the analysis of Sherrington (1909b). Reflex tonus, he declares, is postural contraction and has the following characteristics: (1) the low degree of tension it usually develops; (2) the long periods for which it is very commonly maintained without obvious fatigue; (3) the difficulty of obtaining by artificial—e.g., electrical—stimulation reflex contraction at all closely simulating the postural contraction produced by the natural stimuli (whatever these may be); (4) ease of reflex inhibition; (5) lengthening and (6) shortening reactions.

Sherrington concludes that the tonus of the striated muscles which he tested depends upon the afferent fibers of the

tonic muscle itself, and that "in the decerebrate cat preparation no other afferent fibers than those of the tonic muscle
itself are actually essential for the exhibition of the tonus."
This point of Sherrington is important for the method of
progressive relaxation; for it makes clear how relaxation of a
muscle, by diminishing the afferent impulses to the same
muscle, results in furthering continued relaxation. Of similar
significance is his comment:

> The postural action of muscles and nerves . . . . is a main outcome
> of the functioning of the proprioceptive part of the nervous system, at
> least it is so as regards the skeletal musculature, perhaps as regards the
> visceral and vascular musculature also. Reflex maintenance and ad-
> justment of posture is a chief portion of the reflex work of the propriocep-
> tive system, just as sensation of and perception of posture is a chief portion
> of the psychical output of that system.

More recently Liddell and Sherrington (1924, 1925)
describe the myotatic reflex. They find that a stretch applied to a muscle, whether relatively gradual or sudden,
evokes contraction in it. This reflex response to stretch is
said to be the basic element of postural tone; for the reflex
follows with fine variations in grade and time and locality
the changes in the muscles during movement as well as in
sustained posture. When the rectus muscle of a decerebrate
cat is detached and freed from tension the muscle ceases to
be rigid and no action currents are led off; but if tension of
several hundred grams is applied, small rhythmic oscillations are seen in the galvanometer record (Fulton and Liddell, 1925). It had earlier been shown that these cease upon
stimulating an inhibitory nerve (Dusser de Barenne, 1911;
Einthoven, 1918). Therefore the adequate stimulus for the
development of tonic rigidity in the decerebrate animal is
presumed to be the slight stretch occasioned by the normal
attachments of the muscle. The knee-jerk and clonus are

considered to be fractional forms of the myotatic reflex; and when the tonus is sufficiently raised, muscular spasm appears (Cobb, 1918*b*; Viets, 1920). From this various evidence, postural tonus as well as kinetic movements are explained on the basis of one neuromuscular mechanism: as in the case of the knee-jerk, stretch plays a universal rôle.

The property of a skeletal muscle with normal attachments to remain in a tonic state at various lengths has been called the "plastic" element of muscular activity and this is analyzable into certain reflexes, namely, the so-called "lengthening" and "shortening" reactions. If the freely hanging hind limb of a decerebrate cat is extended at the knee-joint, thus shortening the quadriceps muscle, the latter will then offer considerable resistance to forcible flexion. This is the "shortening reaction." If, nevertheless, the knee is forcibly bent, the resistance suddenly gives way and the knee can be flexed with little difficulty, whereupon the knee tends to remain flexed at approximately the angle to which it has been carried by the manipulation (Sherrington, 1908*b*, 1909*b*, 1915). This is the "lengthening reaction." To explain the shortening reaction, Fulton recalls that the leg, when released by the manipulator after full extension, often drops a little before becoming "fixed" in a relatively extended position. There thus occurs a slight stretch of the vasticrureus muscle, stimulating it to the tonic contraction which is called the "shortening" reaction. On the other hand, the lengthening reaction appears to be a relaxation of the extensor muscles produced by the stimulation of inhibitory nerves arising in the same muscles. Motion of the limb in stepping can be analyzed into a rhythmic series of such shortening and lengthening reactions, each phase automatically stimulating the next.

A full understanding of the relation between "stretch"

and the tonus of smooth and cardiac muscle must await further investigation. This is likewise true of the tonus of skeletal muscle in the intact animal, including man. In these instances we do not know whether the slight stretch occasioned by the normal attachments of skeletal muscles is an adequate stimulus for the development of tonus, in which event complete relaxation would require the intervention of inhibitory nerve impulses, which seems unlikely; or whether, as seems more probable, the particular slight stretch mentioned is an element whose presence is a *sine qua non* for tonus but yet is not sufficient in itself without additional stretch or other stimulus to excite tonus. Pending the solution of this and various other problems, it would seem premature to attempt to relate matters above discussed to the present investigation in a final manner. In man, under normal conditions (chap. ix), prolonged stretch of the quadriceps produced by the limb hanging flexed at an obtuse angle is not an adequate stimulus to produce tonus in all instances, as tested by the knee-jerk; for at least in some individuals complete relaxation is attended by the absence of a reflex, that is, by the apparent absence of tonus.

The traditional definition that tonus is the slight contraction present in intact or healthy muscle even during resting states is founded chiefly on the interpretation of postures or other gross phenomena rather than on direct measurement. Extremely sensitive voltage-testing equipment now makes it possible to go into this question directly in terms of action-potentials, as will be related in chapter xvi. We can assume that the presence of muscular contraction, however slight, will be denoted by action-potentials (Fulton, 1925; Jacobson, 1930; Davis and Davis, 1932). Accordingly, anticipating the results stated in chapter xvi, measurements in striped muscle of man, the dog and the frog fail to sustain the traditional view. *During quiet states in these organisms practically no evidence of contraction is detected for periods of minutes at a time.* These measurements are so delicate that we can assume that action-potentials are absent or physiologically negligible. On the other hand, when the resting individual or animal appears quiet to the naked eye, marked action-potentials in various muscles often disclose that the inactivity is far from complete. It is evident that rest commonly varies in degree and in locality.

Tonus, according to Fenn and Garvey (1934), is no *single* property of muscle but includes elasticity, viscosity, contractility and the many nervous processes controlling the muscle reflexes. Admitting that the term is used in various senses, I find that common usage means *moderate, sustained contraction.*

According to the traditional view, tonus in striped muscle favors speed of response to stimuli. Tests on reaction times in man disclose an increased speed if there is preliminary slight contraction rather than complete relaxation in the reacting muscle, which accords with the traditional view; but this increase is strikingly small: the completely relaxed muscle can, nevertheless, respond promptly (Jacobson, 1936).

The use of the term "tonus" in the traditional sense for striped muscle does not seem warranted. It is suggested that *"tonus in striped muscle" be redefined as meaning a state of slight contraction, more or less constant or irregular, often present but sometimes absent in health.*

Before leaving the topic of tonus and relaxation, mention should be made of several further matters. According to Fulton, the curve of relaxation following an isometric twitch in an unfatigued frog muscle is concave (upward) in outline, suggesting that a passive process is involved in its production. He believes that in his experiments the curve of relaxation is in reality no more than the curve of a "viscous" body stretched by a decremental force. Quantitative studies of the effects of heat on contraction as compared with relaxation (Fulton, 1926) reveal that the rate of relaxation is relatively less influenced by temperature changes, and the curves likewise indicate that relaxation is a passive physical process rather than an active chemical one. The conception of relaxation as a passive process in the foregoing sense, if confirmed, is fundamental to our understanding of certain problems of the present investigation.

Relaxation occurs after the cessation of impulses along motor nerve fibers extending to a muscle, but quite another phenomenon has been called "inhibitory relaxation." Sherrington (1893) discovered that the knee-jerk in the "spinal" animal can be inhibited by repeated stimulation of an ipsolateral nerve. He noticed that the period of inhibition considerably outlasted the duration of the inhibitory stimulus and also that after recovery there followed a period of enhanced excitability during which the knee-jerks were of larger size than usual. The after-effects of inhibitory stimuli were analyzed further by Ballif, Fulton and Liddell (1925). They find the knee-jerk in the decerebrate animal ten to twenty times less susceptible to inhibition than in the "spinal" animal; moreover, the condition of inhibition in the "spinal" animal continues for only 0.1 second after application of a strong ipsolateral stimulus. A single, strong break-shock may, after a brief latent period, lead to an abrupt relaxation, the speed of which corresponds with the speed of cessation of activity of a motor-nerve tetanus. Studies of the effects of repeatedly applying a weak stimulus during sustained extensor contraction (the "stimulation plateau," Liddell and Sherrington, 1925) reveal a gradual relaxation. It is clear that inhibition of this character does not abolish motor discharge simultaneously in the total group of motor neurones affected. Evidence indicates that the gradual decrease in tension is due to suppression one by one of progressively large numbers of active motor neurones.

In chapter xvi experiments are described which lead to the conclusion that the tonus of visceral muscle is at times reflexly influenced by the tonus of striated muscle. Is the reverse also true? Carlson (1913) found that the tonus of skeletal muscles, as tested by the knee-jerk, increased with gastric hunger contractions—a result recently confirmed (Johnson and Carlson, 1927). Likewise, Daniélopolu and Carniol (1922) have dwelt upon the rôle of the vegetative nervous system in causing hypertonus of the skeletal muscles. This seems to help us understand our clinical observations that patients with chronically spastic viscera, including the esophagus, pylorus and colon generally require a particularly long period of training before they learn to relax the skeletal muscles.

It is evident from general considerations as well as from the foregoing discussions in this chapter that muscular activities are an index of the activities of the nervous system. For tonus or contraction in a skeletal muscle generally occurs only as the result of stimulating impulses from efferent nerves which lead to the muscle; and this can take place only if nerve centers and their afferent nerves become active. There is thus opened an approach to the clinical studies of the nervous conditions which will later be discussed.

Do Striated Muscles Have a Sympathetic Innervation?

As contraction in striated muscles diminishes toward the zero point, a prolonged period may be required before the advanced stages of relaxation are reached. The individual very readily relaxes the arm or leg so far as the large visible contractions are concerned, but a minute fraction of contraction, called residual tension in a previous chapter, often tends to persist. This, we have seen, is what calls for a finely cultivated method. We are led to ask whether this persistent remnant is not possibly of different origin from the grossly visible contraction; whether it does not possibly arise from the vegetative nervous system. We therefore come to the discussion whether muscle tone has a sympathetic origin.

In 1904 Mosso suggested that tonus is the function of nerve cells, especially the sympathetic. He quoted Bremer (1882) as the first to discover the sympathetic in addition to the somatic innervation of muscle fibers (in *Lacerta muralis*) and Grabower (1902) as finding non-myelinated fibers in addition to the myelinated in human muscle.

Much argument has centered about the innervation of the small striated fibers of the muscle-spindles discovered by Kühne in 1863. Medulated nerve-endings running to these fibers were observed by Cajal in 1888, by Huber and De Witt in 1897 and by others, including Hines in 1927, who by destroying the corresponding dorsal root ganglia in the frog, proved that they are sensory. Kerschner in 1888 had observed myelinated nerve fibers in man losing their myelin sheaths and winding themselves about the muscle fibers. This was followed by Ruffini's (1893) finding three types of nerve-endings in the spindles of the cat and of man, which he regarded as sensory (although he described also a motor type). As pointed out by Hines, whose recent review, like that of Cobb and Coman, is of marked interest in the present connection, physiologic proof must be added to histologic observation before the function of nerve-endings can be definitely ascertained. Using the cat and monkey in 1894, Sherrington cut certain ventral nerve roots as well as the dorsal roots between the cord and the dorsal ganglia. He observed medulated nerve fibers in the spindles with normal appearance 190 days after the operation. Cutting the sciatic nerve likewise failed to produce atrophy of the muscle fibers of the spindle. This proved that the spindle muscle fibers do not depend upon somatic motor nerves.

That non-medulated nerve fibers supply the spindle muscle fibers has been frequently observed: by Tschiriew in 1879, who described what he called *terminaisons en grappe;* by Giacomini in 1898, in birds; by Huber and De Witt the same year, in the spindle capsule of birds; and by Perroncito in 1901. Kulchitsky in 1924, from histologic studies of the python, concluded that the non-medulated nerve fibers were of sympathetic origin; a conclusion in harmony with the above-quoted theory of Mosso and the findings of Sherrington.

Physiologic evidence of the sympathetic innervation of some of the spindle muscle fibers is found in a few investigations. Hines concludes that the motor innervation in the frog is sympathetic, presumably because she found that the non-medulated fibers disappeared when the abdominal sympathetic chain was removed. Sherrington's experiments suggest that the innervation is sympathetic. Agduhr (1919*b*) cut the spinal nerves in cats between the root ganglia and the point of divergence of the white *rami communicantes.* This was followed by complete degeneration of all medulated nerves; but many non-medulated nerves remained intact and could be traced to the spindles as well as to other muscle fibers. Kuntz and Kerper (1924) confirmed these results.

Leaving aside the spindles, various histologists have collected evidence that striated muscle fibers may be innervated not only by medulated nerve fibers but also by the non-medulated variety. Bremer in 1882 found the latter in what he called the motor end-plate; Perroncito in 1901 in the lizard; Botezat in 1906 in birds; Boeke in 1913 in reptiles, birds and mammals; Kulchitsky in 1924 in the python; and Hines in 1927 in the frog. Boeke has emphasized that the fibers he considers sympathetic are distinguished not merely by their lack of myelin but also by their small caliber, peculiar structure, fine endings and distribution in loose plexuses —all in contrast with the non-myelinated somatic nerve-endings. Physiologic proof has been added that the non-medulated fibers under discussion are at least in certain instances of sympathetic origin. Boeke and Dusser de Barenne (1919) cut the anterior and posterior roots of the sixth to the ninth thoracic nerves and extirpated the corresponding spinal ganglia. After degeneration, no myelinated nerves, no motor end-plates and no sensory end-organs were found, but the muscles were still plentifully supplied with non-myelinated fibers and grapelike endings, indicating that sympathetic fibers were present. Agduhr had similar results, as mentioned above. Kuntz and Kerper (1924) confirmed Agduhr and also removed the stellate ganglion, which was followed by degeneration of the non-medulated fibers. Dusser de Barenne (1917) had an analogous finding with the hind-leg muscles. Contrary evidence is cited by Ranson (1926), who states that Hinsey has found that many or most of the non-myelinated fibers are not of sympathetic origin, since they persist in the terminal branches of the motor nerve after degeneration of all of the sympathetic fibers. Nevertheless, as Cobb and Hines indicate, the weight of evidence in favor of a sympathetic innervation of some striated muscles practically amounts to proof. Cobb (1925) reports a personal communication from Bielchovsky that the end-organs described by Boeke are most numerous in the external ocular muscles, less numerous in the facial muscles, and only occasional in the muscles of the extremities. If it should be confirmed that these end-organs are rare in the extremities, we may dismiss the matter as of diminished importance for the method of relaxation, since it would not explain residual tension in these parts.

If we assume that instances of striated muscle fibers being innervated by sympathetic nerves have been demonstrated, the question still remains whether the latter are sensory or motor. In 1913, De Boer cut the *rami communicantes* of the abdominal sympathetic nerves in frogs and reported a loss of tone in the ipsolateral leg muscles. But Beritoff promptly reported the contrary for frogs and cats in 1914 and Cobb found likewise in 1918. Langelaan (1915) agreed with De Boer, and maintained that the capacity of a muscle to retain a particular form and length ("plastic tonus") is due to the sympathetic innervation, while contractile tonus is due to the somatic nerves. This distinction has failed to meet with wide acceptance, for it seems possible to explain all contractions by gradual increase in the number of fibers involved.

Direct electrical stimulation of the sympathetic efferent (?) fibers has frequently failed to excite tonic contraction (Bottazzi, 1897; Beritoff, 1914; Cobb, 1918; Deicke, 1922; Uyeno, 1922). In the exceptions to this rule, the effects could generally be ascribed to circulatory changes produced by such stimulation. Orbeli (1923), using a bloodless frog preparation, believed that sympathetic stimulation added to somatic stimulation at the stage of fatigue increased the force and extent of contraction, which he considered a "trophic" effect; but Wastl (1925) could not confirm this with the frog and the anesthetized cat, while Varades (1926), also, secured negative results.

Later Maibach (1928), as well as Labhart (1929), found

positive evidence but particularly Baetjer (1930). In the cat with intact circulation, sympathetic excitation often provoked a notable augmentation in contraction, accompanied by no marked increase in afflux of blood. Diminished contraction, when produced, could be ascribed to vasoconstriction. Similarly, Jaschwili (1928) noted that excitation of *rami communicantes* lowers the rheobase and the motor chronaxie of the sciatic nerve. Most recently (1936) Esser agrees that stimulation of sympathetic fibers does not in itself arouse skeletal muscle contraction. However, if the gastrocnemius muscle in the frog is stimulated by infra-maximal or threshold shocks applied to its motor nerve, contraction appears or is augmented upon stimulating the sympathetic fibers. This occurs even in the absence of local blood supply and is accompanied by vasoconstriction in the interdigital membrane. The view is suggested that activity in the sympathetic nervous system, by augmenting the excitability of muscular units, renders efficacious nervous influences not previously operating overtly.

Esser finds a rise in the base line, showing increased tonus, during stimulation of the sympathetic; but this does not occur if time is allowed for relaxation to become complete between the volleys of discharges or if the frequency of stimuli to the sciatic nerve is sufficiently decreased. In certain respects, then, action of the sympathetic fibers tends to augment tonus.

In the effort to relieve certain conditions of spastic paralysis in man, Royle has cut the sympathetic branches to the affected limb (1924). He asserts that he has produced marked decrease of tone. Coman (1926), as well as Forbes and associates (1927), failed to confirm these results on animals. Royle's method has not proved popular among American surgeons but is still used sometimes (Delchef and

Roudil, 1934). His results in a large series of cases are open to the criticism that quantitative measurements are generally lacking. Carlson examined children operated in Chicago and concluded that suggestion plays an important rôle.

Without seeking to draw conclusions from a single instance, the results with one patient having Little's disease, treated with progressive relaxation, seem worthy to report, as supporting the view of Carlson and of Franz (1923) that to some extent it is possible to influence spastic paralysis by functional means. A child of twelve showed spastic paralysis, more accurately paresis, particularly of the right limbs, resulting from Little's disease. The peroneal muscles were relatively undeveloped. She dragged the limb in walking, while the feet, particularly the right, were moved in the valgus position. Proper walking therefore seemed impossible, and she had never run or engaged in the special games of children. Dr. Josephine Young found the intelligence quotient of the child to be 53–57 in three tests according to Terman's modification of the Binet-Simon scale. The patient was given tri-weekly treatments in relaxation, with the co-operation of Miss C. Kelly, graduate student in psychology, over a period of 9 months, with special periods of daily practice, including attempted relaxation while using her limbs. At the end of this time there apparently was much decrease of spastic resistance to passive flexion at the joints. Objective signs of diminished spasticity were the improved gait, since the shuffle had very largely disappeared, and the ability to run and to play for the first time in her life and even to use roller-skates.

The foregoing considerations suggest that we must perhaps take functional factors into account in order to explain Royle's results. Davis and Kanavel have reported negative results following operations on human beings as well as on cats (1926). After the operation, Hunter maintained that tonus progressively diminished during the following weeks. As Forbes and Cobb and others have pointed out, effects taking place after such a long interval could not be due to the mere withdrawal of efferent sympathetic impulses; for impulses cease to flow from the cut surface of a nerve within a few seconds after this section is made. We may therefore tentatively conclude that, while the results from human operation are still uncertain, there is evidence from animal studies that sympathetic fibers ending in striated muscle may exert an augmentative influence toward maintaining tonus.

THE NERVE IMPULSE

It is convenient to review certain portions of the physiology of nervous conduction and to seek to interpret our observations on relaxation in the light of recent conclusions. Spontaneous activity of nerve trunks has never been demonstrated; nervous activity, as observed, occurs only after stimulation. Even cerebral centers, it has been believed, do not remain active in the absence of stimulation. This view has been modified since Berger (1929) placed electrodes adjacent to the skull and, using an oscillograph, discovered voltage changes or waves evidently occurring spontaneously. These proceed from brain regions in various rhythms and rates, and to certain of them he attributes (1937) interrelation with psychic activity. Investigations of "brain waves" are under way in many laboratories.

It is generally agreed that at least in peripheral nerves, energy does not flow in steady streams but in series of impulses. When local excitation takes place, there pass propagated disturbances along nerve cells and fibers (cf. Adrian and Lucas, 1912). As is well known, Helmholtz, using frogs, was the first to measure the speed of such waves. In man, this was found to be about 117–125 meters per second (Piper,

1908), but Munnich (1915) found the rate to be about 30–71 meters per second. Rates vary in different fibers (Erlanger and Gasser, 1930). A wave of potential, negative to inactive portions of the tissue, distinguishes the nerve impulse (Ostwald, 1890). This passes not only over the motor nerves but over the sensory as well (Einthoven and Jolly, 1908; Hartline, 1925; Adrian and associates, 1926). The duration of a single wave of negative electrical potential along a nerve fiber is on the order of 0.7–1.6 sigmata in the phrenic nerve of the dog (Erlanger, Bishop, and Gasser, 1926).

It was formerly believed that nerve fibers carry stronger or weaker impulses depending upon the strength of stimuli and causing a corresponding degree of muscular contraction. Now it is generally held that the nerve, like the muscle fiber, at any moment responds as fully as it then can or not at all ("all-or-none law"). Gotch (1902) first threw doubt upon the old view, suggesting that the submaximal contraction of a muscle or the submaximal electrical response of a nerve in answer to a weak stimulus resembles the effect produced by a maximal excitation of a few of the fibers, since the time relations of the submaximal effect do not differ from those of the maximal effect. The response of muscle fibers is all or none (Pratt, 1917, 1925). Exceptions are ascribed to injury (Steiman, 1937). That the same holds likewise for nerve fibers follows from a large body of investigations (Symes and Veley, 1910; Vészi, 1912; Lucas, 1912; Lodholz, 1913; Adrian, 1912, 1913, 1914, 1924; Kato, 1924; Davis, Forbes, Brunswick and McHopkins, 1926).

Lucas points out that the nerve impulse is a disturbance which depends for its progression on the local energy supply in the nerve itself. He compares it in this respect with the firing of a train of gunpowder, and this has been practically proved by Adrian (1912). Transmission of excitation from nerve to muscle perhaps depends upon the liberation of minute quantities of substances like acetylcholine and "sympathin." A very small amount of acetylcholine injected into the empty blood vessels of a normal mammalian voluntary muscle occasions a short tetanus (Brown, Dale and Feldberg, 1936; Brown, 1937; Bacq and Brown, 1937). Transmission of impulses through sympathetic ganglia also appears to be due to acetylcholine (Cannon and Rosenblueth, 1937). Although sympathetic nerve impulses probably are usually mediated by the adrenin-like sympathin, some may act through liberation of acetylcholine (Cannon and Rosenblueth, 1933; Dale and Feldberg, 1934). Investigations on chemical transmission have been reviewed by Kuntz (1934), Brown (1937), Rosenblueth (1937) and Eccles (1937).

Nerve impulses probably depend for transmission on the supply of energy along the course of the nerve (Lucas); for they use O_2 (Bayer, 1903; Fillié, 1908; Thorner, 1908, 1909; Gerard, 1927) and give off CO_2 (Tashiro, 1913) as well as a very slight amount of heat (Hill, 1912; Downing, Gerard and Hill, 1926). The initial energy liberation of the nerve, judging from the duration of the rising phase of the action-current, is very short, perhaps about 0.75 sigmata. About 90 per cent of the total heat is liberated after the stimulus is over. These writers find that the observations of Parker on the extra CO_2 output resulting from the passage of impulses along the nerve agree well with the total heat observed, on the hypothesis of the oxidation of some ordinary foodstuff. In frog's nerve the initial heat for a single maximal impulse has been calculated as about one ten-millionth of a calorie per gram, and the total heat about ten times as much (Gerard, Hill and Zotterman, 1927). This is only about one eight-thousandth of the amount of total heat given off by an equal weight of muscle produced by a single twitch.

As nerves are stimulated at increasing rates per second, the heat per impulse falls off; in other words, the nerve apparently has insufficient time for a full return of energy. If the interval falls below certain limits, energy will be insufficient for wave propagation, which explains the absolute refractory period. Hartree and Hill (1921) found the return of heat-liberating power in a frog's muscle at 10° C. to be complete in about 0.2 seconds, which is just about the time taken in complete relaxation at that temperature. This "recovery" in muscle has therefore seemed to investigators to be associated in some direct way with relaxation; and since relaxation is known to be accompanied by heat production, it was suggested that in nerve the return to its initial condition will also be accompanied by a production of heat. This has led to the belief that in nerve, as in muscle, there may be three phases of liberation of heat; namely, an initial very rapid phase corresponding with the rise of action current, a second phase complete in about 50 sigmata concerned with the restoration of excitability and conductivity and a third phase lasting about 10 minutes during which eight-ninths of the total energy is liberated. In later investigations (1927) Gerard has been able to increase the accuracy of measurement of the two phases of heat production of nerve and has gathered evidence that nervous activity involves increased oxygen consumption and carbon-dioxide production rather than a glycogen-lactic acid system like that described by Hill for muscle.

The presence of heat in nerve subsequent to nervous activity for a period of about 10 minutes and associated with delayed oxygen consumption permits interesting speculations. What was called nervous hypertension in the earlier chapters of this volume, a condition of hyperirritation and hyperexcitation which I believed could be traced to an excessive number of neuromuscular reactions (lack of relaxation), may eventually be found to depend upon an altered physicochemical state of nerve substance, in some measure a function of delayed heat production and perhaps delayed oxygen consumption. The assumption that chronic excess of nervous activity in man fails to permit adequate "recovery" of nerve to its initial condition, locally or generally, may help to explain the pathological physiology of occupational and other neuroses.

It seems at least of passing interest that the experiments described in the early chapters of this volume also bring out that the character of the response of the whole organism, like that of the single nerve, depends not merely upon the stimulus but upon the properties of the part stimulated. If we accept the "all-or-none" point of view, it is suggested that as relaxation progresses in a muscle-group, the number of impulses per second per fiber undergoes a diminution (Wachholder), perhaps even to zero in some sets of fibers earlier than in others. In consequence, the muscle-group, when tonus has disappeared, will be inexcitable by stimuli which in the previous state of tone would have been adequate to produce a response. Further experiments to test this view are desirable. Theoretically the effects of progressive relaxation can readily be explained in terms of the so-called "all-or-nothing" law.

Since motor and sensory nervous activities consist of series of impulses, it is but to be expected that only the first of these impulses travels along the resting nerve (Piper, 1912; Vészi, 1913; Dittler and Günther, 1914). That there is a refractory period following the nerve impulse during which the nerve fails to respond to stimulation was first demonstrated by Gotch and Burch (1899). Lucas states that the following roughly holds for the frog's sciatic nerve at 15° C: After a previous impulse, conduction is impossible from 0.0 to 0.003 seconds, is impaired from 0.003 to 0.015 seconds, and is supernormal from 0.015 to 0.1 seconds. Sherrington (1904, with Sowton, 1915) investigated the refractory period of reflexes and found that in the flexion reflex it was 0.76. Recent investigators find that this period varies from about 0.6 to 3.6 (Forbes, Ray and Griffith, 1923).

Hoffmann (1922) has found that the refractory period of a tendon reflex elicited by an induction shock (extensor muscles of the foot) exceeds 0.2 seconds. Decrease in the extent of reflexes occurring at the rate of 5 per second could be prevented by voluntary extension of the foot (*Bahnung*) during the delivery of the stimuli; at the rate of 50 reflexes per second decrease could still be prevented, even with moderate voluntary contraction and a weak electrical stimulus; but as the rate of reflexes was increased up to 120 per second, not alone was it necessary to increase the stimulus to the reflex but also much stronger voluntary contraction was required. Hoffman feels warranted in concluding that voluntary contraction tends to shorten the effective refractory period of the reflex while voluntary relaxation tends to lengthen it.

Further investigations are needed on the relation of voluntary relaxation to the refractory period following various reflexes. It is suggested that the disorder present in various functional neuroses is often due to the excess and diversity of neuromuscular processes present at a particular moment. Failure of response might also be due to the frequently observed tendencies of central nervous processes to inhibit other simultaneous ones (Setschenow, 1863; Herzen, 1864; Freusberg, 1874; Bubnoff and Heidenhain, 1881; Oddi, 1895; Hering, 1898; Wedensky [cited by Bechterew, 1908]; Vészi, 1910). In agreement therewith, conscious processes (sensations) often tend to inhibit each other (Jacobson, 1911, 1912).

From this point of view it seems plausible that one effect of relaxation is to promote order through the reduction of superfluous activities and the consequent elimination of various inhibitions.

Binswanger points out that the old theory of nervousness as irritability failed to take into account the many inhibitions that characterize neurosis and therefore gave way before the more comprehensive conception of Beard. As shown above, the objection made by Binswanger does not apply to the present conception of nervousness as nervous hypertension which was outlined in chapters ii and iii.

Sherrington in 1906 summarized the differences between conduction in the nerve trunk and the reflex arc, adopting the theory that they are due to the presence of the synapse in the latter. Among these he included (a) slower speed as judged by latency of response; (b) after-discharge, i.e., persistence of response after stimulation has ceased, often for several seconds; (c) summation; (d) fatigue on continued stimulation, in contrast with the nerve trunk which exhibits extraordinary resistance to fatigue; (e) mutual relations between allied or antagonistic reflex arcs, manifesting themselves as reinforcement or inhibition. Assuming the correctness of his views, we might explain the observed effects of relaxation in slowing reflex times as due to diminution of the impulses whose cumulative energy is needed to overcome synaptic resistance. Another effect frequently observed in man when voluntary relaxation is quickly performed is the sudden termination of reflexes such as the scratch reflex. It seems plausible that this is due, at least in part, to lessening of the after-discharge at the synapses produced by the diminution of afferent impulses from the muscles during relaxation. Similarly, this diminution of afferent impulses may be partly accountable for the diminished fatigue which follows

relaxation, as well as for the diminished production of reinforcements and of excitations which lead to inhibition. In this
way the observations on relaxation readily can be explained
in terms of current views of the functions of the synapse.

THE CONTROL OF MUSCULAR CONTRACTION
BY THE CENTRAL NERVOUS SYSTEM

It is beyond the scope of the present volume to cover more
than the salient features of the central nervous regulation of
tonus. Adequate reviews of this subject have recently appeared (F. R. Miller, 1926; Langworthy, 1928).

That the tonus of skeletal muscles depends upon a more
or less continuous activity of their motor nerves is generally
agreed. Also, various centers of the cord and brain are commonly said to be in a state of tonic activity, including the
vasoconstrictor center, the center for the sphincter muscle of
the iris, and the centers for the sphincter muscles of the bladder and the anus. As is well known, the respiratory centers
probably respond directly to chemical changes in the blood.
Certain brain potentials, the 10 per second or alpha waves
of Berger, become more irregular and lower in amplitude
upon sensory stimulation, but increase again in regularity
and amplitude upon the return of moderate relaxation. Their
significance as related to muscular control is not yet clear.

The Cerebellum

Numerous investigations have revealed that the cerebellum plays an important part in the maintenance of muscle
tone. From the proprioceptive sense organs of muscles, as
well as possibly from those of joints and tendons (Marburg,
1904; Bing, 1906), impulses travel through the dorsal and
ventral spinocerebellar tracts, which enter the cerebellum
respectively via the lower and superior peduncles. From the
cerebral cortex (association centers and motor areas) im-

pulses are relayed to pontine nuclei (respectively by corticopontine tracts and by collaterals from the corticospinal tracts), thence via the middle peduncle to the median and lateral parts of the cerebellum. From three cerebellar nuclei axons pass over the superior peduncle to end in the red nucleus of the midbrain. This is connected through the rubro-spinal tract with motor cells of the anterior horn of the cord; but there are other connections to the thalamus and cerebral cortex. Important connections exist (e.g., from the nucleus fastigii) to the nucleus of Deiters. Various descending pathways from these pontobulbar centers finally lead to the spinal motor neurones which maintain postural tone.

Most important among early studies is Flourens' (1824–42) account of the pigeon, which deprived of the cerebellum, cannot jump, fly or walk. Equally well known are Luciani's (1891, 1915) investigations of dogs and monkeys with the entire cerebellum or certain parts removed. He found that violent disturbances of movement slowly disappeared after the operation and concluded that they were due to surgical irritation. But later investigators attribute them to removal of inhibitory influence exerted by the cerebellum on muscle tone (Jackson, 1884; Sherrington, 1898; Head, 1921). After 8 or 10 days, the characteristic defects appeared to him to be loss of force in muscular contractions (asthenia), loss of tone (atonia) and loss of steadiness (astasia). This condition of atonia may be compared to the loss of tone in the hind-leg muscles of the frog produced by cutting the dorsal roots in Brondgeest's (1860) experiment. His theory, therefore, was that the cerebellum is an augmentary organ for activity of the neuromuscular apparatus.

Luciani observed that the totally decerebellate dog can swim before it has recovered the ability to stand or walk, which led him to deny that the cerebellum is an organ of equilibration, as has been recently contended by Ingvar and others. Likewise, the decerebellate dog can make progressive movements (Laughton, 1924) which confirms Luciani's view that the organ is not a mechanism for the co-ordination of movements. Muscle sense was proved intact in the semidecerebellate dog by the equal promptness with which abnormal positions were corrected on both sides (Luciani). This conclusion, as well as the characteristic presence of hypotonia, asthenia and astasia were confirmed in Gordon Holmes' (1917, 1922) investigations on men with cerebellar injuries sustained during the great war. A conspicuous symptom was delay in commencing and in relaxing muscular contractions, resulting in movements often overshooting the mark. Holmes believes that the cerebellum influences postural tone by a continuous discharge which tends to maintain attitudes, but also by a discontinuous discharge that regulates the tone accompanying muscular relaxations and contractions.

The effects of electrical stimulation of the cerebellum as well as of ablation experiments of Rossi (1912) likewise indicate that the cerebellum has a facilitating influence on the motor mechanism of the cerebrospinal system; but he suggests (1920) that the symptoms of astasia and asthenia observed by Luciani should be imputed at least in part to insufficient postural tonus. This view has been further developed and analyzed by Noïca (1921), Simonelli (1921), Walshe (1922) and Hunt (1921), in the direction that cerebellar insufficiency is in some sense equivalent to insufficiency of postural tonus.

That the cerebellum has not only an excitatory but also an inhibitory function in the regulation of postural tone is demonstrated by the observations of Löwenthal and Horsley (1897) and of Sherrington (1898) upon stimulating the superior vermis.

Contrary to the foregoing, Magnus and Rademaker (1923) argue that the cerebellum has no part in the production of extensor tone and that decerebrate rigidity may develop very markedly after the entire cerebellum has been removed. No dysfunction occurs upon middle-lobe decortication (Keller *et al.*, 1937). Postural responses follow stimulating certain regions within the vermis and hemispheres (Hare *et al.*, 1936). Results vary, but most agree that the cerebellum takes part in regulating the tonus of somatic muscles.

Mesencephalic Centers

The extirpation experiments of Weed (1914), using cats, led him to believe that the caudad portion of the midbrain must be intact if extensor rigidity is present. Accordingly he gave attention to the red nucleus, since this is the largest group of motor cells in the midbrain region. This nucleus makes connections with the *nucleus dentatus* and the *nucleus pallidus* of the cerebellum, links the *nucleus globosus* and the *nucleus emboliformis* with the descending rubrospinal tract and lies in the pathway of fibers arising in the frontal cortex and going to the *tegmentum* of the midbrain and *formatio reticularis* of the pons, from which they are presumed to send relays to the spinal cord. Decerebration in cats above the level of the red nucleus produces rigidity of strong and constant character (Cobb, Bailey and Holtz, 1917); but if below the red nucleus, the extensor hypertonus is more variable and depends on afferent stimuli being applied to the limbs. These conclusions are challenged by Bazett and Penfield (1922), since in many of their decerebrate-cat preparations, in which post-mortem studies made certain that the red nucleus had been destroyed, extensor rigidity was nevertheless present for days after the operation. However, Magnus and Rademaker (1923) reported that section anterior to the red nucleus, i.e., leaving the midbrain intact, produced no rigidity; but if the red nucleus was injured, rigidity resulted along with abolition of the labyrinthine postural reflexes. Injury of the red nuclei in cats and rabbits by stab wounds in intact animals and in "thalamus animals" characteristically produced rigidity, and the same result followed injury of the rubrospinal tract at Forel's decussation. Accordingly these investigators believe that the red nucleus is part of the mechanism that maintains tonus normally, but they ascribe to it no excitatory but only an inhibitory function. Mussen (1926) produced lesions in portions of the red nucleus and obtained practically negative results as to effects on tonus, apart from some disturbances of righting reflexes. Nevertheless, electrical stimulation of the red nucleus produces marked extension of the contralateral forelimb, and the balance of evidence favors the view that red nucleus cells have an excitatory influence on postural tone (Graham Brown, 1913; Weed, 1914; Cobb, Bailey and Holtz, 1917; Langworthy, 1926); in man (van Gehuchten, 1933; Steblov, 1935).

Vestibular and Other Centers

As is well known, Magnus and De Kleijn (1912, 1915) showed that in some reflexes postural tonus is regulated by the vestibular system. They sectioned the cerebral peduncles in cats, producing decerebrate rigidity. Different positions of the head now brought definite changes in the posture of the extremities. These changes were due to labyrinthine reflexes, since they disappeared upon destruction of both labyrinths. Ewald (1892) had previously demonstrated that impulses from the labyrinth tend to maintain the tone of voluntary muscles, for he found that the latter became flabby when the labyrinth on both sides is destroyed.

Evidence that Deiter's nuclei are perhaps the chief reflex

centers concerned in decerebrate rigidity is therefore considerable. Other centers to which the function of tonic innervation have been assigned include the inferior olives (Zylberlast-Zand, 1925), the large cells in the *substantia reticularis* (Bernis and Spiegel, 1925), the optic thalamus in an inhibitory direction (Thiele, 1905), the *corpora striata* (Wilson, 1924*b*, 1925), the temporal lobe (Bernis and Spiegel, 1925), but most often the frontal lobe, which will now be discussed. (Regarding the diencephalon, see p. 215.)

The Cerebrum

Since Fritsch and Hitzig first disclosed (1870) the location of a cortical region in the dog which upon stimulation yielded definite movements, their results have been confirmed in practically every physiological laboratory. As is well known, the motor area has been more closely localized in the monkey in the anterior central convolution (Grünbaum and Sherrington, 1902, 1903; Leyton and Sherrington, 1916). From this area arise the pyramidal tracts, and clinical experience has shown that lesions in this part of the cortex are accompanied by paralysis of the muscles of the other side. Thus clinical, pathological, and physiological investigations combine to trace the course of pyramidal fibers through the *corona radiata*, the internal capsule, the cerebral peduncle into the pons, where some pass to the cerebellum, others cross to motor cranial nerve nuclei, and still others continue downward through the anterior pyramidal tracts and the decussation, passing via the anterior and lateral fasciculi in the cord to end in the anterior horn cells.

Another pathway to the muscles, the extra-pyramidal tract, probably commences in the premotor cortex, involving also the optic thalami and the *corpus striatum*. From the latter they travel partly to the red nuclei, the olives, *substantia nigra*, or to pontile nuclei, descending then in the cord.

TONUS AND NERVOUS REGULATION

According to Dusser de Barenne (1920), the cerebral cortex participates in the control of postural reflexes. Removing the *area frontalis* plus the electrically responsive motor area occasions extensor hypertonus (Warner and Olmsted, 1923; Olmsted and Logan, 1925). In cats, King (1927) removed what corresponds with the "intermediate motor area" of Campbell (1925), resulting in contralateral extensor hypertonus. The premotor area in primates probably integrates voluntary motor activity and co-ordinates postural adjustments (Fulton and Kennard, 1934), but involvement in man may or may not give rise to spasticity (Davison and Bieber, 1934). In man, conditions of spasticity resulting from injury to motor tracts by cerebral hemorrhage are well known (apoplexy and Little's disease). Wilson and Walshe (1914) reported cases of injury to the "intermediate motor area" with slight spasticity of the affected legs; following voluntary innervation of these parts, there was marked inability to relax (see also Barris, 1937).

It is very clear from the foregoing that the evidence concerning extensor rigidity conflicts at many points. At best only a beginning has been made toward the adequate understanding of the infinite variations of tonus of which the intact animal, particularly man, is capable. A hundred questions may be asked for one that may be answered. As a reviewer concludes, almost all portions of the central nervous system appear to have a part in the nice control of postural reflexes and with the progressive development of the brain, this control tends to pass to higher levels. In harmony with this point of view are the observations during the present investigation of the effects of general relaxation (man) initiated from cerebral sources The postural reflex as a rule, regardless of what particular area of the central nervous system controls its action, diminishes or disappears with extremely advanced relaxation, illustrating that in man the control has been assumed by higher levels.

Activity in the Intact Organism

a) *Reciprocal innervation.*—Since Sherrington published his first observations concerning the reciprocal innervation of antagonistic muscles (1893), it has been customary to assume that flexor muscles relax while corresponding extensors contract and vice versa. Fulton lists fourteen contributions from Sherrington developing this point of view (published respectively in 1893, 1893, 1897, 1898, 1899, 1900, 1905, 1905, 1906, 1907, 1908, 1908, 1908, 1909) and adds a discussion of the physiology which it is unnecessary here to repeat. Our interest arises from the fact that it is a frequent practice (chap. v) to request the individual who is learning to relax to contract all the muscle groups of a limb simultaneously, producing a rigidity of the member with little or no movement. After he does this with progressive intensity, he ceases to contract, but again progressively—that is, he is instructed to contract "not quite so much" and so on. It scarcely appears necessary to secure mechanical or electrical records of the simultaneous contraction of flexors and extensors under such conditions; palpation unmistakably discloses the marked rigidity of the antagonistic muscle groups. However, electromyograms during certain simple voluntary movements have been secured by several investigators (Pfahl, 1921; Wachholder, 1923, 1925; Golla and Hettwer, 1924; Wagener, 1925). They indicate that synchronous co-contraction can and often does take place. Admitting this, we are not obliged to deny that reciprocal innervation is the rule in the cord reflexes as studied by Sherrington. Obviously, as Fulton remarks, both types of phenomena are observed facts and not merely possible explanations. Indeed, reciprocal innervation commonly occurs in intact human subjects (Beevor, 1904). Such are the observed phenomena, which need not be disputed; the problem remains unsolved how voluntary innervation can sometimes make use of the mechanism of reciprocal innervation yet at other times just as readily produce synchronous co-contraction. Physiological theories of the mechanism of reciprocal innervation will need to take this possible shift into account.

b) The "common path."—It is generally agreed that the efferent paths involved in cerebral control of muscular contraction are chiefly the pyramidal tracts. In explanation of the mechanism that enables one particular reflex or voluntary act to be carried out at any particular moment to the exclusion of other possible ones, Sherrington (1904*b*, 1906) advanced the "principle of the common path." There is but one motor fiber to five sensory ones in the nervous system (Ingbert, 1903; Sherrington, 1906, p. 145). "The receptor system bears, therefore, to the efferent paths the relation of the wide ingress of a funnel to the narrow egress. Further, each receptor stands in connection not with one efferent only but with many—perhaps with all, though as to some of these only through synapses of high resistance." When two or more reflexes can occur simultaneously using the same final common path, reinforcing each other, they are said to be "allied excitatory reflexes." But if certain other points be so stimulated as to produce a depression of activity of the final common path, the reflex exerting this influence is called "inhibitory," and two or more such reflexes are termed "allied." Accordingly, during normal activity there occurs a competition between the afferent paths to such excitatory and inhibitory reflexes by mechanisms which Sherrington considers at length, and the resultant reflexes make up what is known as "reflex co-ordination."

"Alliance" may also exist between voluntary activity and a reflex, since Hoffmann (1918) finds that voluntary contraction of any muscle increases the reflex excited by tapping the tendon of that muscle. However, in this instance the alliance is not quite mutual, for following the tendon reflex which has been favored by the occurrence of such mild voluntary contraction, there occurs an inhibition of the latter for a brief period, about 0.1 second (Hoffman, 1922), which is perhaps a refractory period. It is obvious that amid such complicated relationships we still have far to go to attain an adequate understanding of the mechanisms of reflex co-ordination.

c) Tonus as possibly a function of higher centers.—At the commencement of every reflex-arc is a receptive neurone extending from the receptive surface to the central nervous organ. Sherrington comments that the "simple reflex," in the sense of an isolable and isolated mechanism, is a convenient but artificial abstraction. The nervous system, he indicates, functions as a whole. Physiological and histological analysis finds it connected throughout its entire extent. Donaldson (1900) opens his description of it with the remark: "A group of nerve-cells disconnected from the other nerve tissues of the body, as muscles and glands are disconnected from each other, would be without physiological significance." Bearing these considerations in mind, we now come to the question whether in man the normal mechanism of tonus essentially involves the action of higher brain centers. At first one is inclined to answer negatively on the well-known ground that tonus exists in decerebrate preparations. But a little reflection reveals the possibility that under these conditions of trenchant mutilation, a short-circuiting may occur in the absence of the cerebrum, which does not occur in the intact animal. In other words, conclusions based upon what happens in mutilated preparations may be suggestive, but they are not final proof. They indicate possibilities, but additional and direct evidence is needed if we are really to know and not merely to infer what goes on under normal conditions.

Accordingly there has been a growing conviction on the part of many physiologists that in the intact organism tonus is occasioned by afferent impulses passing not directly to anterior horn cells but via higher nerve centers. This has been called "long-circuiting." As Pike (1912) remarks, beginning with Rosenthal (1873) and Bastian (1890) and later with Crocq (1901), the idea has grown up that the reflex tonus of the skeletal muscles and most skin reflexes in adult human subjects are dependent upon reflex-arcs which pass through the cerebral cortex, and that the arcs for the tendon reflexes lie through the brain-stem; while in newborn animals and in the lower forms generally, these reflexes may depend upon short arcs through the spinal cord alone. If the path through higher centers is cut, conduction can then take place directly through spinal centers, a pathway which is probably phylogenetically the older. Likewise Langworthy (1928), reviewing investigations on reflex-arcs concerned with the postural reflex, concludes that they "undoubtedly form long circuits." A similar view is expressed by Fulton (1926), partly based on a certain observation during decerebrate rigidity:

It is obvious that the process of long-circuiting of impulses up the cord is of fundamental importance for the integrative processes occurring in the higher centers. It would appear, moreover, that so long as a given controlling higher center is present, the majority of afferent impulses are long-circuited in this way. When, however, the higher centers are cut away, the majority of impulses pursue the phylogenetically older paths across the cord.

As an example of this variety of long-circuiting, he quotes an observation by Amsler (1923). In lightly anesthetized animals, stimulation of the central end of the cut sciatic nerve causes cardiac inhibition and a reflex cry. These phenomena diminish and finally disappear under deep ether or morphine anesthesia; but if the superficial layers of the cerebral cortex are removed, even though the animal remains under deep anesthesia, the cardiac inhibition and the cry immediately return. Amsler concludes that so long as the cerebral cortex remains intact, the pain impulses are long-circuited through this region. Obviously the welfare of the organism is subserved if pain impulses are conveyed through the higher integrative levels of the nervous system, thus favoring cerebral regulation.

Hoffmann (1921; Hansen, Hoffmann and von Weizäcker, 1922) goes so far as to believe that the lowest grade of voluntary excitation accomplishes what we call "physiological muscle tonus."

But there is a discrepancy in the evidence: Jolly (1910) found that the interval between the tap on the patellar tendon and the beginning of the muscle action-current of the knee-jerk may be as brief as 6 sigmata. He subtracted the time required, according to Piper's (1912) figures, for the nerve impulses to traverse the distance to the cord and back again to the muscle, and from this obtained a figure representing the time consumed in passing through the central nervous system alone. This amounts to but 2 sigmata, which is not long enough to permit impulses to pass to the brain. Still less time is consumed in the central nervous system than even Jolly calculates, if Hoffmann (1922, p. 57) is right in correcting Jolly's figures on the ground of temperature differences. According to either set of figures, we must conclude with Jolly that during the knee-jerk the brevity of time proves that the impulses do not pass to the brain but only to the cord and back again to the muscle. If, therefore, we regard the myotatic reflex as "tonus in the making," we have definite evidence that this element of tonus is a purely spinal reflex (see also Herren *et al.*, 1936).

However, since there is evidence on both sides of the question, it seems possible that in some instances studied tonus depends upon reflex-arcs which pass through higher nervous centers, while in others it does not. This double possibility seems perfectly compatible with what we know of the complexity of mechanism frequently found in the nervous system. In the present investigation we find some evidence that afferent impulses aroused by muscle tone arrive at higher centers, for sensations from muscular contraction can readily be perceived. Furthermore, we find that upon extreme voluntary relaxation skeletal muscles can be reduced to a toneless state. Assuming that this is not due to cerebral inhibition of the anterior horn cells whose activity results in the impulses that produce muscle tone, the foregoing is evidence that cerebral (i.e., conscious) centers intervene in some measure in the production of tonus. Finally, we have seen in three investigations (chap. vii and ix) that during extreme muscular relaxation the knee-jerk and the flexion reflex are diminished or fail to occur upon appropriate stimulation, again indicating the part played by the voluntary element in the production of tonus. Indeed my subjects report during general relaxation a certain inertness of voluntary activity, an unreadiness to start to move; that is, voluntary activity, like the reflexes, apparently suffers diminution.

These findings agree with those of Hoffmann during studies of tendon reflexes (*Eigenreflexe*) elicited by electrical stimulation of the mixed nerve near its point of entry into the muscle. While he was not directly interested in problems of relaxation, his investigations have an important bearing on that subject. He finds that voluntary contraction effects an

[257]

increase of reflex excitability. With one exception (the extensor reflex of the foot), he observed that the tendon reflexes could not be electrically produced if the limb tested was in a state of relaxation. Only when the subject was requested to voluntarily contract the muscle involved, could the reflex be elicited upon such stimulation. In a general way, therefore, his studies also bring out the diminution of tendon reflexes elicited electrically with relaxation, as compared with tension.

Of similar import are his observations on reflexes produced with the vibration apparatus (Hansen and Hoffmann, 1922) and on extensor foot clonus in normal persons. If the forepart of the ball of the foot is pressed on the floor and tremor of the extensors is permitted to begin, it continues involuntarily in many individuals. As soon as voluntary innervation diminishes the refractory period of the reflex, clonus takes place. In the relaxed muscle, clonus cannot take place, because the refractory phase of the reflex is too long. I have often observed during clonus that upon voluntary relaxation, although a favorable position is retained, the reflex ceases.

Light is thrown by these considerations upon the probable mechanism when, during the present investigations, the subject is given certain directions which according to our experience tend to help him to become relaxed. As previously related (chap. v) he contracts a muscle-group, such as the flexors of the forearm, and receives the instruction, "This is you, doing! What is desired is the reverse of this—just not doing." This simple instruction followed by some of the others mentioned may lead him to cease voluntary contractions in a progressive manner. In consequence, tonus may in some instances diminish to the zero stage. If we assume, with various of the foregoing investigators, that tonus depends for its maintenance on reflex-arcs which pass through cerebral centers, we need no further explanation of what substantially occurs: the cerebral portion of these reflex-arcs progressively ceases to function as voluntary relaxation advances, and tonus must therefore diminish and eventually disappear. If we make the more moderate assumption that in some instances tonus at least partially depends upon such cerebral arcs, we are still provided with an adequate explanation of the phenomena described in the present investigation: the subject encouraged to relax by so doing removes a measure of muscle tone from the muscle in which a reflex is being stimulated, so that the reflex suffers diminution or does not occur at all. In the light of present knowledge, the latter assumption and explanation would appear most reasonable; but other alternatives remain to be considered (chaps. xiv and xv).

d) Rates of nervous and muscular discharge during voluntary activity.—A very brief review of some of the observations and conclusions on the rate of action-currents during voluntary activity will suffice for present purposes. As is well known, Piper (1907, 1908, 1914) stated that action-currents during voluntary contraction occur with a fairly regular rate of about 42–100 per second, averaging about 50. Most observers agree that they are irregular and of higher rate than attributed by Piper, namely about 100–200 per second, averaging about 160. Whether these rates are peripheral (muscular) or central (nervous) in origin has aroused much discussion; Piper took the latter view, but Buchanan (1908) defended the first, finding larger waves (3 to 14 per second), which she attributed to the nervous elements. Accurate counting with the string galvanometer requires that the string be critically damped, a fact ignored by many. Counts made with the oscillograph and Adrian-Bronk electrodes during voluntary contraction show rates from 5 to 20

(Smith, 1934) or 3 to 50 per second (Lindsley, 1935). Wach-holder (1923) found that muscles when not quite completely relaxed give off action-currents, 5–10 per second. With increase of muscle tension, more fibers participate and at higher rates.

e) Inhibitory action.—During various decades evidence has been gathered that higher centers, particularly the cerebrum, exert an inhibitory influence over functions of the cord. This may be conveniently grouped as follows: (1) In 1863 Setschenow made his classic observation that if the optic lobes or medulla of the frog are stimulated with salt crystals or with the induced current, the foot is withdrawn much more slowly from acidulated water than normally—that is, reflex time is increased. (2) That removal of the hemispheres augments spinal reactions is a familiar observation, made by Marshall Hall in frogs as early as 1833. (3) Bubnoff and Heidenhain (1881) demonstrated that during morphine narcosis in the dog there may be produced certain slight muscular contractures in an extremity which may be made to cease upon slight stimulation of certain points of the cortex or upon stimulation of the skin or sense organs. They believed that these contractures were of spinal origin. Results of similar import were had by Exner (1882) on the rabbit, by Oddi (1895) and by Hering (1902). (4) Sherrington (1893–94, 1905) upon stimulation of the cortex demonstrated inhibitory effects on the extra-ocular muscles as well as on the extensors of the elbow and knee in monkeys (1897, 1898). The cortical area responsible for inhibition of the active extensors also produced contraction of the flexors, while a different area produced contraction of the extensors. Weed (1914) traces the inhibitory pathways from the cerebral cortex via the pontine nuclei to the cerebellum, particularly the superior vermis. Stimulation of the premotor area in cats, dogs and monkeys often inhibits bilateral motor responses in hind limbs. Results vary, but it is possible to evoke excitatory or inhibitory influences independently (Rioch and Rosenbleuth, 1935). (5) It is well known that reflexes which can be elicited with machine-like regularity in the decerebrate or spinal animal are uncertain and unpredictable if the animal is intact.

The modern conception of "levels of function" in the central nervous system is foreshadowed by Charles Bell, according to Thompson (1926) and Fulton; but physiologists commonly attribute it to Hughlings Jackson (1884). Higher levels are presumed to have a twofold influence, namely, excitatory as well as inhibitory. It was Jackson who first argued that hemorrhage in the human brain, by throwing this higher level out of function, released from inhibition the spinal centers, whose activity being then unrestrained produced spasticity. Experimental evidence in favor of this view was furnished by Sherrington's (1897a) discovery that removal of the cerebrum in cats and dogs produced rigidity.

Hoffmann, on the ground of observations recounted in the following chapter, has suggested that voluntary contraction appears to be little more than a series of augmented tendon reflexes. If we adopt Hoffmann's suggestion, according to Fulton (1926, p. 489), "voluntary contraction may in fact consist almost entirely in release of the lower motor neurones from higher inhibitory control." Here we have an apotheosis of inhibition which needs cautious consideration, especially when we recall the history of the subject.

In the sixties and seventies of the last century, a lively debate arose whether centers exist whose special function is to produce inhibitions in the brain and cord. Setschenow took the affirmative on the ground of his observations, including those above mentioned, and of the increased excitability of reflexes after decapitation. But Herzen (1864), on the contrary, showed that every marked mechanical and chemical stimulation of the lower region of the cord tends to produce a diminution of the reflexes of the upper region of the cord, and that, even when the whole brain and cord are removed, inhibitions can still be demonstrated. Hereupon, Setschenow was obliged to assume that inhibitory centers exist anywhere and everywhere in the cord and brain. This notion, however, was said to complicate matters rather than to clear them up, since it seemed too much to assume that every sensory stimulus occasions not alone a corresponding reflex but also simultaneously excites inhibitory centers scattered throughout the entire central nervous system; hence it was combated, especially by Goltz and Freusberg (1874). The latter drew attention to the fact that in dogs with the cord cut above the lumbar level, there occurred, upon lifting the hind leg, a rhythmic extension and flexion which could be inhibited by stimulating (stroking or pinching) such divers places on the periphery as the back and tail. Therefore he concluded that there are no "definitely bounded centers of inhibition either in the cord or brain, but rather that any sensory stimulus in any part of the cord may work an inhibition." These conclusions came to be generally accepted by physiologists and evidently are worth bearing in mind today.

Observations on the inhibition of spinal reflexes produced by stimulating the cerebral cortex lead some authors to go so far as to regard the pyramidal tracts as purely inhibitory paths. But Sherrington (1906, pp. 283, 284) found that inhibition (cessation of contraction) of a muscle-group could be produced by stimulating a definite area in the cortex or internal capsule but was accompanied by simultaneous contraction of their antagonists. This indicates, as Fulton comments, that excitatory as well as inhibitory fibers exist in the pyramidal tracts. Here, then, is another point against the view that voluntary activity is chiefly a release phenomenon.

We have previously recounted the effects of sections of the brain-stem at various levels, namely an increased spasticity or tonus until the level of section excludes the vestibular nuclei, when tonus becomes less than normal. If we explain this effect on the basis of release of lower centers from inhibition, we should expect the animal deprived of his whole brain, the spinal animal, to show at least as marked extensor hypertonus as the animal deprived of his cerebrum alone, the decerebrate animal. But such is not the case. For although the knee-jerk in the spinal animal is of more ample duration than in the Magnus preparation or than in the normal (Fulton, 1926), which is evidence for the foregoing explanation, yet the spinal animal shows no rigidity but far more a hypotonia. In man, where the spinal cord was completely divided due to wounds of war, Riddoch (1917) observed that a stage of reflex activity followed a stage of muscular flaccidity, with knee-jerks and ankle-jerks appearing 21–53 days after injury; but that no extensor rigidity developed at any time, and hypotonia (as compared with normal man) was persistent. Furthermore the knee-jerk appeared to be briefer in duration than in normal man.

No satisfactory explanation has yet been advanced to explain relative hypotonia in the spinal animal and in man. The possible existence of inhibitory pathways has been suggested which become extensively "released" after exclusion of the vestibular nuclei, with the result that active myotatic contraction cannot be maintained in a muscle owing to the increased effectiveness of autogenous inhibition. But such presumed increase of autogenous inhibition obviously is contradicted by the fact that the knee-jerk in the spinal animal is of more ample duration than in animals sectioned at higher levels. Clearly, further investigations are needed to complete our understanding of the origin of rigidity in the decerebrate animal and of relative hypotonia in the spinal animal. Evidence that higher centers exert a predominantly inhibitory rather than excitatory influence on activities of the cord is as yet inconclusive.

On the other hand, there are positive grounds for rejecting the view that voluntary activity consists almost entirely in release of the lower motor neurones from higher inhibitory control. These grounds are not obvious when the subjects of laboratory study are the relatively simple reflexes of decerebrate or other preparations whose range of activity perforce is limited by the conditions of the experiment. Such reflexes perhaps can with little stretch of the imagination be analyzed into series of tendon reflexes released from moment to moment from higher inhibitory control. But in this respect the step from the narrow range of reaction of the decerebrate animal to the vast possibilities of intact man is apparently too great to permit both to be included in so simple a formula. If, for example, we consider the number of reaction combinations possible for a normal man seated at a typewriter, it is evidently not fitting to compare the physiological conditions with the arousal of the knee-jerk at a certain time. For in order to explain voluntary activity in typewriting on the theory that it is a release phenomenon, we should have to assume that the subject is stimulated simultaneously to strike every key and that his lower centers tend to provoke every such reaction but are restrained by the cerebrum except in respect to the one reaction which actually takes place. Evidence is lacking for the positive existence of such manifold reflex reactions which would appear but for restraint imposed upon them. Indeed, a consideration of the possible reactions of intact man in common daily situations reveals that the range is so vast, perhaps at times unlimited, as to rule out the possibility of accounting for voluntary activity as a form of "release phenomenon." It is far simpler to assume that the one reaction made by the individual out of many at any particular moment is the resultant of excitatory and inhibitory cerebral activities leading in that direction than to assume that reflex tendencies exist in him to make all possible reactions but that all are inhibited save one. We must conclude, therefore, that analyses into elements of tendon reflexes alone are not likely to completely solve problems of voluntary activity; we must await methods that permit investigations on the intact animal or human subject, controlled with the aid of autosensory observation and checked at all stages by objective methods.

Résumé

Owing to the wide variety of topics discussed in the present chapter, nothing more will be here attempted than a review of some of the leading statements and conclusions.

The muscular contractions dealt with in the present investigations in man are tetanic in type. There is evidence that during the tetanic contraction of decerebrate rigidity the O_2 consumption in skeletal muscle is increased 10–25 per cent (as compared with atonia) but that during tonic contraction of smooth muscle there is little increase of metabolism. The energy for muscular contraction possibly arises from hydrolysis of phosphocreatine, suggested by Lundsgaard's finding that contraction can occur in muscle poisoned with monoiodoacetic acid, when no lactic acid is formed. Breakdown of reactive muscle sugar finally yields lactic acid. During oxidative recovery occur resynthesis of phosphagen, removal of lactic acid and restoration of glycogen. The onset of fatigue follows the accumulation of lactic acid. Creatinin is increased in the urine of man after prolonged voluntary contraction but not after walking. However, critics point out that as yet no satisfactory evidence has been adduced that creatin metabolism is in any way associated with tonic muscular contraction.

By tonus is generally meant the state of slight sustained contraction which formerly was held to be generally characteristic of all healthy muscle. It disappears in skeletal muscle when the nervous elements are stripped off, but whether this holds also for smooth muscle is not yet settled. It is generally accepted that muscular tonus is reflexly maintained through afferent nerves. If so, we are aided to understand the mechanism of progressive relaxation, where there occurs a diminution of afferent impulses in connection with a diminution of tonus.

Sherrington (1915) gives evidence that a slight steady contraction is not universally characteristic of healthy muscle: depending upon the posture of an animal, it may be present or absent in any particular group of muscles. He concludes that no other afferent fibers than those of the tonic muscle itself are actually essential for the exhibition of the tonus. If this is true, it is clear how progressive voluntary relaxation of a muscle in man, by diminishing the afferent impulses from and to the same muscle, results in furthering continued relaxation.

The property of a skeletal muscle with normal attachments to remain in a tonic state at various lengths has been called "plastic tonus" and has been analyzed into simpler reactions, including the reaction to "stretch." But "stretch" due to normal attachments does not unaided evoke tonic contraction, as is abundantly evidenced in the present investigations.

The type of relaxation in which we are chiefly interested

is probably brought about by the cessation of impulses along motor nerves extending to a muscle and probably is to be distinguished from the mechanism of the relaxation due to stimulation of an ipsolateral nerve called "inhibitory relaxation." Evidence has been gathered that the presence of tonus is the skeletal muscular system tends to evoke tonus in the visceral muscular system and vice versa. Accordingly we are prepared to understand the presence of a "vicious circle" in divers conditions of neuromuscular hypertension, explaining why patients with chronically spastic viscera generally require a particularly long period of training before they become habitually relaxed.

A review of the literature leads to the conclusion that an augmentative influence toward maintaining tonus probably is exerted by sympathetic fibers ending in striated muscle. The evidence therefor from animal studies has been growing.

According to various investigators, activity of nerve trunks and nerve centers depends for its onset and continuance upon the occurrence of sensory nerve stimulation. This is in harmony with the explanation that the mechanism of progressive relaxation involves the diminution of proprioceptive muscle sensation and is of interest in connection with the cultivation of the muscle-sense described in chapter v.

In harmony with the "all-or-none" principle, we may assume that as relaxation progresses in a muscle, there is a diminution of the number of nerve and muscle fibers which are responding as fully as they can. During almost complete relaxation, according to Wachholder, the number of impulses per second is markedly decreased. (See chap. xvi.)

The rôle of various portions of the central nervous system in the regulation of muscular contraction was discussed at length. It seems probable that the cerebellum has not only an excitatory but also an inhibitory function in the regulation of postural tone. Likewise the balance of evidence favors the view that the red nucleus has an excitatory, perhaps also an inhibitory, influence on postural tone.

In some reflexes, postural tonus has appeared to be regulated by the vestibular system, and certain investigators have assigned a function of tonic innervation to various other centers, including the inferior olives, the *substantia reticularis*, the optic thalamus, the striated bodies, the temporal lobe, but most often the frontal lobe, particularly in the anterior central convolution. In brief, almost all of the chief divisions of the central nervous system appear to have a part in the nice control of postural reflexes, and with the progressive development of the brain the superordinated control tends to pass to higher levels. In harmony with this view are our present observations that diminution of postural reflexes in man can be effected by voluntary relaxation, that is, can be induced from cerebral sources.

Concerning the intact organism, various matters are discussed: (*a*) Voluntary contraction of flexor and extensor muscles can take place simultaneously, often producing a condition of rigidity, as of a limb. (*b*) Since there is but one motor fiber to five sensory ones in the central nervous system, the motor fiber has been called "the common path." Reflexes may be "allied" or "inhibitory" as regards their involving a "common path." (*c*) Evidence is discussed that tonus is occasioned by afferent impulses passing not directly to anterior horn cells (as in decerebrate or "spinal" animals) but via higher centers, including the cerebrum, before impulses arrive in the motor nerves to the muscle. In support of this view, the present investigations suggest that afferent impulses aroused by muscular contraction arrive at cerebral centers, for sensations from muscular contraction can readily be perceived. Assuming that voluntary relaxation is not effected through cerebral inhibition of lower centers, including anterior horn cells, the phenomena of voluntary relaxation afford evidence that cerebral (i.e., conscious) centers intervene in some measure in the production of tonus. The corresponding explanation of the mechanism of voluntary relaxation would be that the cerebral portion of the reflex arcs involved in the maintainance of tonus progressively ceases to function as voluntary relaxation advances, and tonus must therefore diminish and eventually disappear. (*d*) As previously said, the rate of nervous and muscular action voltages per second per fiber is diminished as relaxation proceeds. (*e*) The higher centers, particularly the cerebrum, exert an inhibitory influence over functions of the cord. But this must not be interpreted to imply that the pyramidal tracts are chiefly inhibitory in function and that voluntary action is chiefly a release of lower centers from inhibition. This view appears untenable.

ADDENDUM

Cortical influence upon autonomic activities.—Stimulation of points in the cortex in cats, near the motor centers, specifically evokes rises or falls in blood pressure or heart-rate. The evidence furnished by earlier investigators has been rendered conclusive by Hoff and Green (1936). Responses of the nictitating membrane in cats to cortical motor and inhibitory points have been found by Morison and Rioch (1937).

These findings suggest that cortical pathways may be involved during voluntary relaxation in man, when we observe diminution of contraction in smooth muscle. Examples include relaxation of the esophagus (chap. xix) and reduction of blood pressure (in press).

CHAPTER XIV
AUGMENTATION AND RELATED PHENOMENA

OF PARTICULAR importance for the present volume is the influence on response produced by an active state in muscle or in the nervous system, as compared with rest. Many instances are on record where stimulation has produced a stronger response if the part was previously active than if at rest. Again, if two simultaneous submaximal stimuli affect different afferent nerves leading to the same effector, their joint effect may exceed that produced by either stimulus alone (Vészi, 1910). Such instances are called summation, augmentation, facilitation, *Bahnung*, or reinforcement depending upon the context and author, as will be illustrated in the following paragraphs; but some authors, including Sherrington, use these terms interchangeably.

The importance of summation in the activity of the reflex arc was first disclosed by Setschenow in 1863. After removing the cerebrum from the frog, he found that a single induction shock failed to produce a reflex response but that frequent repetition of the stimulus succeeded. In 1875 Stirling observed that the response followed more promptly within certain limits as the stimulus was made stronger or was more frequently repeated. A single stimulus, even if very strong, failed to take effect; and this suggests, as Lucas remarks, that summation in the reflex-arc depends on failure of the first nervous impulse to pass through the center, while making it possible for a subsequent impulse to succeed.

Exner applied the term *Bahnung*, usually translated "augmentation," to the phenomena which he studied (1877, 1882,

1894). He discovered that the motor excitation of a limb in rabbits from the cortex augments the reflex caused by direct excitation of the skin of this part, and vice versa. Augmentations apparently in spinal centers were produced by stimuli to the higher sense organs. Contractions of the muscles of the paw were strengthened by auditory stimuli. The effect of a stimulus to the cortex was to augment that of another which followed soon afterward. In these various instances stimuli which were too weak to be effective when they occurred singly nevertheless produced a reaction when they occurred together or shortly after one another.

Augmentation resulting from previous activity in skeletal muscle was recognized by Ranke in 1865 and by Marey in 1866, but Bowditch first put the matter clearly before physiologists in 1871. Using a constant stimulus, he observed that the early contractions in heart muscle (sometimes excepting the first three or four) show a steplike increase, which he named *Treppe*, the German word for staircase. Various workers confirmed his findings for other tissues: Tiegel (1875), for frog's skeletal muscle; Rossbach and Harteneck (1877), for the skeletal muscle of various mammals; Romanes (1876), for medusa; Waller (1896), for frog's nerve; Stirling (1875), as above mentioned, for the central nervous system of the frog; Sherrington (1900), for that of the dog. The staircase effect has recently been demonstrated for the knee-jerk (Brown and Tuttle, 1926). Evidence is given by Lee (1907) that summation and the staircase effect are physiologically similar processes. Fulton (1925*b*) emphasizes the importance of summation most clearly by the statement that prolonged contractions of skeletal muscles are made possible probably entirely by summation.

Reinforcement of the knee-jerk was first observed by Jendrássik (1883), produced by clenching the hands and by other violent movements. Mitchell and Lewis (1886*a*) confirmed his findings, but found reinforcement consequent also upon volitional acts directed to other parts of the body as well as upon painful pinching and upon exposure to sudden heat, cold, light and electric stimulation. They stated that in order to reinforce the jerk, the muscular action must precede the tap by a period which was as yet undetermined. Subsequently Bowditch and Warren studied these time relations (1890) and noted when inhibitions took place instead of reinforcements. According to their observations, the interval at which the effect changes from positive to negative varies with different subjects from 0.22 to 0.6 seconds. After an interval of 1.7–2.5 seconds, the extent of the jerk returns to its normal value. In two subjects the effect of muscular contractions mentioned above was wholly positive. Visual and auditory stimuli used on their three subjects were almost wholly positive in effect. Increase of the knee-jerk due to divers forms of stimulation has been noted by many investigators, including Lombard (1887), Russell (1893), and Luckhardt and Johnson (1928). Russell observed such an effect during asphyxia, hemorrhage, anemia (produced acutely by compressing the abdominal aorta), and following the administration of nitrogen, nitrous oxide, oxygen, ether, chloroform and strychnine in certain dosages and manners. Lombard's results are generally conceded to indicate very clearly that *the irritability of the spinal cord varies with almost every marked change in mental activity.*

Sherrington (1906*b*) dwells on the importance of summation of subliminal stimuli in reflex-arc conduction. Here it is found to an extent not known for vertebrate skeletal muscle or for the nerve trunk. He secured records where the scratch reflex in the decerebrate dog appeared only after the fortieth successive double shock (at rate of 11.3 per second) and others which indicated that a momentary stimulus shows its facilitating influence on a subsequent stimulus applied even 1.4 seconds later.

That the interval for augmentation in the intact animal, as compared with the nerve or nerve-muscle preparation, is vastly increased is illustrated by Brown's experiments (1915) on the motor cortex of monkeys. He found that a second stimulus following upon a first within 20 seconds evoked an increased reaction. The increase was greater according as the stimulus was stronger or more prolonged and as the interval was shorter. Subliminal stimuli, when sufficiently repeated, produced a reaction. Repeated stimulation producing fatigue tended to shorten the state of raised excitability. Brown gives evidence that facilitation may take place within the cerebral cortex itself. The importance of these points for the present investigation will be obvious to the reader.

We are not to believe that all of the preceding phenomena are fundamentally the same type of physiologic process. Lucas (1917) clearly distinguishes one type of summation due to the persistence of a residuum of excitation at the spot where the electrodes are applied to the nerve. Accordingly, the increase is not observed if the second stimulus is applied at a different site. We can dismiss this type of summation as having little to do with the subject of relaxation. Another type of summation, according to Lucas, is produced in a nerve-muscle preparation provided that the second stimulus follows the first within a certain range of time intervals (Samajloff, 1908). Adrian and Lucas (1912) found that the effect was absent if the stimuli fell directly on the muscle. The effect was not altered by warming or cooling the site of the second stimulus, showing that it was not due to an alteration of excitability there. It was therefore clear that the increase must be due to better conduction of the second im-

[273]

pulse. Since the occurrence of summation is favored by fatigue, mild curarization, removal of calcium and treatment with acids (conditions known to affect particularly the myoneural junction of the nerve-muscle preparation), Lucas concludes that this type of summation is dependent upon conditions at the myoneural junction. Related conclusions are drawn from the work of Lapicque (1912) on reflex contractions in the frog, elicited on stimulating the central end of the cut (contralateral) sciatic nerve. He confirmed Stirling's results and determined the frequency of stimuli needed to produce a response with a minimal stimulus. He showed that the frequency varied with the temperature of the cord but not with that of the tissue stimulated. This virtually demonstrated that the site of augmentation in these instances lay in the nerve center. It seems reasonable to infer that this would apply also to the observations of Sherrington mentioned above on summation in reflex-arc conduction, as well as to those of Setschenow, most of those of Exner, and those of Brown.

What part, if any, does fatigue play in the production of summation? As we have seen, Lucas shows that conditions of fatigue at the myoneural junction favor but are not prerequisite for summation. The important observation has been made by Rollet (1896) and others that with increasing fatigue a muscle requires a prolonged time for relaxation. Accordingly, Fröhlich (1905) has sought to explain the staircase effect or summation by the development of fatigue products; and Lee (1907), carrying the matter farther, has demonstrated that perfusion of a preparation with fatigue products tends to cause increase of contraction. Thörner (1908), using the capillary electrometer with medullated nerves, found evidence of summation in consequence of delayed restitution—that is, a form of fatigue. But Beritoff

(1913) observes weak stimuli calling forth maximum effects soon after the first three or four stimuli, a condition of summation evidently too early to be called fatigue in the ordinary sense; and Fulton (1925) finds that the staircase effect is not necessarily due to fatigue.

Ishikawa (1910) calls it "false *Bahnung*" when increase of response is due to fatigue in nerve centers. Verworn (1913), following studies by Borrutau and Fröhlich (1904) and Fröhlich (1909), identifies fatigue in part with the refractory period. Facilitation not due to fatigue is studied by Lorente de Nó (1935); he applies two shocks to affect motor neurones of the third nucleus and records responses of the left internal rectus muscle with an oscillograph. The period of facilitation begins from 0.4 to 1.10 milliseconds after the conditioning shock, independently of the absolute refractory periods, and may last from 1.5 to 80 milliseconds. It is absent if the reticular substance in the pons and medulla is destroyed. He ascribes it to activation of internuncial neurones arranged in closed self-re-exciting chains capable of sustained rhythmic activity. Synaptic delays in these chains account for prolonged facilitation. Evidence is accumulating that divers internuncial neurones are the locus of inhibitory (Hughes, McCouch and Stewart, 1937), as well as facilitatory effects. The mechanism suggested by Lorente de Nó may be involved in the nervous start, as well as in other phenomena discussed in this volume.

If fatigue favors summation, there is a possible clinical bearing. It is commonly agreed that (chronically) fatigued individuals manifest symptoms of undue irritability. There is often an increase of reflexes as well as of subjective "sensitiveness" to stimulation. The explanation appears plausible that this is a summation effect, due to fatigue. So far we are in agreement with the old conception of "neurasthenia"; but the reader is urged to recall the reasons for believing (chap. ii) that the fatigue effect is not primary (but is only the result of nervous overaction exceeding the capacity of the individual) which led us to prefer the clinical term "nervous hypertension."

In the foregoing illustrations, the additive effect of repetitive impulses (at certain time-intervals under each set of conditions) in muscle and nerve fibers, nerve centers and (probably) myoneural junctions is amply illustrated. The existence of this type of summation as a frequent fundamental event in the physiology of the organism can be regarded as established. Without doubt, then, such summation effects commonly occur in consequence of the numerous afferent impulses from the many and varied muscular contractions that characterize the activities of the intact organism. This would obviously imply that the diminution of such muscular contractions and afferent impulses, as effected by voluntary or cultivated relaxation, should be expected to produce a corresponding diminution of summation effects. Furthermore, general progressive relaxation, involving diminished activity of higher centers, probably has, in addition to the foregoing, an analogous central effect: when higher centers cease to be active, presumably excitatory impulses from them dwindle in number, and so diminish the frequency of phenomena of summation in the nerve tissues.

It seems justifiable, therefore, to infer that the foregoing type of nullification of phenomena of summation frequently takes place during general relaxation. But is this, so far as augmentation is concerned, the complete explanation? Putting this question in terms of the knee-jerk, which has been shown to be a typical spinal reflex (Jolly, 1911; Dodge, 1910; Snyder, 1910), how shall we explain reinforcement produced by immediately preceding voluntary contraction of other parts? Lombard (1887) concluded from his many observations that "in general the knee-jerk is increased or diminished by whatever increases or diminishes the activity of the central nervous system as a whole." This can be expressed very simply in present-day terms: Increased activity in the cord or brain, present during the reinforcing contraction, results in excitatory impulses passing down, "overflowing," or "irradiating" to the anterior horn cells of the quadriceps group, where their effects are added to those of afferent impulses aroused by tapping the patellar tendon. The result is an increased jerk. The point of view adopted by Lombard is, according to Howell (1925), shared by most physiologists today.

Jendrássik, however, was not so simple in his conceptions. He states that contraction of muscles supplied by the crural nerve diminishes or prevents the jerk, while the contraction of muscles supplied by the ischiadicus nerve does not hinder the jerk. In evidence he affirms that in man, if one thigh is crossed over the other in the usual manner, the jerk can be elicited; but that it is diminished if the lower thigh is removed, making it necessary for the individual to hold the tested limb in the air while the leg hangs free. In this event the quadriceps muscle is not contracting and is therefore not the cause of the failure. But if we let the thigh hang loose while the leg is bent at the knee, the jerk is present although the antagonists of the quadriceps are in action. Also, with weak contraction of the quadriceps the knee-jerk is diminished or absent, for instance, when the thigh is supported and the foot touches a nearby object. However, if the muscles are allowed to relax, the contraction can again be felt. The opposite condition is present during activity of the muscles on the posterior side of the thigh. If the thigh is voluntarily pressed down upon its support, or if the heel is voluntarily pressed down upon an object, the jerk is augmented.

[277]

There are individuals, he adds, who cannot readily relax but who hold the muscles supplied by the crural nerve tense, wherefore the knee-jerk may be small or absent. In this event, the individual is requested to cross his thighs and press down the upper one, in order to relax the quadriceps muscles when he contracts the flexors. In the same manner, I presume, Jendrássik would explain the reinforcing effect of the arms pulling against interlocked fingers: this leads to relaxation of the quadriceps muscle and so frees the jerk from inhibition. At any rate, such is the view of Kroner (1907) who states that "the principle of reinforcement of Jendrássik is through distraction and avoiding of inhibitory activities which prevent the relaxation of the quadriceps necessary for the knee-jerk." Kroner is careful to add, however, that in addition to phenomena which produce augmentation in the foregoing manner, there are others which do not involve a release of the knee-jerk centers from inhibition but which depend upon the arrival of positive excitatory impulses at the knee-jerk center, an example of which is reinforcement following walking.

We have hitherto assumed (p. 128) that when a tense patient, who has no nervous structural lesion, fails to reveal a knee-jerk, it is because he is contracting his flexor muscles so as to prevent the kick. This, it is believed, is the prevalent view among experienced clinicians today. In fact, when they suspect that a patient is thus holding his limbs rigid, they customarily test for rigidity by attempting to extend and flex the leg, whereupon they meet with resistance and confirm their suspicions. Clinicians find that the individual can and commonly does contract any set of flexor and corresponding extensor muscles at the same time. As previously said, it is now admitted by physiologists that during voluntary activity, simultaneous contraction of flexors and extensors commonly occurs (Graham Brown, 1915; Golla and Hettwer, 1922, 1924; Tilney and Pike, 1925). We can readily understand that as his attention is distracted by clenching his hands, by other strong voluntary activity, or by a sensory stimulus, he ceases the act of holding his tested limb rigid, and so permits the jerk to occur. But is this the full explanation or do other mechanisms, notably that espoused by Jendrássik and Kroner, also take part? There is an apparent contradiction between Jendrássik's assertion that contraction of the quadriceps muscle inhibits the knee-jerk and the commonly accepted belief of many investigators that the extent of the knee-jerk is a measure of the degree of tonus existing in the extensor muscles. Sherrington (1906), for example, finds that the "knee-jerk varies *pari passu* with the reflex spinal tonus of the *vastus medialis* and *crureus*"; he speaks of the reflex tonus of these extensor muscles as "admittedly a *conditio sine qua non* of the jerk"; and repeats that whether or not we regard the knee-jerk as a reflex, it is none the less "an index of the reflex tonus of the extensor muscles." Fulton (1926) states that under most circumstances it is undoubtedly true that the knee-jerk is a measure of the degree of tonus existing in the extensor muscles. He reminds us that in flaccid paralysis (e.g., caused by lesions of the anterior horn cells) tonus is absent, and so is the knee-jerk; and that, on the other hand, the knee-jerk usually increases in briskness, as many clinicians have found, with increasing spasticity. But, he adds, one is not wholly justified in believing that the knee-jerk is always an index of tonus, for ample jerks are to be obtained in spinal preparations in which tonus is largely absent. However, Fulton evidently does not take the doubt he thus expresses very seriously, for he forthwith quotes and agrees with Viets' (previously mentioned) finding that if the tonus of a muscle is augmented,

[279]

even in a "spinal preparation," by contralateral stimulation, the knee-jerks, within certain limits, are augmented. We must conclude that extensor-muscle tonus is but one of the variables upon which the briskness or extent of the knee-jerk depends: under one set of conditions constantly maintained, a given degree of tonus may permit a greater jerk than would another set of conditions; so in a "spinal" animal (perhaps because of the removal of inhibiting higher centers) a relatively low degree of tonus permits the presence of a brisk jerk such as would not be secured with equal tonus in an intact or decerebrate animal; but under any particular set of conditions, otherwise kept constant, variation of degree of tonus, within limits, is attended by corresponding variation of degree of jerk.

At this point of the discussion of augmentation, we may profitably pause to pass in review various characters of the knee-jerk reflex, particularly the mechanisms by which the reaction is aroused or inhibited. Westphal pinched, pricked and irritated the skin in various ways without eliciting the knee-jerk; when a fold of skin was lifted away from the tendon and subjected to blows with a hammer, the result was negative. Freezing the skin did not lessen the jerk. Jendrássik dissected out the patellar tendon, freeing it from the skin and cutting it away from the tibia. A properly directed blow still elicited the jerk; but only when the tendon was in a state of passive tension. Tschiriew (1879) is generally presumed to be the first observer who discovered nerve fibers, without myeline, which terminate in the neighborhood of tendons, which he looked on as the organs of transmission of the muscular sense. Sherrington (1893a) discovered that the muscular portion of the mechanism of the knee-jerk consists mainly of the *vastus internus* and part of the *crureus* divisions of the great extensor muscles of the thigh. In man the spinal center is at the third and fourth lumbar segments. The efferent path in the Rhesus monkey is in the anterior roots of lumbar nerves 4 and 5, proceeding via the anterior crural nerve to the muscles. Afferent fibers arise in the foregoing muscles and proceed in lumbar nerve 5; severing this root or freezing it, or applying CO_2 vapor or cocain, abolishes the jerk. These afferent fibers, therefore, condition the presence of the knee-jerk and are the only afferent fibers that do so. But while the activity of these proprioceptive muscle fibers arising from the extensor muscles is required for the jerk to occur, other afferent fibers from the same source can produce by activity an inhibition of the jerk. Sherrington (1898, 1924) proved this for decerebrate preparations with his discovery that if the cut central end of one of the four nerves which supply the quadriceps muscle is stimulated, the remaining three portions of the muscle are reflexly inhibited. He gives reasons for believing that at least some of these inhibitory fibers carry the sense of pain. Fulton (1926) finds suggestive evidence for traces of such autogenous inhibition (which is largely concealed by the dominance of the positive contraction) in the crossed extensor reflexes of muscles with intact efferent nerves. Mitchell and Lewis (1886) also found that strong electrical innervation inhibits the knee-jerk while weak innervation increases the reaction. On the contrary, Hoffmann finds in man that the strongest stimulation he uses on the femoral nerve-point increases the knee-jerk; he fails to secure the inhibitions described for animal preparations by Sherrington and infers that the latter must have used stimuli in animals of such strength that the pain caused in man would prohibit their use.

That the sensory end-organs whose excitation causes the jerk reside in the muscle-substance was concluded by Sherrington upon cutting away the patellar tendon and nevertheless eliciting the reaction in response to stretch. These end-organs, according to Fulton and Pi-Suñer (1928), are perhaps the muscle-spindles.

The views of Jendrássik have been tested with modern galvanometric methods by Hoffmann (1918), with results of such consequence that it is necessary to discuss them at length. No doubt exists, Hoffmann admits, that tendon reflexes are increased by the hand-gripping method of Jendrássik or other simultaneous contraction of a large, distant muscle-group. But almost every neurologist has his own method to produce the desired reinforcement: one has the subject count, another figure, a third relies on coughing. Admittedly, it is common experience that distraction of the attention or contraction of a distant muscle-group has an augmenting action; but Hoffmann seeks to disprove Jendrássik's assertion that the knee-jerk is inhibited by contraction of the very muscles employed in the reflex, namely, the extensors of the leg. It is by no means self-evident, as Sternberg points out, that a contracting muscle is incapable of making a reflex response to a stimulus by further contraction. Such further response would be impossible only if all of the fibers of the muscle were already engaged in maximal contraction. But this is not the case. Accordingly, the fibers not in a state of contraction could very well participate in the reflex, according to the investigations of Gad.

Therefore, Hoffmann studies the action-currents given off by various skeletal muscles in man upon stimulating the mixed nerve near its point of entry, particularly the extensor muscles of the foot. Following every induction shock the record shows two action-currents: "A" produced by the impulse that passes directly to the muscle from the site of stimulation; "B" occurring a little later in consequence of the impulse that passes to the cord and back to the muscle through the motor fibers of the mixed nerve. "A" need not be considered further, for the reflex, which is of present interest, is represented by "B" alone.

Upon applying this method to the extensor muscles of the foot, the usual reflexes appear as shown by deflections "A" and "B" provided that the muscle-group is held in the usual state of relaxation. As will later be seen, this "usual state of relaxation" is in reality a state of slight contraction. But if during the stimulation the antagonists (flexors) are contracted, the reflex at once disappears (Fig. 52). However, if instead the extensors of the foot are contracted, the height of the reflex is enormously increased (Fig. 53). Under this condition, the induction shock can be made so weak that "A" fails to appear, yet "B" remains evident. In short, voluntary contraction of the muscle-group certainly produces no inhibition of the reflex, but rather an intense augmentation. The stronger the contraction, the greater the augmentation.

In Hoffmann's experience with induction shocks as means of eliciting tendon reflexes, only the extensors of the foot give a reaction during what he calls the "usual relaxed state." In the case of all the other muscles, such shocks produce no reflex reaction unless there is simultaneous augmentation produced by voluntary contraction of the muscle-group involved. This is especially true of the extensors of the leg, where the experiment is found relatively difficult to perform. It is impossible to test whether contraction of the flexors inhibits the knee-jerk, because no deflections "B" appear when the quadriceps muscles are maintained in the "usual relaxed state"; but Hoffmann presumes that the evidence secured in the case of the extensors of the foot holds by analogy for the extensors of the leg.

Hoffmann (1922) finds under his conditions no signs of ef-

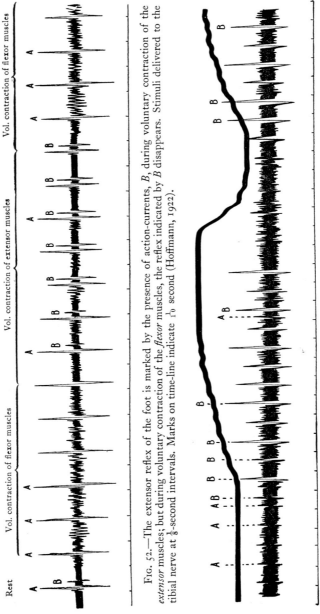

FIG. 52.—The extensor reflex of the foot is marked by the presence of action-currents, B, during voluntary contraction of the *extensor* muscles; but during voluntary contraction of the *flexor* muscles, the reflex indicated by B disappears. Stimuli delivered to the tibial nerve at $\frac{1}{8}$-second intervals. Marks on time-line indicate $\frac{1}{10}$ second (Hoffmann, 1922).

FIG. 53.—During marked voluntary contraction of the extensor muscles of the foot, the action-currents, B, are seen to be greatly increased in intensity. Upper tracing made with lever attached to foot, lifting 50 kilograms; rises in this curve denote voluntary contraction. Middle line shows action-currents. A may be neglected, being due to the impulse which passes directly from the site of nerve stimulation to the muscle. B indicates the reflex. It will be seen that B is present, but not high, before voluntary contraction begins. Lower line denotes time intervals, possibly about $\frac{1}{5}$ (?) second (Hoffmann, 1922).

fects of summation of tendon reflexes upon each other. For a stimulus can be selected too weak to produce a contraction, except with augmentation such as voluntary contraction; thereupon, if voluntary contraction is omitted and electrical stimuli are delivered in series with very brief intervals, no evidence of summation appears; that is, no reflex is produced. On this point Hoffmann appears to disagree with investigators previously quoted. However, he agrees that what he terms *Fremdreflexe*, i.e., reflexes other than tendon reflexes, are like voluntary contraction in that they have augmenting tendencies.

At first sight, the results cited above, indicating that voluntary contraction increases the tendon reflexes, appear to conflict with the experience of every doctor that the tendon reflexes can be elicited only if the involved muscles are held relaxed. Hoffmann believes that the apparent contradiction is resolved according to the following considerations: (1) In customary examination of the tendon reflexes, the mechanical contraction of the muscle and the movement of the limb are taken as positive signs. But during voluntary muscular contraction, consisting of 50–200 impulses per second, the addition of one more—due to the tendon reflex—will pass unnoticed. (2) "When I say that the reflex disappears during complete relaxation of the muscle, I do not mean the usual relaxed posture of the muscle, but the maximal relaxation, which appears only upon contraction of the antagonists of the particular muscle. The muscle which is held relaxed, as has long been known, is persistently in a certain slight condition of contraction which is commonly called tonus." It is of considerable interest that Hoffmann's investigations on tendon reflexes confirm the previously stated conclusions about relaxation (chaps. iv and ix). He agrees that what is customarily called "relaxation" in reality falls short of this state to a measurable extent; and that in extreme relaxation, the electrically elicited tendon reflex disappears. To be sure this appears as a by-product of his studies, for he is not interested primarily in relaxation. This state is produced under his conditions only by contraction of antagonists, and not by the simultaneous cessation of both flexors and extensors as in the present investigations.

Finally Hoffmann reminds us that studies of the tendon reflexes reveal that during every moment of waking life various nervous activities exert virtually countless augmentatory effects. Practically every mental and bodily activity can act in this way. Such observations indicate that our bodily and mental activities do not remain limited to those regions essential to carry them out. Since we know that such excitations must also influence muscle tonus, it is readily seen that in such cases there comes to the muscle an imperceptible increase of tonus. The tonus itself is not perceived by us but can be recognized by increase of tendon reflexes.

Thus Hoffmann arrives at conclusions regarding the existence of augmentatory effects which Lombard early emphasized and upon which various intervening investigators of tendon reflexes are in agreement. Such conclusions permit us to gain an insight of the mechanism of neuromuscular tension and hypertension in the life of the average individual. They remind us of the added augmentatory effects that can occur in excitable individuals and in the disease conditions discussed in chapter iii. It lies near at hand to assume that the effects of progressive relaxation are to counteract various phenomena of augmentation.

Realizing how complex and manifold are the observations and possible interpretations of augmentations and inhibitions of the knee-jerk up to date, we now return to the question, how voluntary muscular contraction in other parts produces reinforcement of the knee-jerk. In summary, there are at least five theoretical possibilities, of which the first three have been discussed above: (1) Voluntary contraction, as of the arms during reinforcement, distracts the attention like any strong sensory stimulus and leads the tense individual to cease holding flexor muscles of the tested limb so rigid as to prevent the jerk by (*a*) mechanical restraint as well as by (*b*) the mechanism called "reciprocal innervation." Evidence was given that this explanation actually holds for at least some instances. (2) We recall that the tonus of skeletal muscles admittedly varies with and is an index of the presence of excitation processes in the corresponding anterior horn cells; for, as present-day investigators agree, no specific inhibitory fibers pass from those cells to their muscles, unlike the conditions that govern the action of smooth muscles. Accordingly, in completely or fairly relaxed individuals, when reinforcement is sufficiently strong to restore or to increase the knee-jerk, it is because excitatory impulses pass from the centers active in the voluntary contraction to the anterior horn cells which arouse the extensor muscles, producing an augmentatory effect with the impulses aroused by tapping the tendon. Substantially the same explanation would be that the foregoing excitatory impulses produce or increase the extensor-muscle tonus requisite for the jerk. The preceding two explanations, taken jointly, are probably most in acceptance by physiologists today, and the writer feels that they reasonably account for known facts. (3) According to Jendrássik and Kroner, contraction of quadriceps-extensor muscles hinders the knee-jerk; therefore augmentation is due to whatever factors prevent contraction of the extensor muscles. They evidently presume that voluntary hand-gripping reinforcement is such a factor. But Hoffmann's direct examination of action-currents, indicating that quadriceps-extensor contraction augments the knee-jerk, virtually disproves the views of Jendrássik and Kroner. (4) Diminished cerebral or spinal inhibition of the knee-jerk center produced by voluntary contraction would perhaps explain the reinforcement. We know, both from clinical and from laboratory sources, that removal of cerebral influence on spinal centers results in increased spinal reflexes (chap. xiii). Perhaps reinforcing activities produce their effect by removal of cerebral influence, e.g., by diminution of impulses along pyramidal inhibitory fibers to local excitatory fibers. Fulton favors this point of view, but Professor Luckhardt (personal communication) finds evidence against it. By having his subject fully extend one arm supporting a weight during a prolonged period in which the tendon is tapped at regular intervals, he secures a series of jerks of ascending heights. If the mechanism of such reinforcement is the irradiation of impulses from the spinal centers of the arm to the knee-jerk center, thereby producing an additive effect (rather than a release of that center from inhibition), increasing the tonus of the extensor muscles should produce a corresponding increase in jerk. His records show striking positive results, which he interprets as ruling out the theory of inhibitory release. The reader will recall similar increase of reflex response produced by heightened muscular tonus in the experiments described in chapters vii, viii and ix, as well as the decrease in response produced by relaxation. Fulton suggests, as an additional mechanism, that (5) since fibers arise in the extensor muscles whose action as above described from Sherrington's observations, is ultimately inhibitory of the knee-jerk, the action of these fibers may be checked at their central end-organs by impulses from the cerebrum arriving along pyramidal inhibitory fibers. Augmentation then results from inhibition of an inhibition.

But as said above, Hoffmann fails to find evidence for the occurrence of such "autogenous inhibition" under ordinary conditions in man.

None of the foregoing five explanations accounts clearly for the observations of Bowditch and Warren that a brief voluntary contraction changed in effect from augmentation to depression of the knee-jerk at an interval between the signal to contract and the blow varying from 0.22 to 0.6 seconds, with a return to normal value at an interval from 1.7 to 2.5 seconds. We need not believe that only one of the five mentioned factors will fully account for augmentation of the jerk; for, as Fulton remarks, it would be in keeping with the known complexity of nervous control to assume that more than one factor is at work. Further investigation is needed before we can be certain at all points of the precise mechanism of reinforcement. In the meantime, if we would select among the five possibilities, the first obviously holds as mentioned, the second appears to me probable as at least a partial factor, the third seems ruled out by Hoffmann's, Luckhardt's, and my own observations, the fourth seems theoretically possible but lacking in definite experimental evidence, while the fifth appears very highly theoretical and must await further tests.

Whatever be the mechanism by which voluntary contraction reinforces the knee-jerk, it is obvious that extreme voluntary relaxation will imply the diminution or absence of that mechanism and therefore the absence of reinforcement. This applies no less to factor (4) above, if we assume this is the true explanation, than to factor (2).

The observations of Sherrington and Viets above-discussed prove that the presence of tonus in the quadriceps group contributes to subsequent reaction of the knee-jerk upon adequate stimulation. In the complete absence of tonus, the stimulus fails to produce the reflex. I am inclined to believe that extreme voluntary relaxation produces a low or absent jerk principally because of reduced quadriceps tonus: for these muscles become yielding and apparently relaxed to experienced palpation, while the experimenter can flex or extend the leg at the knee without detecting the slightest resistance, and the subject, experienced at discerning the presence even of slight contraction in a muscle, as judged by various tests, reports a total absence of the sensation of tonus or contraction. I am inclined to explain similarly the frequent failure of voluntary contraction, just after there has been extreme general relaxation, to produce a knee-jerk as great in degree as that elicited with such reinforcement before relaxation (chap. ix). Likewise Miller's experiments (chap. viii), which disclosed that the flexion reflex occurred after a given stimulus providing that a certain tonic condition was maintained, admits of the same possible explanation. The occurrence of this reflex might also in a sense be called a summation effect. Flexion diminished or failed to occur under conditions of extreme relaxation. In general we are led to question whether any reflex will follow its appropriate stimulus (within limits of moderate intensity) in the absence of peripheral-muscle tonus as produced by extreme relaxation.

By way of recapitulation, these instances and considerations suggest that general relaxation is in effect the negative of various phenomena of augmentation or reinforcement. Certain experimental evidence in favor of this view was advanced in chapter ix. In general, we have seen that the arousal of nerve impulses in any portion of the nervous system not alone tends as a rule to excite a peripheral response but also renders conditions favorable for a time for a summation or augmentation effect, that is, for an increased subse-

quent response. If this double effect or tendency is characteristic of each excitation process, it seems evident that relaxation, in so far as it is the negative of such excitation process, will tend to undo this double effect. We have taken a
step, at least, toward explaining how previous general or
local relaxation may diminish a particular reflex response;
or how habitual relaxation may, in clinical terms, make the
individual less irritable.

CHAPTER XV
EXPLANATORY PRINCIPLES OF RELAXATION

ASSUMING that general neuromuscular relaxation tends to produce a quieting effect upon the nervous system or local portions, we now inquire how this comes about. In part the answer may be sought in the foregoing pages: Experimental evidence was offered that general relaxation of striated muscle-groups results in a diminution of various reflexes. Examples are the diminution of tone in the quadriceps femoris, which effected a decrease in the knee-jerk; the diminution of tone in the arm-muscles, manifested by the decreased flexion reflex; and the diminution of the general skeletal muscle tone, revealed by a decreased involuntary start. Furthermore it appeared that relaxation of the skeletal muscles might reflexly produce a relaxation of visceral muscle, at least in the case of the esophagus. On the other hand observations were quoted from other workers indicating that increased visceral or vegetative nervous activity tends to produce increased tonus of skeletal muscle. Thus two chains of evidence apparently unite in a circle, suggesting that decrease of tonus in either the cerebrospinal or the autonomic sphere may have a corresponding effect upon the other. A consideration of the probable mechanisms by which tonus is maintained, increased, or decreased by laboratory methods in animal preparations clarified our understanding at various points of the mechanisms by which general relaxation produces its effects in man. Studies of tendon reflexes made it apparent that owing to the change and variety of proprioceptive and exteroceptive stimuli in the waking life of man, impulses aroused in one portion of the nervous system tend to have overflow effects in other relatively distant parts, so that almost every point of the nervous system may be in a continually changing state of flux. Various excessive stimuli, including those in pathological conditions, may produce an exaggerated state of this normal condition. Such phenomena of augmentation and the resultant nervous hypertension seemed to result from conditions opposite in nature from those which existed during neuromuscular relaxation, so that the predominance of the latter apparently accounted for the diminution of the former.

Some light also seemed to be thrown upon the nature of hypersensitivity. Under controlled experimental conditions where pain was produced by an induced current, it was found that when the only noteworthy variable was the degree of tension in the motor portion of the reflex-arc, the intensity of the sensation as reported by the subjects varied accordingly. This variation in sensitivity ran parallel to variation in the extent and reaction-time of the reflex. It was suggested that a tense or spastic organ tended *ipso facto* to be hypersensitive. Our general experience with relaxation has furnished many scattered instances that various types of sensations lose in intensity as relaxation progresses; but conclusions on so important a topic require further proof, and careful quantitative experimentation in this direction is needed. In my clinical experience what is commonly called "sensitiveness," meaning proneness to arousal of anger, resentment, disgust, embarrassment and other disagreeable or painful emotions, has seemed to diminish in the course of the months in which the individual has acquired habits of relaxation. This has taken place in instances where hints or suggestions of such effect have not been made to the subject. Experimental tests of this matter are now possible, but in the meantime it is tentatively suggested that "sensitiveness" varies in some relation with the tonus of the muscular portion of the reflex-arc.

Relaxation leading to decreased reflexes was obviously associated with diminution of proprioceptive sensory impulses. In particular, the subjects reported that as they relaxed there was a corresponding diminution of muscular sensations. They were trained to observe these sensations in order to guide their relaxation. As previously said, since the time of Brondgeest and those who soon after him confirmed his conclusions, it has been commonly agreed among physiologists that skeletal muscular tone varies with the presence of afferent or proprioceptive impulses.

Sherrington (1894a) has shown that in the so-called motor nerve supplying the limb muscles from one-third to one-half of the fibers are afferent. He and others have often dwelt upon the great physiological importance of impulses arising from these intramuscular receptors in the co-ordination of reflex and voluntary muscular activity and in regulating and modifying muscular action. They have discovered (Sherrington, 1908b) "additional proof of the existence of afferent impulses from muscle in the so-called 'lengthening and shortening' reactions which Sherrington found in the condition of partial rigidity which he designates 'plastic tonus' and in the 'crossed reflex III' of Phillipson cited by Sherrington." In these experiments there can scarcely be any doubt of the muscular origin of these nerve impulses, since the skin areas were rendered anesthetic and the afferent twig supplying the only joint involved was sectioned. In "de-afferented" preparations, all these phenomena are quite absent.

Forbes, Campbell and Williams (1924), emphasizing the foregoing points, performed a critical experiment. They dissected out the peroneal branch of the sciatic nerve, cut it at the hip and stimulated the central end of the nerve. Their work was well controlled. They also stimulated the muscles of the animal by pulling the leg. During muscular contraction they secured records both of the motor nerve impulses and of the returning proprioceptive impulses.

Accordingly, there seems little room for doubt that muscular contraction excites proprioceptive impulses which lead to, or at least contribute to the maintenance of muscular tone, and therefore that relaxation, the negative of contraction, leads in the reverse direction. What bearing has this upon mental and emotional activity? Evidence has been given in the preceding pages which suggests that an essential part of mental and emotional activities consists of neuromuscular patterns; the latter are not just "expressions" of emotion, as was formerly believed; in physical terms, the energy expended in a neuromuscular pattern is identical with, and not a transformation of, the energy of the corresponding mental and emotional activity. If this is true, it follows that the extreme relaxation of a muscular pattern essential to a particular mental or emotional process must bring with it the diminution of that process.

At this point it seems well to summarize the foregoing points and to add others in order to bring to focus the explanation how general neuromuscular relaxation tends to do away with motor unrest and to calm emotional and mental activity. There are, of course, certain fundamental questions about the nature and function of relaxation, like that of contraction, which cannot yet be answered; but in the light of present-day knowledge the following considerations (which are not all mutually exclusive and independent) are suggested as explaining the sedative effects of relaxation:

1. Relaxation of a muscle-group is of course physiologically incompatible with contraction of that same group. It

is therefore obvious how cultivated relaxation in a muscle-group or set of muscle-groups does away with motor unrest in those groups. Voluntary relaxation is obviously the direct negative of motor unrest and is in effect apparently contrary to various phenomena of augmentation (chap. xiv).

2. It is well known that the occurrence of a reflex depends upon the presence of tone in the muscles that take part. Sherrington (1909b) and Viets (1920), using the vasti-crureus muscle of the spinal cat, have clearly proved that it is not possible to obtain a knee-jerk in an absolutely toneless muscle. But cultivated relaxation has been shown in this volume to diminish tone, resulting in the diminution of the knee-jerk and of other reflexes in man. This indicates the mechanism by which relaxation can diminish reflexes even when their appropriate stimuli are present.

3. As relaxation progresses and reflex contractions diminish, there is consequent diminished production of proprioceptive impulses, tending thus toward a progressive decrease in the production of further reflexes.

4. The effects of relaxation can be explained on the basis of reduced excitation or of irritability of nerve centers. Evidently, general relaxation with closed eyes diminishes visual as well as muscle and joint sensation and so diminishes the excitation of nerve centers. Certain autonomic functions probably are represented by localities in the cortex (Hoff and Green). Voluntary relaxation perhaps includes diminished impulses to these localities from motor centers or other cortical sources.

We think of two ways in which the irritability and excitability of nerve centers can be physiologically reduced. The one which has been most discussed in the foregoing text is by diminution of excitatory impulses arriving at that center. But just as the nerve centers of various viscera can become less irritable and excitable either owing to diminution of excitatory impulses along one set of nerves or owing to inhibitory impulses along another set, so also this dual possibility is to be considered for the nerve centers governing skeletal muscle reflexes. It is conceivable that when the subject relaxes, the diminution of reflexes is due, in part at least, to inhibitory impulses which proceed from the cerebrum to lower centers. According to this view, diminution of the knee-jerk produced by relaxation would be due to depression of the knee-jerk center. Sleep would involve similar depressant influences from the cerebrum on the "sleep center" (Hess, 1932). Contrary evidence as to the knee-jerk was recounted in chapter xiv. As for natural sleep in man, my experience likewise leads away from the inhibitory hypothesis of the Pavlow school. For sleep eventuates as relaxation progresses to an advanced state, in particular when the eyes become fully relaxed.

However, investigations to date do not permit us to decide finally whether the process of complete relaxation in man involves the passage of inhibitory nerve impulses from cortical to lower centers. That the cortex can exercise an inhibitory influence upon neuromuscular activity is shown by the observations of Sherrington and Weed and by other evidence discussed in chapter xiii. But in their instances of relaxation produced by cortical stimulation, a concomitant contraction of the antagonist was each time produced, apparently by the mechanism called "reciprocal innervation." This mechanism does not apply to the present investigations, where relaxation of a muscle-group and its antagonists are observed to occur simultaneously. Our conditions are such as seem calculated to produce diminished reflex activity and therefore diminished cortical stimulation and activity. What takes place cortically at the moment when an individual sets out to relax a muscle-group is of course unknown, and it is safest to avoid speculation.

Fröhlich and Meyer (1912) have found the skeletal muscles during tetanus toxemia in a shortened yet relaxed state, and have been led to regard the relaxation of mammalian muscle as being in some way directly under control of the central nervous system. However, the significance of their observations still remains to be finally determined. The conception of Pavlow (1923) that sleep results from "inner inhibition" rests upon no immediate examination of chemical processes in the brain and must be regarded as highly speculative.

Several writers, notably Nikolaïdes and Dontas (1907-8), have made observations which they have interpreted as pointing to the existence of specific inhibitory fibers to skeletal muscles. But certain oversights in their work were found by Woolley (1907); while Hoffmann (1904) and Sherrington (1906*b*, p. 100) likewise failed to find satisfactory evidence of the existence of such fibers. Moreover, Verworn (1900*b*) has shown that during inhibitory relaxation of skeletal muscle, the excitability of the relaxed muscle and its motor nerve to electrical stimuli remains undiminished. Accordingly, it is now commonly agreed that no specific inhibitory fibers of the foregoing type have been demonstrated.

5. Evidence has been given that increased tonus or relaxation in either the cerebrospinal or the autonomic sphere may reflexly have a corresponding effect upon the other.

6. Reasons were given (chap. xiii) which support the belief that general relaxation tends to diminish the inhibitions in the nervous system of the neurotic individual. Clinical studies (chaps. xvi, xvii, xviii) suggest that failures and irregularities of nervous response are frequent in such patients.

7. The importance of what is called "habit formation" in

the nervous system is clear. Any learning-process of course depends upon habit. So only can we explain increase of ability to relax voluntarily and increase of the apparently relaxed state of the individual who practices from month to month. It is a matter of nervous re-education—the cultivation of relatively passive responses.

8. Progressive relaxation of one part in our experience tends to effect a similar condition of other parts. This would accord with "Pflüger's laws," restated in terms of relaxation. One can with care observe that an individual who is successfully relaxing one member is also as a rule at the same time relaxing others. It sometimes happens that a beginner striving only to relax one arm becomes generally relaxed and may even fall asleep. Apparently we are dealing with "conditioned reflexes."

In addition to the preceding considerations, others may be advanced on the psychological side to explain the calming effect of relaxation on mental and emotional activity:

9. Evidence was offered in chapter x that mental and emotional activity always involve a motor element. By decreasing this motor element, relaxation apparently diminishes such activity. Nervous individuals tend to rehearse their griefs, difficulties and problems, considering incessantly and perhaps inco-ordinately what to do about them; and this emotional reflection evidently is a fountainhead of nervous hypertension, which relaxation mechanically shuts off.

10. The patient has the will to relax, or to put the matter technically, he has what has been known since the work of Ach in 1905 as a "mental set" or "determining tendencies." He and later workers found under experimental conditions that subjects instructed to press certain keys according as certain colors were presented had a definite psychological setting, which has recently been somewhat further analyzed

(Watt, 1905; Bentley, 1927). The influence of "mental set" and the physiological processes which it arouses will not be confused with suggestion by those familiar with the literature.

11. Suggestion apparently plays a minor rôle. Of course an element of suggestion enters into almost every normal act; for instance, going to bed at night is suggestive of sleep; but the physician avoids suggestion in a technical sense.

12. In some instances repose seems to be slightly aided by virtue of the attention of the worried or emotional individual being diverted to his muscles. It is common knowledge that the one who becomes utterly engrossed in an emotional experience is greatly moved. The exciting issue is alone in the focus of attention. Training in relaxation appears to cultivate the tendency or habit of turning the attention away from the undesirably exciting issue or mental content to the state of the organism. Thereupon the issue may come to seem less important to the subject, as he attends to something else, and he is less prone to be emotionally disturbed about it.

13. Following psychological laws of "association," we may expect states of agitation to be succeeded by other states of agitation and states of repose to be succeeded by other states of repose. Practically this is a repetition of what was said under 7 and 8 above.

14. That reasoning may play a part is suggested by the observation of a patient, "Instead of giving way to excitement, you stop to reason if you are relaxed!"

It will be evident that voluntary relaxation of an undesired mental activity differs in mechanism from what psychologists and psychiatrists are accustomed to call "suppression" or "repression." The latter activities are commonly attended with effort, which is the reverse of relaxation.

EXPLANATORY PRINCIPLES

SUGGESTION

We come now to the *bête noire* of progress in nervous medicine. We must meet the defense which some physicians offer against investigative progress in any therapeutic direction—their claim that any new method is "suggestion." Until recently it was a current fashion to devote little attention to new methods of physiotherapy and of psychotherapy, since it was more convincing to dismiss them with an authoritative wave of the hand and a statement that "they are simply another form of suggestion." In 1913 when electrotherapeutics were still classed as dark magic by sundry practitioners of medicine, Sommer of the University of Giessen, well known for his experimental precision, found it necessary to come to the defense of the new subject with arguments that the effect of the electric current on the blood and other tissues was not identical with suggestion but could be distinguished therefrom by chemical and other tests. Unfortunately, those who contend that new methods are "suggestion" rarely undertake to define what they mean by that word. They fail to realize that criticism has as much need to have a scientific foundation as has the matter criticized. Every student wishes to know what is meant from a physiological standpoint when a new method is called "suggestion"; else the criticism is empty. On the other hand, phenomena of voluntary relaxation are amply described in various physiological monographs and texts quoted in the present volume, and it seems safe to say that no one familiar with the literature would be inclined to identify them with what psychiatrists call "suggestion."

Without doubt the best answer to any individual who inclines to believe that the present method is a suggestive procedure is to have him take a course in relaxation. All arguments then become unnecessary. He sees for himself how complete is the contrast between physiological relaxation and suggestive procedure. No one of my associates or subjects in the present work—physicians, physiologists, or psychologists —has ever maintained after completing the course that progressive relaxation is a suggestive procedure. It would be begging the question to assert that the individual who is learning to relax receives "suggestions" to do so. Suggestion is an ambiguous term: it is often used in the sense of "instructions," while at other times it has a technical meaning in the sense of hypnotic suggestion or the like. Of course instructions have to be given to the patient in order to teach him to relax. But in the same way he may be instructed how to carry out a physiological experiment or how to solve a complicated mathematical problem. This is not suggestion in the sense applied by critics to new therapeutic methods. Here they mean that the new method accomplishes its results only because it instils into the patient the belief that he will derive benefit. It is suggestion in this sense alone that we need discuss further.

A physician may prescribe digitalis or bromides in the hope of appropriate physiologic effects, yet find it expedient to tell the patient that he will derive benefit. As is well known, such an element of suggestion may accompany any procedure in the practice of medicine and in some cases apparently add to its effects. It can be proved by precise methods that bromides or digitalis have an effect which is essentially independent of "suggestion." In the same way experiments of the present volume are offered to show that relaxation is a fundamental physiologic occurrence, and is therefore not to be explained in terms of a word called "suggestion" whose physiology remains unknown, yet apparently distinct and different. A suggestive element can be added to the use of relaxation deliberately or unconsciously just as this may be done with digitalis or bromides; but experience leaves no doubt that relaxation is essentially different from suggestion.

The physician or investigator who is being trained to use the present method is required to become familiar with suggestive procedures in order that he may strictly avoid them when applying the present method. He who gives such suggestions has no proper sense of the technic of progressive relaxation. It is of interest to note the following list of thirty-two points of difference between relaxation and suggestion:

1. In the method of relaxation, no technical suggestions are given. For instance, the physician would never suggest, "Now your arm is becoming limp!" or "You will feel better after this treatment!" or "This will help you to be quiet!" He simply directs the patient in the same manner as when prescribing diet or exercise.

2. The relaxing patient shows no sign of increased suggestibility. None of the familiar tests of suggestibility gives a positive response. It is not possible to arouse during progressive relaxation suggestive anesthesia, paralysis, illusions, or delusions. His state of mind is shown to be in every way normal.

3. There is nothing in connection with progressive relaxation to correspond with post-hypnotic suggestion.

4. When learning to relax, the patient may be skeptical about the whole procedure and nevertheless do very well. No attempt is made or should be made to convince him that he will be "cured." Objective evidence may show his progress. As is well known, skepticism completely defeats suggestive procedures (Bernheim, 1896).

5. No *rapport* is established between physician and patient, like that in suggestion or hypnosis. Patients often state that they are disturbed by the physician's presence dur-

ing progressive relaxation and do better when alone. They are of course aided by his criticisms and instructions as they would be in learning any other procedure. The patient remains normally responsive to statements made by a third person present, whereas in hypnosis the comments of a third person are likely to be ignored.

6. The physician does not hesitate to interrupt the patient at a moment when he is failing in the attempt to relax and to criticize him vigorously. Such practice is foreign to all suggestive procedures.

7. Hypnotic suggestion is difficult to develop in neurasthenia, according to Möbius, and in children, according to Bostroem. In our experience, relaxation readily applies to both.

8. Relaxation has to be learned step by step with various details of success and failure. It is a learning process by the method of trial and error. It requires the cultivation of the observation and skill of the patient, which largely depends upon practice apart from the physician. Hypnosis does not have to be learned at all. The most susceptible subjects may give the best results with hypnosis or suggestive procedure upon the very first session. Hysterical anesthesias and paralyses may yield to a single treatment. This never happens with progressive relaxation, which requires time and patience like golf.

9. The therapeutic effects of suggestion and hypnosis, while brilliant, are likely to be ephemeral. Symptoms disappear to be succeeded by other ones, for the basic nervous hypertension is not treated. On the other hand, re-educative treatment by relaxation tends to be relatively thoroughgoing and permanent in effect.

10. The skill and habits acquired through progressive relaxation are educational, in the sense that the patient on his

own initiative often applies them to other pursuits like golf or pianoforte, while those of hypnosis are not.

11. Progressive relaxation leads to independence of manner and attitude. The longer the training, the greater on the whole becomes the independence. An advanced subject may even discharge himself as sufficiently skilled for his purposes. On the other hand, a patient who has been subjected to hypnosis for a prolonged period becomes increasingly dependent and manifests a submissive attitude and appearance.

12. In "autosuggestion" the patient imposes states of mind upon himself, imitating the acts of the hypnotist. Nothing like this is done in the present method. Such procedures would be tensions and would have to be eliminated as interferences with relaxation.

13. Headache or other distress often follows stimulating or exciting hypnosis. This is commonly avoided by suggesting to the patient that this after-effect will not take place. Nothing like this occurs after relaxation and no suggestion is given nor needed. A feeling of well-being accompanies successful relaxation.

14. There is nothing connected with progressive relaxation that corresponds with catalepsy of hypnosis.

15. The mode of onset is essentially different. In hypnosis a characteristic method is to have the patient stare at some object, often held high above his head. The trance state is then introduced through persistent unvarying tension of certain parts while others are partially relaxed. Other means of initiating hypnosis likewise represent persistent tensions of the sense organs during attention. Each patient generally concentrates his attention when hypnosis is introduced. On the contrary, such tensions play no part in progressive relaxation, which proceeds like natural sleep with general relaxation of the muscles.

16. Persistent tension in at least one muscle-group is characteristic of hypnotism, even if others are relaxed. Frequently it appears about the eyes. Tension is absent in relaxation.

17. In extreme relaxation the knee-jerk is diminished. (chap. ix). The same holds for the extensor reflex of the foot as electrically elicited (Hoffmann, 1922). Hypnosis has no such influence upon the reflexes (Bostroem, 1925).

18. It is generally admitted that a considerable percentage of individuals cannot be hypnotized, including many who are ready and willing to submit. In contrast with this, barring conditions of unwillingness and of severe pain and distress, relaxation can be learned to a noteworthy extent and degree almost universally. One patient reported that repeated attempts to hypnotize her had failed; however, she learned to relax with average readiness.

19. Persons of suggestible disposition as a rule are tardy in learning to relax. Generally they tend to be unobserving and wait for the physician to do something for them. They lack the independent spirit which is the greatest aid to speedy success with relaxation.

20. Patients accustomed to hypnosis may tend to go into a trance-like state when they lie down. This renders them poor and difficult subjects for the present method and makes it well to abandon the attempt to develop relaxation.

21. The first subjective experience in passing into a trance state of hypnosis is strange indeed. On the other hand, general muscular relaxation is familiar from the outset to everyone who is accustomed to natural sleep.

22. Patients who have learned to relax and to observe signs of neuromuscular hypertension often teach their children or others something of the practice. Obviously there is no strict parallel to this in suggestive procedures.

23. Disturbing noises tend to interfere with a period of relaxation as they do with natural sleep. This effect is likely to be lacking in hypnosis, after it has been induced.

24. The individual is generally awakened from hypnosis by some suggestive signal. Occasionally the awakening proves difficult. In contrast with this, there is no difficulty and no special signal of arousal from profound relaxation. If the individual has fallen asleep, he awakes in every respect as from a natural sleep.

25. Amnesia generally follows deep hypnosis. There is no parallel with this during extreme relaxation if the individual does not sleep. If he falls asleep, the conditions and after-effects are those of natural sleep.

26. There is nothing in hypnosis or in any other suggestive method to correspond with differential relaxation. If a critic were to state that differential relaxation is a suggestive procedure, it would be equally true to allege that the singing or swimming instructor is mistaken in believing that he teaches relaxation, but in reality only uses suggestion.

27. As previously stated, examinations of the conditions of relaxation by competent physiologists and psychologists have agreed that the matter is not suggestion.

28. Progressive relaxation is simply cultivated natural relaxation. Untaught persons who are naturally clever at the matter can secure the same sort of results in lesser degree, although they do not show the requisite control at all times. A physiologist (Professor A. J. Carlson) with no special training other than familiarity with the literature gave the same character of results during experiments on the knee-jerk as did trained patients. That suggestion was absent here seems beyond doubt.

29. In the investigations on general relaxation in man described in chapters vii, viii, ix and xvi suggestive elements

were carefully eliminated. The results clearly reveal the character of relaxation.

30. As will be seen from chapters xiii and xiv, relaxation is an elementary phenomenon, with a clear-cut physiological basis. This can occur even in nerve-muscle preparations. It would appear absurd to attempt to explain such an elementary phenomenon in terms of a more complex one like suggestion.

31. A simple exercise will bring out the distinction clearly. If the reader will raise his hand, then passively let it fall, he will have an experience of relaxation which evidently differs from suggestion.

32. In suggestion, the doctor designates the symptoms which are to disappear. After learning to relax, the patient often reports disappearance of symptoms never before mentioned.

33. It seems unlikely that anyone whose laboratory and general training has included the observation of muscular tensions would maintain that the effects of relaxation depend upon suggestion. Patients often report spontaneously upon the very first period that they have previously observed tenseness in themselves at the time of emotional or nervous excitement. Physicians likewise may observe muscular tensions in patients clinically or with laboratory methods. If muscular tension thus accompanies nervous or emotional excitement it is evident that its negative, relaxation, is not simply a suggestive phenomenon. The present method follows Darwin, Spencer, Lange, James and various experimental observers cited in chapter xii in their observations of muscular activities correlated with emotion. It seems safe to assume that anyone who has thoroughly read those writers will be unlikely to confuse the present method with" suggestion."

CHAPTER XVI

ELECTRICAL MEASUREMENTS OF MUSCLE CONTRACTION AND RELAXATION

THAT states of nervous irritation and excitement in man are marked by increased contraction in specific muscles has been discussed in previous chapters. This accords with the well-known fact that a considerable proportion of the peripheral nerves of the body (about half, according to Sherrington's estimate) supply the skeletal muscle system. Various evidence has been accumulated, as presented in this volume, that relaxation can be cultivated in man to counteract nervous excitement. What was most needed was a new, precise method of measuring muscle tension and, conversely, relaxation.

The problem of measuring muscle tension in intact man is no easy one. Apparatus must be sufficiently sensitive to detect and measure contraction in the muscles of a part that appears to be perfectly quiet. No one has succeeded in measuring slight muscle contraction in intact man by mechanical means, not merely because of the lack of sufficient sensitivity and the impossibility to apply mechanical connections directly to muscle fibers through the skin, but also because of the absence of means to discriminate between a very slight, steady contraction and no contraction at all. Because of these difficulties, I was led in 1922 to investigate the possibilities of electrical measurements. Since the time of Mattucci, 1842, it has been known that, when a frog's muscle contracts, a current of electricity passes along the muscle with the wave of contraction. In 1907, Peiper, using a string galvanometer, first secured electrical records of marked muscular contractions in man. During my earliest observations it was evident that the string galvanometer is not sufficiently sensitive for the present purpose. Accordingly, I set out to develop or assemble apparatus which would magnify the response from that instrument. In 1921 Forbes and Thatcher recorded human muscular contraction with a string galvanometer and amplifying equipment. Since that time other investigators had used various types of similar equipment but never, I believed, sufficiently sensitive for the present studies. Difficulties arise because, upon adding extra tubes, the recording wire becomes increasingly disturbed and irregular due to internal instability as well as to external factors.

The history of the development of the instrument used in the present studies and of the reduction and elimination of electrical disturbances has been told elsewhere. I have been, at all times, indebted to the Bell Telephone Laboratories for aid in this philanthropic endeavor; in particular, to Messrs. H. A. Frederick and D. G. Blattner, and to the late Mr. H. D. Arnold for his kindly interest. It has proved possible to devise and assemble equipment that can record voltage changes in muscles as low as a major fraction of one-millionth of a volt. This has been done accurately and reliably but as yet only within a certain frequency range.

For those not familiar with the string galvanometer, we recall that voltage changes (action-potentials) proceeding from the subject are impressed upon the terminals of a wire, approximately one ten-thousandth of an inch in diameter, extending in the field of a very powerful magnet, 20,000 gauss. The shadow of this wire, magnified six hundred times, appears on a screen, where its vibrations are recorded on a moving photographic film (see Fig. 54).

To record the activity in a muscle—for example, the muscles that flex the right forearm (see Fig. 55)—actual contact is made with the muscle fibers. Wires of platinum irid-

FIG. 54.—Photograph showing the string shadow (indicated by arrow) at rest

ium, so fine that even highly nervous persons are not irritated by them and do not object to their presence, are inserted through the skin. For the very occasional patient who objects to skin puncture, surface electrodes can be substituted.

In general, the greater the contraction of muscle fibers near the wire electrodes at any instant, the greater the vibra-

[311]

tion of the string at that instant. The extent of the vibration, of course, is the resultant of the voltages in the neigh-

Fig. 55.—Wires connected with muscles in the arm of this subject, here fairly relaxed, lead to the instruments which record the muscle potentials.

borhood of the wire. A photograph which illustrates the kind
of records secured from a tense muscle is shown in Figure 56,

and this may be contrasted with the record from a relaxed muscle, as shown in Figure 57. It has been my custom to take records for a half-hour or hour period. During this time the

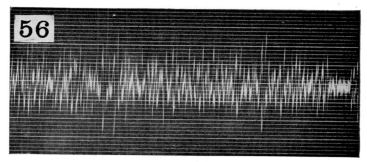

Fig. 56.—Tenseness in the muscle is disclosed here. Marked vibrations of the string shadow produce long approximately vertical lines, the length of which depends upon the voltages in the electrodes in the muscle. These vibrations vary with the extent to which the muscle is contracting.

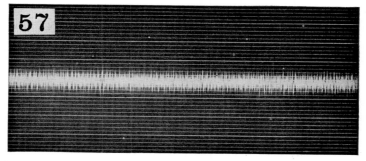

Fig. 57.—During complete relaxation of a muscle, the shadow of the recording wire is quiet, except for slight vibrations due to the instrument itself. Cf. Fig. 58.

photographic paper is exposed at the rate of $1\frac{1}{2}$ inches per second. For economy, automatic devices have been set up so as to limit each exposure to about 3 seconds, followed by a 17-second interval. In this way a photograph is secured, about 2.4 inches wide and perhaps about 20 feet long, containing ninety sections per half-hour.

PROGRESSIVE RELAXATION

Photographic tracings now serve as invaluable tests of the patient's state before treatment is begun, and are repeated from time to time to furnish records concerning progress. In addition they are useful to detect and to measure residual tension and to test speed of relaxation, which needs to be cultivated in therapy.

Results can be put in graphic form when desired by measuring the length of the vertical lines representing the string vibrations. Each point that is plotted is the maximum peak voltage (V_m) per unit of time, commonly 0.2 second. This is derived from the length of the longest vertical line found in a given unit of time, but subtracting therefrom the length of the longest line in a similar unit of time during a period when a short circuit exists across the input of the amplifier. We are required to subtract from the string vibrations, when the subject is in line, the string vibrations that take place when he is excluded from the circuit. The resultant can be translated into microvolts (millionths of a volt) and plotted accordingly. The vibration of the string for one microvolt is determined during each period of operation by a standardizing device which supplies these voltages at 57 cycles per second. Technical descriptions of the apparatus and of methods of procedure have been published elsewhere (Jacobson, 1934).

Thousands of records have been taken up to date (1937). As previously stated (p. 229), they fail to confirm the traditional view that healthy muscles, even in the resting state, are always in a state of slight sustained contraction, commonly called "tonus." They show clearly that rest in healthy man and in healthy animals may be complete in a muscle or nerve. Of course, this does not apply to the essential heart and respiratory muscles. The term "tonus" has been used in many senses; and we are obliged, as stated, to give up the traditional one. However, a common usage is in the sense of

a slight, more or less sustained contraction. This usage we retain, because we find the state referred to frequently present
in human and animal organisms.

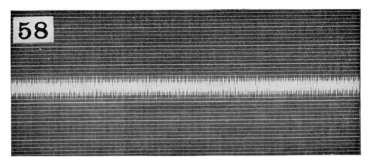

Fig. 58.—A short circuit exists across the input terminals of the amplifier, excluding the subject from the circuit. Slight vertical vibrations of the shadow are
recorded, owing to the instrument itself. They are scarcely less than those in
Figure 57. Evidently the relaxation indicated in Figure 57 is almost, but not quite,
complete. The patient complained of nervous symptoms and had not been trained
to relax. Within 17 seconds after the tracing of Figure 57, she became tense again,
as shown in Figure 56.

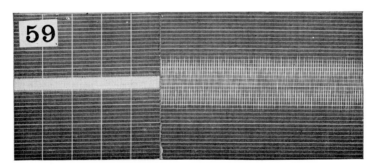

Fig. 59.—Intervals between vertical time lines=0.2 second. On the left, the
string is quiet, for there is no connection with the amplifying portion of the circuit.
On the right, 15 mm. (at 57 cycles) = 1 microvolt.

Methods of electrical measurement enable us to test
whether patients trained by methods of relaxation really
learn to relax and acquire a more relaxed state.

[315]

Students not trained to relax.—For control tests, relaxation in college students was tested in two groups of both sexes, mostly males. Details of this study have been previously published (Jacobson, 1934) and will not be here repeated. Group 1 consisted of ten undergraduate students in a course of elementary physiology who volunteered their services. They ranged in ages from eighteen to twenty-five. Group 2 consisted of six males, including three undergraduates within the above-mentioned age limits and three graduates who were twenty-eight to thirty years old. To illustrate the findings, the results of a second electrical test on Group 2 are presented in Figure 60. In general, we may assume, the more tense the muscle-group during the period of test, the greater the action-potentials (microvoltage). No student achieved or maintained complete relaxation of the arm muscles during the 30-minute period. The closest to this was achieved by one student who showed approximately zero voltage (action-potentials) for 79 per cent of the time. For an additional 14 per cent of the time, the action-potentials were very slight (not in excess of 0.3 microvolts); his relaxation was practically complete for 93 per cent of the time. This subject was a graduate student of sociology, age twenty-nine, who stated that he was very well, excepting that he had noticed chronic constipation and occasional mild backache for about two years. He added that he had trained himself to relax.

For the five other students, relaxation was approximately complete for smaller periods of the time, ranging down to 20.2 per cent. We may assume that the results presented in Figure 60 are a fair specimen for an average group of persons not trained to relax but trying to do so while lying down. More recently a third group of unselected university students presented considerably lower voltage curves. Various students, not trained to relax, showed complete relaxation of muscle-groups such as in the arm, in the leg or in the abdominal wall for many minutes at a time.

Patients before training to relax.—In the present investigation one or more records were made of each of fifty individuals with disorders classified as neuromuscular hypertension, chronic or intermittent, or other nervous or fatigue states with or without marked chronic colitis or vascular hypertension. Complete sets of records before, during and after training were made in twelve instances; partial sets of records were made in other instances, single records were made at one stage or another in the remaining instances. Concerning the central problem of this chapter, whether relaxation can be cultivated in man, the results in general are uniform.

Records for five subjects made before training to relax are represented in Figure 61. These five were selected because their action-potentials during the test averaged higher than those for other nervous subjects for whom complete sets of records had been made.

As is strikingly evident, the individuals with nervous symptoms whose records are represented in Figure 61 exhibit an almost complete failure to relax. This point becomes clearer by comparison with the preceding Figure 60, for the normal subjects. The "nervous" individuals show not alone higher peak muscle potentials but also greater and more irregular variations than do the "normal" students.

As will be recalled, one student achieved approximately complete relaxation during 93 per cent of the period. In contrast therewith, no individual in this "nervous" group achieved approximately complete relaxation in the arm muscles for as much as one-third of the period. One subject failed to become relaxed for even a fraction of a second. The

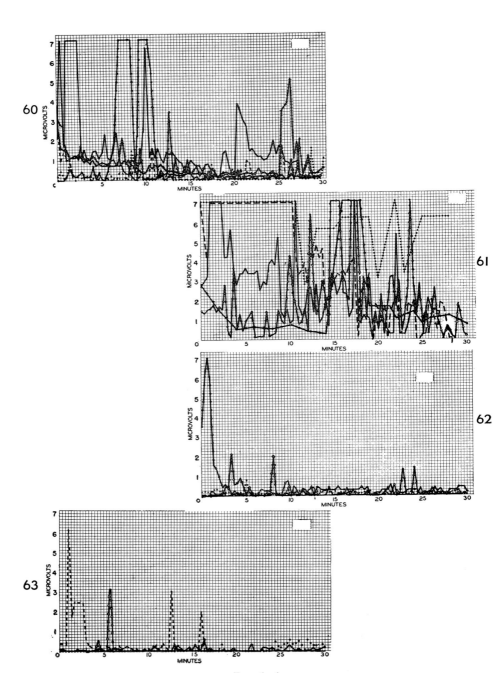

FIGS. 60-63

longest period during which any one of these five subjects remained continuously relaxed in the muscle-group tested was about 4 minutes.

The "nervous" subjects under consideration frequently became so tense and the voltages in consequence so high that the string shadow shot off the camera face with the sensitivity generally used, making it necessary to reduce the sensitivity of the assembly in order to obtain a record.

After training to relax.—Treatment of the groups of patients herein described was, in practically all instances, limited strictly to methods of relaxation. Instruction was given generally twice a week for periods of an hour. In addition the subject was requested to practice during each day for 1 or 2 hours.

Records made after training to relax (generally 4–9 months) are represented in Figure 62, which is to be compared with Figure 61 for the same five subjects before training. It is obvious that the microvoltages (action-potentials) in these subjects after training are much smaller than before training.

The results for a sixth subject are not included in Figure 62 because they yield a curve more irregular and somewhat

Fig. 60.—Curves for six "normal" students, not trained to relax. Microvoltage (V_m per cm.) is plotted against time. The instruction given is to relax as fully as possible. (In this and the following figures, values $\geqq 7.1$ microvolts are plotted as 7.1.)

Fig. 61.—Curves for five patients with certain neuromuscular disorders before training to relax. As will be seen from comparison with Figure 60, the microvoltages for this group are considerably higher than those for the "normal" students.

Fig. 62.—Curves for the same subjects as shown in Figure 61, but after training to relax. The microvoltages are considerably lower than before training. They are lower also than those for the students shown in Figure 60.

Fig. 63.—Curves for five other patients after training to relax. No electrical records had been made with these subjects previously. Since the curves on the whole are as low as those in Figure 62, it is evident that repetition of tests is not the cause of the progressive decrease in microvoltage generally observed during training to relax.

higher in microvoltage than the five curves in that graph, so that, if included, it would obscure the curves for the other subjects. However, considering this patient alone, the curve of voltages after training is distinctly lower, as a whole, than that before training.

The possibility must be considered that the microvoltages represented in Figure 62 are low in consequence of repetition of tests, that is, because the subject gets accustomed to the environment and not because of the results of training. Repetition of basal metabolic tests tends toward lower values, at least with some subjects; and this result is generally attributed to increase in muscular relaxation.

Five subjects were available who had clear histories of severe nervous symptoms but who had been trained to relax. Before treatment one had exhibited chronic spastic esophagus with a background of moderate cyclothymic depression; the second had feared jumping out of the window and had other nervous difficulties in practicing his profession; the third stammered severely, especially when under stress; the fourth had frequent loose bowel movements with mucus and passing of flatus, interfering with normal social activities; and the fifth gave a general history of nervous maladjustments and deported himself rigidly. All originally showed outward evidence of restlessness and inability to sit still with limbs persistently relaxed. The symptoms mentioned above had mostly disappeared during the course of training to relax, but no electrical records had previously been taken. Here, therefore, the factor of becoming accustomed to the tests does not enter.

Figure 63 represents the first records for these subjects, with one exception, where it seemed proper to retake the record, owing to noises in the laboratory occurring during the last 12 minutes of the period, and where the first record during quiet was essentially the same. Evidently the micro-voltages for these five subjects, for whom no question of habituation enters, are even somewhat lower than those represented in Figure 62.

The five subjects whose records before training are represented in Figure 61 and after training in Figure 62 were tested also from time to time during training. As a rule, records during training are intermediate in average microvoltage and in the number of moments of high microvoltage. The curves, on the whole, become progressively lower with training until a relatively low level is attained. Thereafter records may continue to show slight improvement or may continue on practically the same level; needless to say, there are instances when they show the reverse of progress. In my opinion, and according to reports of the subjects, various factors influence the results, as, indeed, they do at all stages of training—most commonly intercurrent illness, anxiety or grief.

Restatement of results.—Another way to state results of the electrical measurement of muscular activity for a person is to express it simply as a certain number of microvolts. This number or value represents the average excursions of the string for a unit of time during the period of test. More precisely, what we have here measured is the average mean peak microvoltage per unit of time during the ninety sample exposures per half-hour period of test.

A brief way to sum up the results for an entire record, then, is to state the mean peak microvoltage as a single number (mean $V_m = X$).

In summary, the foregoing tests on two groups of unselected students, age nineteen to thirty, generally showed mean $V_m < 1.4$. For all students in both groups, the average of mean V_m was 0.90. Considering only those students (twelve) for whom mean $V_m < 1.4$ in both tests, the average of means was found to be 0.57 microvolt. It is important to note that, for one student out of each five tested, mean $V_m < 1.4$ in one of the two tests.

Records of the six nervous subjects under consideration exhibited a relatively higher mean V_m, ranging from 1.46 to 8.00. (Similar averages for the students generally fall below 1.30 in value.) The highest of these values is over seventy-two times the lowest value found for any student.

Before training these subjects to relax, the average of mean V_m was 3.63, while after training it was only 0.26. For the fifteen students in both sets of tests the average of mean V_m was 0.90, a figure about three and one-half times as great as that for these trained subjects. Furthermore, before training, all subjects but one had moments when $V_m > 7.1$, while after training, four of the six subjects showed no excursions beyond 2.0 and seldom any as great as 1.4 microvolt. After training, the other two subjects showed some excursions exceeding 7.1, but they were now relatively infrequent. In the second group of patients on whom no tests were made until they had been trained to relax, the average of mean V_m as a group was 0.10. This figure is but one-ninth of the similar value for the fifteen students. As previously stated, at one or more points in the first record of thirteen out of the fifteen students tested, the microvoltage was 7.1 or more. This was never true of the four trained subjects, who were tested only once. The maximal V_m at any moment was respectively 1.1, 2.9, 3.1 and 4.0. These results are presented graphically in Figure 64.

Tense patients.—The first record taken of twenty patients with various disorders as stated above (p. 317) yielded records with microvoltages as low as the so-called "normal" ($V_m < 1.4$) in half the instances. Of the latter, at least five had been characteristically restless and nervously irritable.

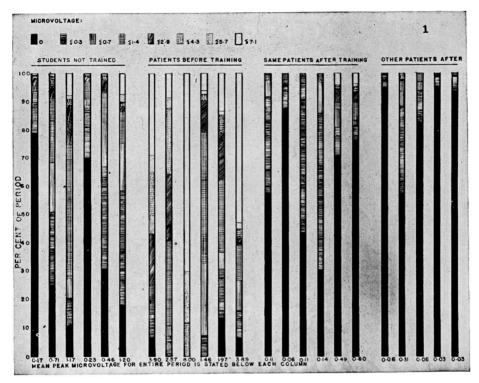

Fig. 64.—In the column for each subject, the various markings indicate during what percentages of the half-hour period the microvoltage (V_m) is less than or equal to certain values (see key). For example, the first column shows that this subject was relaxed 79 per cent of the half-hour and at least approximately relaxed 93 per cent of the half-hour. Below each column is stated the mean peak microvoltage (mean V_m) for the same subject and period.

(These five included instances respectively of cardiac neurosis, fatigue neurosis, atypical facial neuralgia, vascular hypertension and vague anxieties. Three were hypochondriacal to the point of disability.) In the remaining five, restlessness and nervous irritability had been less marked. (This subgroup included one sufferer from recurrent headache, one from chronic colitis and arthritis, one from insomnia, fatigue and mild colitis, and one from restlessness and hyperemotionality.)

Why persons even of pronounced neurotic type may yield a first record similar in microvoltage to those from unselected university students may be explained as follows: (1) Many nervous persons are highly variable in their manifestations —at one time, relatively calm; at another, excited. We should therefore expect such persons at times to give a record fairly "normal." An example in point is the patient in this group who has a cardiac neurosis and whose mean $V_m = 0.9$. Ten days later a second record shows mean $V_m > 7.1$. (2) According to my clinical experience, some individuals relax well lying down but become hypertense during activities that involve stress and strain. That is, they fail to relax differentially. (3) Three of those who show records in which mean $V_m < 1.4$ report that during past years they have, as a rule, rested one or more hours daily. It is conceivable that in these instances the effect of such habits has been to reduce the tension when lying down. But in any event, according to their reports and our clinical observations, they had been restless and over-excitable when sitting and conversing. (4) A record from a particular group, such as the right flexor muscles, does not necessarily represent the state of contraction in most or all other muscles. When other muscles contract, there is likely to occur an increase in the group being tested, as has often been verified. But occasionally in "nervous" patients (and characteristically in subjects who have been trained to relax), such increase is absent.

The direct measurement of "nervousness."—"Nervousness" can be measured directly in electrical terms by methods entirely similar to the foregoing (Jacobson, 1933). In this procedure, platinum-iridium wires are inserted directly in some peripheral nerve, such as the ulnar nerve of either arm. No anesthetic is required since the patient is little disturbed by the very fine wire.

CONCLUSIONS

1. Apparatus and procedure are described which make it possible to measure practically all degrees of muscular contraction in electrical terms. To effect this, transient potential differences (action-potentials) are recorded to fractions of a microvolt. Degrees and variations of muscular contraction over a prolonged period can be represented in graphic form while the mean value during the period can be expressed as a certain number of microvolts.

2. In health, during resting states, the traditional view that human muscle is always in a state of slight contraction is not confirmed. In intact man, in the frog and in the dog, persistent complete relaxation in a muscle-group is found during rest, at times, as tested by electrical methods.

3. Apparently healthy persons (college students) attempting to relax completely commonly fail to do so to some extent. From time to time their arm muscles contract, at least slightly.

4. Certain patients suffering from "nervous disorders," vascular hypertension or chronic colitis characteristically yield records showing marked inability or failure to relax.

5. Measurements made on these individuals before and after training afford evidence that relaxation can be cultivated. After training, they relaxed more fully, as a group,

than did the untrained college students. The results of daily rests alone, without training, remain to be determined in terms of electrical measurements. However, in spite of frequent rests, patients manifesting nervous hypertension are familiar in clinical practice. Even in bedridden patients, hypertensive states are seen, and their excessive tensions give the usual signs of subsidence upon training.

6. Low microvoltage is found after training to relax, even if no test has been made previously. Therefore repetition of tests evidently is not the cause of the progressive decrease in microvoltage generally observed during training to relax.

7. Some of the untrained subjects with disorders mentioned showed hypertension in some records but fair relaxation in others. They varied considerably until they were trained to relax more nearly habitually.

8. Measurements directly of states in nerves ("nervousness") can be made if the electrodes are inserted into a peripheral nerve such as the ulnar nerve.

CLINICAL USE

Electrical measurements of muscle voltages furnish a direct measure of tension. Photographic records are secured from one or more muscle-groups which represent samples from the individual at the time of the test. This method is used in addition to those enumerated on pages 60–62, inclusive. It furnishes the best indication and measure of residual tension.

Records made from the patient before, from time to time during, and after treatment serve as measures of progress and as guides to therapy. The patient may complain of tension in a particular region, or the physician may suspect its presence there. Thereupon, tests can be made locally and the results of therapy can be followed effectively.

CHAPTER XVII
ELECTROPHYSIOLOGY OF MENTAL
ACTIVITIES

IN CLINICAL and experimental studies, as stated, trained observers reported independently that sensations as from slight or pronounced muscular contractions occur during mental activities. During the past thirty years the writer has repeatedly made the same observations upon himself. We desire to test the possibility thus opened that the muscle sense, if carefully and critically employed, may put us on the track of observations that may be objectively verified for the whole range of our mental experience.

The electrical methods described in the last chapter will apply here. In the studies of mental activities we deal with action-potentials and not merely with resistance changes, such as the galvanic reflex (Jacobson, 1932).

It is most convenient to employ subjects who previously have been trained to relax. Otherwise restless movements are likely to be marked by action-potentials which obscure the readings concerning the phenomena of mental activity. Subjects not trained are equally satisfactory, but only if they relax adequately upon signal during the period of test. Relaxation, then, is required to furnish control tests and, in a sense, to isolate the mental activity investigated.

The subject lies relaxed upon a couch in a darkened, quiet, partially sound-proof room. His eyes are closed. Distraction by the operator is carefully avoided. It is agreed between the operator and the subject that the clicks of a telegraph key are to be signals: the first, to engage in a particular mental activity; the second, occurring soon afterward, to relax any muscle tensions present. By means of a signal magnet, the instant of occurrence of each click is recorded on the photograph.

Imagination of movement.—To secure data for statistical consideration, one instruction is used many times with one subject and repeated with others. The subject is to imagine bending the right arm. An imagined persistent or steady flexion of the right forearm is meant, rather than an imagined brief flexion and extension. One electrode is connected with the right biceps-brachial muscles and the other to a (partially) indifferent point—under the skin in the elbow pit.

During a period of general relaxation, when the galvanometer string with one lead attached to the right biceps is recording an approximately straight line, the signal is given for the subject to imagine what would, in actual performance, involve contractions in the above-mentioned muscle; for instance, to imagine that he is steadily flexing the right forearm. *Generally within a fraction of a second, the string ceases its steady course and engages continually or intermittently in relatively large swings, which cease soon after the signal is given to relax any muscular tensions present* (see Fig. 65). This test has been made upon about twenty subjects; and the results have been in agreement, with modifications later to be stated.

However, under the same conditions, if the first signal is to mean, "Imagine bending the left foot," the string shows no such onset of large swings but generally continues quiet or unaltered in its course (Fig. 66). This type of control test will be called "Control 1." Similarly, the string continues in a steady course even after signals have been given, "Imagine bending the left arm" (Control 2), "Actually bend the left foot one inch" (Control 3), "Actually bend the left arm one inch" (Control 4), "Imagine the right arm perfectly relaxed" or "Imagine the right arm paralyzed" (Control 5), "Imagine extending the right arm" (Control 6) and "When the signal comes, do not bother to imagine" (Control 7).

The negative results noted in the control tests rule out two possibilities: (1) that the string deflections observed during imagination are due to effects of the sound on the subject; (2) that the act of imagination involves action-currents all over the body.

Reaction-time from the beginning of the first signal to the onset of the string vibrations commonly varies from about 0.2 to 0.4 second. Results have been presented in tabular form previously (Jacobson, 1932). That the electrical phenomena observed following instruction to imagine bending the right arm are not artifacts due to training in relaxation has been confirmed with approximately ten subjects not so trained. If the subject, instructed actually to flex his right forearm, moves the wrist a fraction of an inch or a little more, the photograph quite resembles those secured upon imagining or performing the same act; but since the micro-voltage is considerably greater during overt movement, the galvanometer is set to record higher voltages.

Imagination and recollection of various muscular acts.—In a further series of investigations the subject is to imagine lifting a 10-pound weight with the right forearm. Under these conditions, electrical fluctuations are found to occur in the flexor muscles of the right arm during such imagination, but not during the foregoing and succeeding periods, when the subject is instructed to relax. Likewise, they are not found to occur in the right arm when the subject imagines that he is bending his left arm, or in the other control tests previously specified. Again, it seems unnecessary to present results in statistical form, since this has been done previously.

Recently this experiment has been repeated with fifteen

[329]

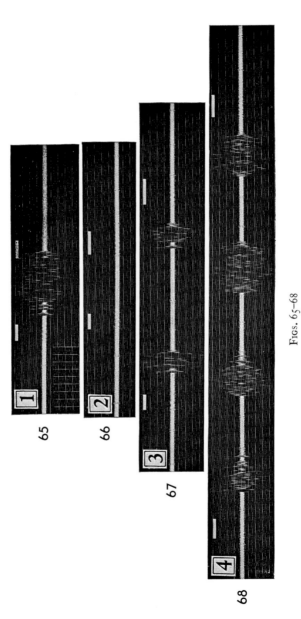

Figs. 65–68

Fig. 65.—Photographic record of string deflections during imagination. The subject, previously trained to relax, has been instructed, "Upon hearing the first signal, imagine lifting a 10-pound weight in the right forearm. Upon hearing the second signal, relax any muscular tensions, if present." One electrode in the biceps muscle, about 6 cm. above the other, which is inserted under the skin in the elbow-pit. Intervals between vertical time lines shown in left corner = 1/5 second. Distance between horizontal lines, approximately,

$$1 \text{ mm.} = 1.3 \times 10^{-6} \text{ volts.}$$

This photograph shows between signal marks (appearing at the top) groups of vertical lines, evidently due to action-potentials, which are absent in the periods of relaxation before the first signal and which subside soon after the onset of the second signal, Flexor muscle-fibers in the right arm contract as the subject imagines lifting with the right forearm.

FIG. 66.—Photographic record of control test. Conditions the same as in Figure 65, excepting that here the instruction has been, "Upon hearing the first signal, imagine lifting the weight with the left forearm." No evidence of action-potentials from the right biceps here appears. Results from the right biceps are negative similarly in tests where the instruction has been, "Do not bother to imagine," or "Imagine bending the right leg" (see text). The latter instruction is characteristically followed by action-potentials in the right leg.

FIG. 67.—Photographic record of string deflections during imagination of a repeated act. Conditions the same as for Figure 65. The instruction has been, "Imagine hitting a nail twice with a hammer held in your right hand." Between signal marks, two series of long vertical lines, indicating action-potentials, are seen, separated by an interval of practically horizontal line, indicating relaxation.

FIG. 68.—Photographic record of string deflections during imagination of a rhythmical act. The instruction has been. "Imagine turning an ice-cream freezer." This photograph shows, between signal marks, four groups of vertical lines, indicating action-potentials from the flexor muscles of the right forearm, separated by approximately equal lengths of practically horizontal line, indicating action-potentials from the muscles that flex the right forearm.

subjects previously trained to relax and with twelve subjects not trained. During imagination of lifting the weight with the right arm, action-potentials were recorded from the right biceps in 54 out of 62 tests with the trained subjects and in 41 out of 63 with the untrained subjects. During imagination of lifting the weight with the left arm, the galvanometer showed no electrical change from relaxation in the right biceps in 24 out of 24 tests with the trained subjects and in 20 out of 25 tests with the untrained subjects.

Imagination and recollection of various other acts commonly performed with the right arm have been studied. The following instructions were selected without consulting the subjects beforehand: "When the signal comes, imagine sweeping a room with a broom. Upon the second signal, relax any muscular tensions present." Other examples of instructions are: "Imagine writing your name"; "Imagine yourself rowing a boat"; "Imagine yourself boxing"; "Imagine scratching your chin"; "Imagine plucking a flower from a bush"; "Imagine combing your hair"; "Imagine playing the piano."

In half of these preliminary tests (14 out of 28) the string excursions were markedly increased during imagination as contrasted with relaxation; but in the other half the contrast was not marked. The subjects sometimes stated that, following the instruction to imagine, they carried out the imagined act with some muscle-group other than the flexors of the forearm. In other instances they stated that they visualized themselves performing the act but failed to have arm-muscle sensations. If these statements are correct, they evidently account for the absence of action-potentials from the biceps region in some of the above-described tests.

If the instructions selected by the investigator fail in some instances to induce the subject to imagine various muscular acts as carried out with his right biceps, but lead him instead to visualize or to carry out the imagination with other muscle-groups, another approach is suggested. The subject is requested to mention various acts which he would naturally perform with the right arm and would naturally imagine as so performed. In making such selection, he will be aided if he has had experience at observation of the muscle-sense. Four subjects were used in the following tests who had such experience. In addition, all except one had been trained to relax.

The instructions selected by them were as follows: "Imagine lifting a cigarette to your mouth"; "Imagine pulling a microscope toward yourself"; "Imagine pulling up your socks"; "Imagine grinding coffee"; "Imagine chinning yourself on a horizontal bar"; "Imagine hugging"; "Imagine shifting the gear of your automobile to first speed"; "Imagine pulling a door open"; "Imagine lifting a glass of milk to your mouth"; "Imagine climbing a rope"; "Imagine throwing a ball"; "Imagine pulling up weeds"; "Imagine pumping a bicycle tire."

Results for these four subjects were positive for action-potentials from the right biceps-brachial region in 97.5 per cent of these tests (159 out of 163). All of the control tests proved negative.

The microvoltage for such imagination, as averaged for all subjects, was about 26. This was much less than the average value found for imagining lifting a 10-pound weight (41 microvolts). No attempt is made to measure the micro-voltages as they occur in the tissues, but only in the attached electrodes. Corrections need to be made for various errors, as previously discussed; therefore the figures presented are not final.

Beautiful records are secured when the subject is requested to imagine some act performed rhythmically, such as "climbing a rope," "pumping a tire" or "turning an ice-cream freezer" (Figs. 67 and 68). Following the signal to imagine, the string shows a brief period of long vibrations, then an interval of rest for a fraction of a second or more, then similar periods of vibrations and rest, occurring in rhythmical succession until the signal to relax is given. On the other hand, instructions to imagine some act that involves only one relatively brief muscular action, such as "throwing a ball," typically are followed by only one brief series of long vibrations. But if the instruction is varied, "When the signal comes, imagine throwing a ball. After an interval, repeat this experience," then the record shows two periods of long vibrations, separated by an interval of rest. Corresponding results follow the instruction to imagine the act three times.

Recollection is tested like imagination. Instructions are to recall such activities as lifting a glass or a cup of tea at dinner, reading a certain paper, removing a shirt, putting on a coat or rubbers, raising the right arm, picking flowers, pulling up dandelions, pulling up an auger, playing the piano, playing the violin, boxing, punching a bag, brushing the teeth, washing dishes, sweeping, running a carpet-sweeper, scrubbing, rowing, hitting a ball at tennis, fishing, landing a fish and slinging a sledge. The subject stated before or after each test that he had previously had the experience denoted by the instruction.

Evidence that specific muscles contract during imagination. —The discovery that action-potentials exist in electrodes connected with specific muscles during the process of imagination or recollection leads to the question whether the fibers involved actually contract. Accordingly, a lever was arranged so that under controlled conditions flexion of the right arm can be magnified about eighty-fold and recorded photo-graphically along with the electrical curve.

The results show that, following the signal for the subject to imagine steadily bending his right arm or to imagine lifting a 10-pound weight in the same manner, the lever records a flexion of the arm, generally of microscopic extent. Soon after the second signal, which signifies that the subject is to relax any muscular tensions present, the lever suddenly returns to the position it had while the arm was at rest. But in control tests, where the subject imagines bending the left arm or imagines lifting the weight with that arm, no microscopic flexion of the right arm and no action-potentials from the biceps region are recorded.

The maximal excursion of the wrist, averaged for each subject, varies from 0.07 mm. to 0.32 mm. In the majority of the records there was a practically uniform velocity in arm bending, which ranged in most instances from 0.02 to 0.05 mm. per second. Considering, also, other evidences in the physiological literature, we conclude that the detection of action-potentials in muscles always signifies the presence of shortening of fibers.

The results of investigations to be recounted later permit us to distinguish two classes of persons: (1) those who characteristically show microscopic contraction in the right-arm musculature when they imagine acting with that arm, and (2) those who do not. The evidence indicates that members of the second class visualize their actions with the arm, or, less commonly, use inner speech or some other muscular activity to represent it to themselves.

Trained subjects of Class 1 all agree that in their experience it is impossible to imagine bending the right arm and to keep it perfectly relaxed simultaneously. Certain tests are devised so as not to depend upon the reports of these subjects: (1) The instruction is given to "cease to imagine" upon hearing the second signal. This effects the same consequences as the instruction to "relax any muscular tensions present in the right arm" upon hearing the second signal (Fig. 65). Evidently this supports the view that the muscular microscopic contraction is here requisite to the act of imagination. (2) If the experienced subject is instructed to relax his right arm completely, but simultaneously to imagine bending the right arm or to imagine lifting with his forearm, he generally reports after repeated attempts that he finds this impossible. He is requested to try again, but, at any rate, to engage in imagination. The results are like those shown in Figure 65. In a second test the subject, finding it impossible to imagine and to relax simultaneously, is requested to try again, but, at any rate, to keep his right arm perfectly relaxed. The results are entirely like those shown in Figure 66. These results indicate, without the intervention of the report of the subject, but entirely in conformity with that report, that it is found impossible to imagine bending the right arm and to keep it perfectly relaxed at the same time. The analogous results apply to imagining lifting a weight with the right arm. The photographic records indicate that the subject imagines—whereupon action-potentials appear along with the microscopic flexion, or else he relaxes the right arm completely—whereupon the electrical and mechanical records are negative. Accordingly, action-potentials, signifying contractions in specific muscular regions, have been found characteristic of imagination and recollection in these investigations.

An objective measurement of introspection.—The present methods make it possible to check on the ability of the subject to report accurately concerning his sensations from muscular contraction. This is accomplished by instructing him to press a button which actuates a signal magnet, producing a record photograph, concerning his perception of the muscular sensations. On the same photograph is recorded the action-potentials from contractions in the same muscle-group. The results indicate a fair correspondence between the periods of indicated perception and of the action-potentials. It is necessary to allow for the reaction-time of the subjects. The procedure just described is, so far as I know, the first instance or record where such objective measurement of introspection has been accomplished.

Relaxation of mental activities.—Following the onset of the second signal, which signifies that the subject is to cease to imagine, there generally occurs with subjects who have been trained to relax a distinct diminution of the large deflections within a time-interval of 0.2 to 0.5 second. A complete disappearance of such deflections generally occurs under the same conditions within 1.5 second. Exceptions to this generally were infrequent with the trained subjects, but they occurred more often with two out of three of the untrained subjects. Most notably, an untrained subject failed to relax, as indicated by the persistence of lengthened deflections, on one occasion for 7 minutes and on another occasion for more than 2 minutes.

The amputated limb.—In the light of the foregoing observations it becomes of interest to inquire what occurs in individuals with amputated limbs who assert that they can do everything in imagination with the lost part that they can do with the intact part.

The left arm of a graduate at the University of Chicago had been amputated above the elbow joint at about the age of eight years. He said that he could imagine doing anything with his left hand that he could with his right. This assertion would harmonize with our results if it meant that, upon imagining acts with his left hand, he merely visualizes or verbalizes the action; but not if it meant that he could imagine acts with the lost hand fully in the same physiological manner as normal subjects.

In brief, the results with this subject indicated that, when he imagines bending the missing left hand, there generally occur action-potentials from the stump-biceps muscle as well as action-potentials from the muscles that flex the right hand. In other words, when this subject engages in mental activity concerning his left hand, certain muscles contract; but these —for example, in imagined flexion—instead of being merely the muscles that flex the left hand, as in intact subjects, are in the stump of the upper arm, or in the intact arm, or in both places. The contraction that takes place is, so to speak, substituted for one that is missing. The subject had not been informed as to the purpose or methods of the investigation. It was therefore of striking interest when, after he had evidently engaged in subjective observation during a number of tests, he suddenly volunteered that he desired to correct his original statement that "he can imagine doing anything with his left hand that he does with his right." He now stated that when he does something with his right hand, the left seems in imagination to duplicate the performance, going through the same experience. But he never has experiences of his left hand's performing any act independently of the right. He added, "My imagination of bending the left hand is but a shadow—a duplicate of what the right hand is imagined to perform." In short, his original statement is ambiguous and he corrects it. He can imagine doing anything with his left hand that his right hand does, but only under one condition, namely, that the right hand, at the moment of imagination, actually engages in the same act or is imagined to engage in that same act. He is obliged to use what we called a "substituted contraction." No independent imagination, such as characterizes intact subjects, exists for this subject's left hand.

In normal subjects, it was found—at least after a little practice—that mental activities involving the muscles of the left hand are generally accompanied by no reaction in the right arm. A certain measure of independence exists for the two sides, particularly after a little practice. But this subject apparently has lost the ability to imagine his left hand acting independently of his right hand.

Visual imagination and recollection.—Visual imagination and recollection are tested with electrodes placed near the eyeballs. The circuit used yields a pattern on the photographic record distinctive for looking in each particular direction. During a period of general relaxation, when the galvanometer string is vibrating slightly and uniformly, recording a fairly horizontal band on the photographic paper, the signal is given which will incite the subject to visual imagination. Generally within a fraction of a second, the string shoots forth and back or alternately for fractions of a second or more, producing deviations from the horizontal on the photographs. These deviations cease soon after the signal is given to relax any muscular tensions present.

If the room is not too dark, the operator can generally observe that the closed eyelids of the subject appear quiet during extreme relaxation, whereas a slight movement or convergence of the eyeballs as a rule can be detected at the moment of visual imagination. Likewise the photographic records for visual imagination show patterns resembling those above-described for eye movements. For instance, the pattern for imagining Eiffel Tower is practically identical with the pattern of the same subject for looking upward. Evidently in imagining the tower, the subject's eyeballs move

[339]

upward, somewhat as they would upon actually seeing a tower. Correspondingly, he reports that in imagining he looks from the base to the top of the tower.

In general during visual imagination and recollection, we find that electrical records are secured from the ocular muscles, producing photographic patterns like those following instructions to look in one direction or another. As shown by these records, eye-movements (or convergence) characteristically occur during visual imagination and recollection.

When the trained subject is requested to relax his eye muscles, the curve becomes relatively horizontal and straight and, according to his report, the visual imagination or recollection disappears. Furthermore, if interrupted at any time when the curve has been relatively horizontal and straight, even if no instruction to relax has been specifically repeated, he reports that visual imagination and recollection have been absent.

Variation of specific muscles contracting during imagination.—On introspective grounds, psychologists have long assumed that the same act or object can be imagined by the subject at different moments through different types of imagery. For instance, lifting a weight with the right arm might be imagined with "muscular imagery" in the right arm, or again with visual imagery. The physiologist is obliged to avoid such assumptions but needs to develop objective tests. These are made in the present instance by attaching a set of electrodes in the ocular region of the subject, while the lever is used simultaneously to record flexion of the right forearm.

When the trained subject is instructed "to visualize bending the right arm" voltage changes in the ocular region, as from eyeball movement, are recorded in almost all instances. But under the same conditions, action-potentials are absent from the arm muscles in almost all instances. This indicates that the matter of bending the right arm may be imagined in at least two physiological manners: (1) in visualization, that is, with contraction occurring in eyeball muscles; and (2) in muscle sensation, that is, with contraction occurring in muscles of the right arm.

Further tests were made with electrical and mechanical recording instruments applied as above set forth, but the instruction merely was to "imagine bending the right arm" or to "imagine lifting a 10-pound weight." While the three subjects employed minutely flex the right forearm in all instances during such imagination, they visualize in some instances but do not in others, as indicated both by the objective records and by their subjective reports.

In different subjects, then, the same muscles do not always contract during the imagination or recollection of a particular act or object. But the results indicate that, during imagination or recollection, muscular contraction, if absent from one region, will be found in another. It is presumed that this principle explains why the instruction to imagine using the right arm is not invariably followed by the occurrence of action-potentials in the right arm: in negative instances, we assume, the subject merely visualizes the act. Further tests in this direction remain to be performed.

The speech musculature.—Investigations also have been carried out on the musculature of speech during imagination, recollection and abstract thinking. Since Plato called thinking "an inner speech," nothing definite has been developed on the physiology of this relationship. A brief abstract of development in this field has been stated previously (Jacobson, 1932).

The present tests are made with electrodes in the muscles of the tongue or underlip. Instructions are chosen to include not alone imagination and recollection but also other types of mental activity. Briefly stated, the instructions include: to imagine counting; to imagine telling your friend the date; to recall certain poems or songs; to multiply certain numbers; to think of abstract matters such as "eternity," "electrical resistance," "Ohm's law," the meaning of the word "incongruous" or "everlasting"; and to make up your mind what you are going to do tomorrow.

When the electrodes are connected in the speech musculature of the trained subject, the string shadow is practically quiet during relaxation. But promptly after the signal is sounded to engage in mental activity involving words or numbers, marked vibrations appear, indicating action-potentials. Soon after the subject hears the signal to relax any muscular tensions present, the vibrations cease, and the string returns to rest.

The series of vibrations occur in patterns evidently corresponding with those present during actual speech. For, if the amplifier rheostat is set for a relatively low sensitivity and the instruction has been, "When the first signal comes, count aloud but as faintly as possible, 'One, two, three,' " three series of vertical lines are generally found on the photograph separated by a horizontal interval. If, now, the amplifier rheostat is changed to a sufficiently higher sensitivity and the instruction is to imagine or to recall counting but not actually to count, a quite similar photograph is secured. Likewise when verbal matters, such as a poem, are imagined or recalled, records of action-potentials are secured resembling those from actual faint speech, but of considerably lower voltage.

For control tests the instructions are: (1) "When the signal comes, do not bother to imagine"; (2) "Imagine Eiffel Tower"; or (3) "Imagine lifting a 10-pound weight with the right forearm." In the tests previously described, (1) was found to be followed by no evidence of muscular contraction, (2) by evidences of contraction in muscles that move the eyes, and (3) by contraction in muscles that flex the right forearm. But in the present tests for contraction in the muscles of speech, these three instructions afford negative results.

Following each of the objective tests here described, re-ports were secured in detail from three subjects who had been trained to introspect. They agreed that during mental ac-tivities involving words or numbers they feel tenseness in the tongue and lips as in saying those words or numbers, except that the feeling is slighter and more fleeting. These reports conform with our objective findings.

Other investigations.—Allers and Scheminsky (1926) used a loud-speaker and amplifying system. With leads from the arm flexor muscles, they heard noises when the subject had the "motor idea" or "motor thought" of doubling his fist or (less loud) of bending his knee. Obviously their subjects did not relax. The significance of the muscle phenomena was not completely "analyzed." A specific relationship was not dem-onstrated as in the present studies. In 1937 Davis and Shaw reported, at the Midwestern Psychological Association, results on imagination involving the arm. Their apparatus was suffi-ciently sensitive; and their findings, according to Professor Davis, tended to confirm those made previously in my labora-tory. Other subsequent investigations with sensitive equip-ment were made by Max (1933–37). Using deaf subjects, but otherwise duplicating in various respects the methods and procedures described above, he has succeeded in recording action-potentials during imagination of various types. In his published accounts he commonly fails to state when he is repeating work that has been done previously. It is evident, also, that he is most interested in theoretical views.

1. When the subject, lying relaxed with eyelids closed, engages in mental activity such as imagination or recollection, contraction (commonly slight and fleeting) occurs in specific muscles. Evidence is thus afforded that the physiology of mental activity is not confined to closed circuits within the brain but that muscular regions participate.

2. During visual imagination or recollection the muscles that move the eyes contract, as if the subject is looking at the imagined object.

3. During what psychologists term "inner speech," muscles in the tongue and lips contract as if to say the words in swift and abbreviated manner.

4. During imagination or recollection of muscular acts or of matters that involve such an act on the part of the subject, contraction occurs in some of the muscle-fibers which would engage in the actual performance of the act. Exceptions are noted when visualization occurs alone, as is characteristic in some subjects.

5. During a particular mental activity the muscles of a quietly lying subject, trained to relax, remain inactive, as a rule, excepting those specifically engaged as above stated.

6. Electrical records, along with subjective reports, indicate that, during general progressive muscular relaxation, imagery and thinking processes dwindle and disappear.

7. Relaxation of the specific muscular contractions present during a particular mental activity brings about the disappearance of that activity. This is accomplished by trained subjects in periods sometimes varying from about 0.2 to 0.5 second as measured by action-potentials.

[344]

8. The action-potentials measured in the present investigations are proved to be very different from the galvanic reflex.

9. A method of measuring neuromuscular processes in mental activities is presented, and the possibility of the development of a branch of study analogous with physical chemistry, and which may appropriately be called "physical psychology," is herewith opened.

CLINICAL APPLICATIONS

The results of electrical measurements agree with and confirm the findings as related in chapter xvi, that relaxation of specific muscular processes *ipso facto* does away with specific mental activities. Physiology thus provides a method which can be turned to clinical use where it is desired to control certain types of imagination or emotion, including worry and excessive mental activity.

CHAPTER XVIII

THE CULTIVATION OF QUICK RELAXATION AND OF SLOW PROGRESSIVE RELAXATION

RELAXATION in a muscle-group can be effected about as quickly as contraction, at least at times, according to Orschansky (1889) and other early investigators. The authors agree on the need of practice to accomplish this. We now have the instrument described in chapter xvi and can determine precisely the time from the beginning of the signal to relax (the click of a telegraph key) to the beginning of relaxation, as signified by the first clearly discernible fall in action-potentials. This interval is comparable with "reaction-time," which means the time from the beginning of the signal to the beginning of reaction. "Relaxation-time" means the time from the beginning of the signal to the termination of action-potentials, signifying relaxation.

Investigations are in order to determine beginning and complete relaxation-times in healthy persons, not trained to relax, as compared with the times in neurotic and hypertense individuals before and after training to relax. For this purpose, fourteen university students not trained to relax were employed. The custom is to request the subject, while lying, to flex the right forearm slightly at the first click of a telegraph key and to relax as quickly and completely as possible at a second, like signal. No preparatory warning or presignal was given, except telling him to be ready just before a set of tests.

Practice tests are made until it appears that instructions are fully understood and that the wrist does not move more than a few centimeters. During this series a mechanical, as well as an electrical, record was taken in a few instances. The interval between the two signals varies, mostly from 1 to 2 seconds, but is almost always less than 3 seconds. Longer time-intervals are not employed because they seem to enable some untrained subjects to relax more promptly, obscuring individual differences, which was contrary to our purposes. It is necessary to vary the interval in order to prevent the subject from expecting the second signal at a given moment and therefore tending to relax at a fairly constant interval after the first signal, in place of reacting to the second signal as desired. In a few records some subjects evidently failed to wait for the second signal to relax, for they showed instances where relaxation began before the second signal and other instances where relaxation began in very brief intervals after the second signal (e.g., 0.04 sec.). Such records were discarded. The subjects who made them were informed of the failure to follow directions and were given practice in proper performance before final records were made.

Generally the photographic record is taken continuously for at least 10 seconds after the second signal. Thereafter, if action-potentials are still above 1.0 microvolt, the camera usually is turned on and off at set intervals for about 2 minutes. In most instances three or more tests were recorded for each subject, but sometimes a single test seemed a fair sample. It should be mentioned that even the trained subjects had no previous experience in relaxing upon signal. The aim is to study in normal and neurotic individuals the course of action-potentials in early tests following the signal to relax rather than to make statistical determinations of intervals, as did earlier investigators.

As a rule, the shunt resistance and the string tension are adjusted so that the excursion of the shadow is 3.5 mm. on either side of the zero-line when a potential difference of 1 microvolt at 57 cycles is impressed upon the input of the amplifier. In consequence, while the arm is bending after the first signal, the string shadow invariably shoots back and forth beyond the limits of the photographic record. The voltage here is undetermined, excepting that it exceeds 7.1 microvolts both positively and negatively. For statistical purposes, as in the previous investigation, we read only on one side of the zero line, and set down excursions beyond the record as 7.1 + microvolts. In averaging, the plus factor is ignored. Where this is done, the averages obviously do not represent absolute values but may be used for purposes of comparison. Toward the end of the series, when the string shadow passed beyond the limits of the photograph for a number of seconds, the operator quickly changed the shunt resistance so as to diminish the string excursions and permit the microvoltages to be determined, whatever their magnitudes.

Results. Gross contractions. Students not trained to relax.— Figures 69 and 70 are illustrations from records of the second set of tests with each subject. One subject shows relaxation not yet attained even 2 minutes after the signal. As a rule, with the fourteen students tested, relaxation was not so delayed as occurs in Figure 69 or so promptly attained as occurs in Figure 70.

In Figure 73 is shown for both groups of students the course of relaxation as measured in action-potentials. For group 1, the curve A represents the results in the first set of tests, while the curve A' represents the results in the tests made a week later.

In the untrained subjects tested, relaxation commonly occurred more or less abruptly and incompletely at first and generally was not well sustained. Occasionally there was a more or less linear decline of action-potentials for several seconds, followed by further irregular declines or by irregular increases. Figure 71 is an illustration of linear progressive relaxation, showing that this can occur with diagram-like precision. Such precision harmonizes with the view held by many investigators that action-potentials are a definite function of the physicochemical process underlying normal muscular contraction.

In summary, thirteen out of fourteen students failed to attain complete relaxation in the arm muscles, following bending, within 1.0 second. At this interval, on the average, V_m fell to about 3.0 microvolts. In some instances failure to relax is prolonged for several minutes or more. Frequently the students failed to maintain complete relaxation while awaiting the first signal.

Patients before training to relax.—The records show that neurotic subjects often exhibit striking failure to relax upon signal. While they await the signal to bend the arm, they often show marked action-potentials. An illustration appears in Figure 72 where, even 2 minutes after the signal to relax, the potentials exceed 10 microvolts.

Of sixteen patients here considered, no more than four relaxed upon signal, even if we disregard the first test recorded for each subject. At 1 second after signal and thereafter, the microvoltage for these four subjects generally was less than 1.4. (Three of these four subjects likewise showed lower mean V_m per half-hour during general rest than did the other twelve subjects in this group.)

In Figure 73 the curves P and $\mathcal{Q}$ represent these sixteen patients, divided into two groups (six and ten). The members of the smaller group, p, exhibited very marked restlessness before training to relax, as judged by clinical observation and

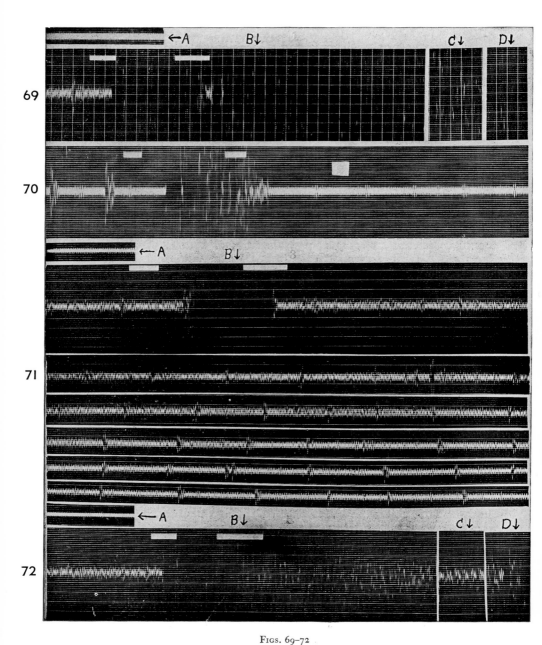

CULTIVATION OF QUICK RELAXATION

by electrical measurements. At any instant during the 5 seconds after the signal to relax, the action-potentials for this group greatly exceeded the action-potentials for either group of students. Likewise during the tests on prolonged rest the action-potentials for this group generally exceeded those for either group of students. The other group showed moderately higher action-potentials than the students, beginning approximately 1.5 seconds after the signal to relax. It seems warranted to conclude that continuing or intermittent high action-potentials (e.g., 5 microvolts, with the present setup) characterize the failure to relax in certain neurotic or hypertense subjects; but, if such action-potentials are moderate, the difference from normal states is not sufficiently great to be distinctive.

Figs. 69, 70, 71, 72.—Intervals between vertical time lines=0.2 second. This time-interval applies to all records, which should be read left to right. Signal marks at the top of records indicate clicks of a telegraph key. The subject is to bend the right forearm slightly when he hears the first signal, and is to relax as quickly and completely as possible when he hears the second signal.

Fig. 69.—Distance between horizontal lines: 1 mm.=0.4 microvolt. *A.* Short circuit across input terminals of amplifier. *B.* Subject, a student not trained to relax. Upon the first signal, he bends the right forearm slightly: action-potentials reach relatively high voltage, so that the excursions of the shadow pass beyond the limits of the photograph. About 0.4 second after the signal to relax, the excursions diminish, but for less than 0.2 second, after which they again pass off the photograph. This failure to relax continues through *C* and *D*, although the intervals between *B* and *C*, and *C* and *D*, are 30 seconds.

Fig. 70.—Excursions on short circuit same as in Fig. 69, *A.* Subject, another untrained student. Note action-potentials, indicating failure to relax as directed, prior to first signal. (In trained subjects, action-potentials are then absent as a rule.) About 0.8 second after the second signal, he evidently relaxes completely, no action-potentials being noted except periodic ones due to arterial pulsation.

Fig. 71.—*A.* Excursions on short circuit. *B.* Illustrating linear progressive relaxation in an untrained student. The original record was cut into six portions: the right end of the first portion is continuous with the left end of the second portion, and so on. Other conditions same as in Figures 69 and 70.

Fig. 72.—1.4 mm.=1 microvolt. *A.* Short circuit. *B.* Failure to relax in a highly neurotic subject. Note higher voltages of action-potentials. Interval between *B* and *C*=46 seconds; between *C* and *D*=105 seconds.

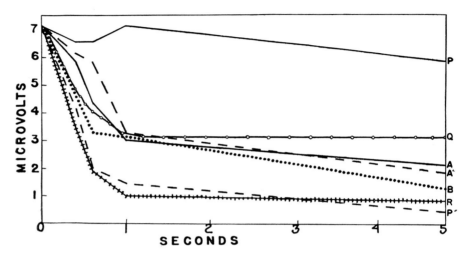

Fig. 73.—Curves showing the course of relaxation in the right biceps-brachial flexors in normal students, contrasted with neurotic and hypertense patients. The onset of the signal to relax occurs at zero (o) on the time scale. The curves *A, A'* and *B* show the mean values for the students tested (see text). *P* indicates the mean values for six highly neurotic and hypertense patients before training to relax, while *P'* shows the difference after training. *R* represents the results from a control group of similar patients, who had been trained to relax but never before tested. Although unaccustomed to the conditions of test, these patients nevertheless show relaxation more promptly attained and better sustained than in the students. *Q* represents another group of patients (see text), not trained to relax.

Patients after training to relax.—Following a period of training which varies for each subject, the results for group *p* are presented in Figure 73 in the curve *P′*. At the end of one second, the microvoltage for each subject is less than 1.5, which is less than half the values for the students at that time, and considerably less than one-fourth of the microvoltage present at that interval when tested before training (*P*).

As previously, it is necessary to consider what part habituation to the conditions of the test may play in this reduction of relaxation times. Accordingly, we employed another set of patients (five) who had been trained to relax but without electrical testing hitherto. These patients, also, exhibited very marked restlessness before training to relax, as judged by clinical observation. The curve for this group, *R*, runs very close to that for the other patients previously mentioned, after training, the average microvoltage at the end of 1 second being approximately 1.0. This is evidence that the more prompt relaxation observed in patients after training is due not to repetition of tests but to training.

Slight contractions.—During activities such as imagining or recalling, as previously stated, slight but specific muscle contractions commonly take place which can be relaxed directly by the trained subject but which in both trained and untrained subjects evidently subside when the particular mental activity ceases. These experiments are here repeated because improved apparatus makes possible a more satisfactory determination of the moment when the action-potentials disappear completely. Subjects are again employed who, upon instruction, can maintain complete relaxation. Otherwise voltage changes due to failure to relax occur irregularly in the record and render it difficult or impossible to identify those changes which mark the act of imagination or recollection. The instruction is to "imagine bending the right arm" or to "imagine lifting a 10-pound weight in the right hand" at the first click of a telegraph key and to relax any muscular tensions present at a second, like signal. An alternative instruction is to cease to imagine at the second signal.

The results for the various subjects are, on the whole, similar to those described following the instruction to bend the arm. Subjects showing marked restlessness, as judged by clinical observations and by electrical measurements during prolonged attempted rest, as a group show longer relaxation-times than the presumably "normal" students; but after training, the relaxation-times are significantly shortened.

SUMMARY AND CONCLUSIONS

Of fourteen students not trained to relax, thirteen failed upon signal to relax the forearm flexors, following flexion, within 1.0 second. At this interval, on the average, V_m fell to about 3.0 microvolts. In some instances failure to relax was prolonged for several minutes or more. Prolonged relaxation may be linear in its progress. Frequently the students failed to maintain complete relaxation while awaiting the signal to bend the arm.

Certain patients were selected for test who before training to relax exhibited very marked restlessness, as judged by clinical observation and as confirmed by electrical measurements of prolonged rest, showing relatively high microvoltages. These patients before training exhibited failure to relax upon signal much more striking than any of the students tested, with one possible exception. Certain other patients definitely neurotic or hypertense at times, according to clinical observations, nevertheless gave curves like those of some of the students. After training, the patients all relaxed while awaiting the signal, and they relaxed more promptly and completely as a rule than did the untrained students (V_m = 1.5 μv at the end of a second). (No practice at relaxing upon signal was included in the course of training.) This affords further confirmation that relaxation can be cultivated in man. Similar confirmatory evidence of shortened relaxation-times in the trained is found upon testing the slight muscular contractions characterizing mental activities.

CLINICAL APPLICATIONS

In many instances the nervous or tense patient presents a relatively brief reaction-time; upon stimulation, he is quick to contract. Evidently, if he is to acquire control over his tensions to any marked extent, it will be necessary for him to learn to relax quite as quickly. That speed in initiating and in completing relaxation can be cultivated is readily confirmed upon electrical tests. Well-trained subjects learn to relax slight tensions to zero in a brief interval such as 0.2 second or a little more. How this pertains to mental activities is typically illustrated in Figure 65. According to clinical experience, patients can be trained to relax promptly the tensions during mental activities, thereby doing away with the mental activities. This experience is shown, by electrical tests, to have a foundation in fact. Failure to relax promptly and completely appears to be responsible for the persistence of outstanding symptoms, including excessive imaginations, reflections, emotions and other mental states in many neurotic or hypertense individuals. Training in relaxation tends to meet these difficulties if the individual practices contracting and relaxing various muscle-groups in response to signals. Present-day methods also include the cultivation of slow progressive relaxation of representative muscle-groups (Fig. 71), as well as of the general musculature. Success in this is marked by the diminution to zero of action-potentials gradually during the course of minutes. In present practice this procedure is particularly applied in the treatment of persistent tensions in smooth muscle, such as are found in colitis, arterial hypertension and other spastic visceral states.

CHAPTER XIX
SPASTIC ESOPHAGUS AND MUCOUS COLITIS
Etiology of Alimentary Spasm

THE cause and treatment of alimentary spasm[1] be-
long among the outstanding problems of medicine today. Even innervation of the alimentary tract is
not yet a closed question. In conflict with theories of "vagotonia" and "sympatheticotonia," perhaps the vagi as well as
the splanchnics have both motor and inhibitory effects on
the lower esophagus, cardia, stomach and small intestine
(Carlson, Boyd and Pearcy, 1922a). Anatomical and functional studies by Kiss (1932) and by Ratkoćzy (1936) lead
them to include the vagus in the sympathetic system, explaining old observations anew. In alimentary hypertensive
states, under varying conditions, phenomena attributable to
excessive vagal or sympathetic activity are encountered (see
also Seifert, 1937), but the following studies suggest that
either type diminishes if relaxation becomes sufficiently
marked and habitual in skeletal musculature.

It is generally admitted that the plexuses of Meissner and
Auerbach control and co-ordinate peristalsis, but there has
been considerable controversy as to what the muscle tissue
itself can do when the nerve elements are stripped away
(Bayliss and Starling, 1899; Carey, 1921; Alvarez, 1922).

Does the hypertonus result from pathological changes in
the mucous membrane, or does it originate in the nerve-
endings, in the intramuscular plexus, in impulses from
the extrinsic nerves, in increased excitations from the motor
ganglions or nerve centers, in decreased excitations from in-

---

[1] The contents of the present chapter have previously been published (Jacobson, 1924a, 1925b, 1927a).

hibitory centers, or from a mixed source? A survey of the clinical literature fails to enable us to decide among these possibilities. We may begin with spasm of the lowermost portion of the esophagus, which often but perhaps not always occurs with spasm elsewhere in the esophagus. Vinson writes of cardiospasm: "The etiology of the condition is unknown, the numerous hypotheses failing to explain all cases." Among other causes he mentions irritative lesions in the vagus nerves, esophagitis, fissures at the cardia, kinking of the esophagus at the hiatus esophagi, extrinsic pressure from the liver, and he recalls the hypothesis of foreign-protein sensitivity (Vinson, 1924). Observers generally admit that many and various local irritations or lesions of the gastro-intestinal tract such as appendicitis, cholecystitis, ulcers, varicosities and tumors may produce spasm reflexly, either in the esophagus or colon or elsewhere. These conditions, it would seem, are due to excessive afferent stimulation. Among other causes of intestinal spasm are mentioned arteriosclerosis, uremia, lead-poisoning and tabes. We know that alimentary spasm can readily be excited by various simple physiologic stimuli to the mucosa. Cardiospasm may be induced by stimulation of the ninth and tenth nerve afferents, as by cold water in the esophagus, or by carbonated water in the esophagus or stomach or by mechanical or chemical irritation of the mucosa in the cardiac region (Kronecker and Meltzer, 1883; von Mikulicz, 1903). We know also that once mucous colitis has been incited with an inflamed mucosa and irritated nerve-endings, foods and other stimuli normally harmless will tend to maintain spasm. But we do not know whether the mucosa changes are primary and the spasm secondary, or vice versa. Steindl (1924) points out that experimental pathologic and clinical observations make it seem probable that cardiospasm (excepting the reflex types) is due in mild cases to vagus neurosis, in severe cases to degenerative changes of the vagus. He follows Meltzer's theory of the origin of cardiospasm from disturbance of vagus fibers, while Rosenheim suggests that the inflamed mucosa causes the onset. He further reminds us of the work of Exner, Heyrovsky, Paltauf and Kraus as indicating that spasm of the esophagus and colon occur as the result of excess activity of the vagus due to perivascular and degenerative changes in the vegetative nervous centers. He cites two cases of spastic ileus in which necropsy revealed lesions in the centers of the vegetative neurons; that is, in the reticular substance of the medulla. His views, I dare say, may require some degree of revision when the distribution and effects of the vagus in man come to be definitely established. Steindl concludes that enterospasm may have a demonstrable pathology in the vegetative nervous system and that disturbances of balance in the entire nervous system, organic or functional, may be in close connection with the occurrence of spasm. That the extrinsic nerves may exert an important influence upon intestinal tonus is suggested by the recent observations of Thomas and Kuntz (1926) upon the action of nicotin.

These various findings and views of investigators, pointing to a variety of causes rather than to a single source of alimentary spasm, are alike in emphasizing the importance of the vegetative nervous system. The possibility is evident that various causes or tendencies, both "organic" and "functional," coexist in one individual, uniting to produce an additive effect. So even a slight or almost healed local irritation or lesion in some portion of the gastrointestinal tract may combine with the reflex effects of neuromuscular hypertension to produce spastic states. Suggestions favoring this hypothesis occur in the investigations to be here reported. It is commonly agreed that the nervous system has much to do

[359]

with the initiation of mucous colitis, spastic colon and spastic esophagus; and the following studies seem to throw light on the mechanisms involved.

THE SPASTIC ESOPHAGUS

Spastic esophagus, according to Brooks (1921), is the most common of gastrointestinal disorders. My discussion will be limited to moderate or incipient conditions, which predominate in number over the severe types. Most typically, according to my experience there occurs a "choking sensation" or feeling of "tightness" or pressure or distension near the lower end of the sternum or in the epigastrium; but distress sometimes seems to be beneath any portion of the sternum or above it in the throat, or in the left upper quadrant, perhaps radiating to the back. At times there may be a dull ache or even intense pangs of pain. The sensation is not generally relieved by taking food, except for a brief period, nor by alkalies, nor altogether by defecation. Belching of air may be frequent and usually brings a partial temporary relief. Distress and a sensation of fulness in the epigastrium may arise from the patient swallowing air, but this condition will not be relieved merely by directing him to discontinue, for the swallowing is apparently an involuntary reflex associated with the spacticity and disappears when the spasticity ceases to be marked. Occurrence of distress does not usually have any characteristic time-relation to meals; occasionally, as when emotional disturbance at mealtime is accompanied by painful acute spasm, the cardiac sphincter may fail to open and food or liquid may be regurgitated. In severe chronic cases atony of the musculature sets in with the well-known effects of dilatation and sacculation, but in mild types this end-result has not taken place.

The character of the distress and particularly its location not seldom suggests a false diagnosis of peptic ulcer or cardiac disease. Dysphagia, the ear-mark of severe esophagospasm, is often present in mild conditions, but is not necessarily a distinguishing symptom. After the history has been carefully recorded, the diagnosis of mild spasm can be made with the Roentgen ray. A single swallow is taken from a teaspoonful of barium paste of such thickness that it almost fails to drip off the down-turned spoon. Under the fluoroscope the lumen of the tense esophagus then appears as an opaque, thin streak which persists as a whole or in parts (particularly where the esophagus is narrowed at the bifurcation of the trachea, the aortic arch and above the cardia) for such prolonged periods as 5–25 minutes or more. At different levels the streak or its portions usually vary considerably in width, and movements can be observed. Normally the esophagus does not present such an appearance except for a brief time: it empties before $1-1\frac{1}{2}$ minutes. Care must be taken to distinguish a thin coating on the walls of the esophagus, which may persist for a long time even under normal conditions, from a complete opaque filling of the lumen at some point or points.

Careful differential diagnosis is required because spasm can result reflexly from various types of inflammation and organic lesions, including peptic ulcer, as well as from nervous hypertension, as will later be discussed. In my experience, emotional individuals display an exaggerated tendency to visceral spasm, so that organic disease, as above cited, in these individuals seems to produce an additive effect. Spastic esophagus or cardiospasm often or usually is accompanied by spasm of the pylorus and colon.

When visceral spasm has been discovered, careful search for underlying pathology that may require surgery is urgent. But the wide prevalence of spastic constipation resting on

[361]

no such foundation is a matter of common knowledge. I have here dwelt on symptoms in the belief that if the picture is differentiated and more commonly borne in mind, the diagnosis will be made more frequently. In peptic ulcer, which, as is known, is frequently accompanied by mild spasm of the esophagus or pylorus, analysis discloses that some of the so-called characteristic symptoms arise from the spasm rather than directly from the ulcerative lesion. However, where no such lesion exists or where healing has taken place, it seems probable that mild spasm of the esophagus often is not recognized and treatment is misdirected because of the tendency to confusion of the symptoms of spasm with those of ulcer. If spastic conditions cause symptoms without leading to correct diagnosis, there is always the danger of unnecessary surgery—an occurrence, I am inclined to believe, which is far too common.

Severe cardiospasm is perhaps most often treated by passing a tube or bag, sometimes with the aid of water pressure. But in the mild cases here under discussion, this is not commonly done; and little has been developed in the line of treatment. Not seldom, if examination for peptic ulcer has proved negative, the patient is informed that he has a "functional disturbance" of unknown character or that "there is nothing the matter with him; he should go home and forget it!"

The esophagus is an interesting organ from various points of view. Evidence will be given for believing that it can respond to our mental operations by contractions and relaxations. In persons of certain nervous types it sometimes goes into a condition of mild spasm that may last hours, perhaps weeks or months. Textbooks describe a condition called *globus hystericus* in which the individual feels something like a "ball" or "lump" or "clutching" in his throat. The explanation of this condition had been uncertain. On examination with proper instruments, no pathology is observed in the throat or speech apparatus. A sound passed down the esophagus may meet no resistance, and the esophagus may present

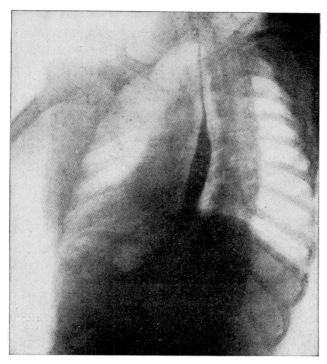

Fig. 74.—(S. L.) Spasm of esophagus in *globus hystericus*, confirmed under the fluoroscope. Later examination, when the distressing sensation was absent, disclosed no spasm.

little more than a little redness or hyperemia. There is now general agreement that *globus* occurs not only in hysterical patients but in many others as well, and the descriptions given of the sensations often vary considerably from the classical expression "ball in the throat." Current literature often refers to *globus* as spasm or as possible spasm, but generally without proof. In 1924 I showed in three cases with Roentgen-ray studies that *globus* is a motor condition of spasm (Figs. 74 and 75). Doubtless an element of hypersensitiveness of the mucosa is also present. It seems reason-

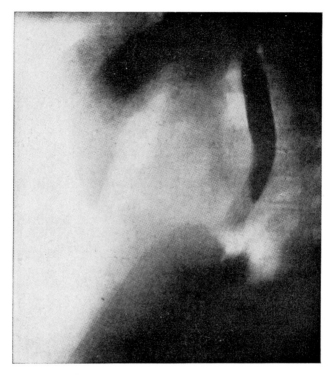

FIG. 75.—(B. K.) Spasm of esophagus in *globus hystericus*, confirmed under the fluoroscope.

able to expect a spastic organ to be hypersensitive if our conclusions (chap. vii and viii) about the nervous start and the flexion reflexes were correct; for there it was seen that the motor reaction which follows a sensory stimulus (and which is the objective criterion we commonly use and mean when we speak of "sensitiveness") depends upon the presence of a motor state, neuromuscular tonus.

[364]

Furthermore, in 1925 there remained to investigate whether the method of cultivated relaxation could affect internal organs like the esophagus. Textbooks generally teach that the esophagus and other portions of the gastro-intestinal tract are removed from voluntary influence. No instance, so far as I know, had ever been reported of voluntary influence on the esophagus, which receives its motor supply in man from the vagus. Here was another opportunity to test the present method, and perhaps to extend its field. Accordingly, physiologic tests were made, as will be indicated in the following abridged case-reports along with such illustrative details of the clinical method as limited space permits:

Case 1 (H.A.)

A university student of nineteen, first seen in January, 1923, complained of severe cramplike or burning pain in the epigastrium which had been continuous for hours each day during the previous three years. There frequently was a feeling of fright which he could not distinguish from the pain. This appeared when under nervous strain, such as during recitation in school, when present at large gatherings, or when in the company of the other sex. He mentioned also frequent feelings of irritability and difficulty in concentration.

According to his recollection of his earliest symptoms, cramplike pain first set in one day in 1920 about $1\frac{1}{2}$ hours after a customary breakfast. Although not severe after lunch, it again became marked after dinner for about 1 hour. On the following morning after breakfast it reappeared for about 1 hour, and remained moderate for the rest of the day. During the next 5 days he remained in bed, awaking each morning in pain. Relief often followed meals for 1–2 hours, after which pain would return more severely than before; but there was no characteristic relationship. However, during the succeeding 2 weeks, taking of food often was followed by pain so severe that he had to cease to eat. No vomiting and no jaundice appeared. After about 3 weeks he had lost 10 pounds and had become somewhat weak.

Three months after the onset he was examined by an internist, whose findings I am permitted to quote. Upon physical examination nothing of note was revealed except a little epigastric tenderness. The Roentgen-ray report was negative except a slight irregularity of the bulbus, along with a little duodenal stasis and a somewhat persistent fleck. Motor meals and fractional examinations showed hyperacidity, but no blood. The feces, blood, urine and Wassermann examinations proved negative. A milk-and-cream diet with periodic alkalies was tried for about 1 month, without adequate improvement. Accordingly he was sent to a ranch in May, 1921, for 1 month and again in September, 1921; but he returned home soon thereafter because of diarrhea and cramplike pain. His work at school was good, in spite of certain worries and experiences which cannot here be discussed. Accordingly, in 1922 a neurologist applied suggestive and persuasive treatment, which cured him of his fears directed toward women but left him still with other fears and with the pain of which he now complained.

Upon general physical and proctoscopic examination in 1923, I found nothing new. He then appeared somewhat undernourished, with remnants of acne on the head and shoulders. An Ewald meal yielded 370 cc., total acidity 60, free hydrochloric acid 52, combined hydrochloric acid 4, combined acids and salts 56, organic acids and salts 4, with a slightly positive Weber and benzidine test. Duodenal aspiration and stools were negative. Detailed study with the fluoroscope as described below plainly revealed esophageal spasm as the principal source of pain. Spasm was also revealed in Roentgen-ray study of the pylorus and colon, and it is therefore possible that some of the pain proceeded from this source. We again found the duodenal fleck. The history of onset along with the finding of hyperacidity and the duodenal fleck with local tenderness suggest that duodenal ulcer may have been present, but this remains in doubt. However, it seemed safe to conclude in 1923 that no active ulceration was now present.

Progressive relaxation was begun January 8, 1923, with tri-weekly treatments and daily practice by the patient. The following excerpts from daily notes indicate that his fear and pain were correlated with increase of neuromuscular tension and that decrease of these symptoms proceeded parallel with relaxation:

1–8–1923. Period 1. Initial instructions given. The patient states that he has often tried to relax in the epigastric region in order to get relief, but has failed. His general appearance in the physician's presence is embarrassed or fearful, with wide-open eyes, occasional wrinkling of forehead, and frequent shifts of position. His head is held stiffly, partly from old habit, partly from a broken jaw healed a month ago.

[366]

1-18-1923. Period 5. Instructions to try to become a little more relaxed each day. Fails repeatedly to recognize tension or contraction of left hamstrings (−2S−3I). His reports indicate that he has engaged in thought, thus hindering observation. This leads him to inquire whether he is to stop thinking when he tries to relax. Reply is made that no instruction is given on this point: he is simply to relax muscles. In practice at home, he admits, he has been suggesting to himself, as for example, "The left leg is relaxed!" It is made clear that this is contrary to instructions: he is not to give himself suggestions but to learn to relax here as he would in dancing.

1-20-1923. Period 6. To date arms and left leg alone have received practice. He reports slight improvement in symptoms. Trunk seems stiffly held.

1-23-1923. Period 7. Physician gives an illustration of general relaxation for imitation. Patient's epigastric region rigid upon palpation.

1-25-1923. Period 8. Reports distinct improvement. A little gain in weight. Special practice with abdominal muscles by the method of diminishing tensions. Various bending motions included. Comments that in the past he has characteristically held the abdomen too tensely when standing erect, and now is learning to act with more relaxation. Musculature of trunk now appears less tense than heretofore. Sleeps today during period.

2-6-1923. Period 11. Neck +1S. Neck receives practice today and following three periods, before the extraordinary stiffness diminishes. He reports that he can now relax without telling himself to do so.

2-8-1923. Period 12. He states that he has been trying too hard to relax, impairing his success.

2-15-1923. Period 15. Pain has diminished, he asserts, during the last 10 days. During the relaxing period his breathing fails to become regular and he sighs. He is reminded to relax his chest further: a sigh indicates that he has been too tense.

2-17-1923. Period 16. After recitation in physics yesterday, he was troubled with "fear-pain" for 2 hours. In addition, a burning epigastric pain began 10 minutes after dinner and lasted 1½ hours. Such pain has been unusual of late.

3-3-1923. Period 21. Appears pale today, with roving eyes and frequent slight frowning. Physician states to him that during the last 10 days he has apparently not improved. He replies that for 5 months previously he has avoided going out alone with the other sex, but for the last 10 days has resumed this practice and has been accordingly disturbed.

3-6-1923. Period 22. As judged by the usual clinical signs, he is observably less tense. Reports that he is more relaxed and that the pain is much diminished.

3-13-1923. Period 25. Detailed practice has been given on thought-processes. He reports that when his eyes are perfectly relaxed, he has (1) no fear and (2) no pain and (3) thought-processes are diminished. (This observation is spontaneous.)

3-17-1923. Period 27. In contrast with his original pallor, shifting eyes and anxious expression, he now appears well, with good color, sure of himself—a "determined look about his eye."

4-17-1923. Period 40. For the first time, the head can be pushed about readily in all directions, showing limpness of neck. He independently reports the best relaxation up to date, and that there were moments of freedom from thought-processes.

4-21-1923. Period 42. Differential relaxation begun.

5-1-1923. Period 46. Arms have been covered. Reports that he is acquiring the habit of walking differentially relaxed.

At this stage I desired to test under the fluoroscope whether the patient was really succeeding in relaxing the esophagus as he seemed to be doing, according to his subjective reports. Since spells of marked pain had become relatively infrequent, he aided the experiment on one occasion (5-22-1923) by exercising vigorously and remaining active almost the entire night, for such excess had generally in his experience been followed by severe distress. On the next day when he swallowed the barium at intervals (we had not yet adopted the single-swallow test), spasm was observed under the fluoroscope for an hour. Then after 20 minutes of voluntary relaxation, barium was again given. This time no spasm was revealed under the fluoroscope or on the film. The esophagus appeared broad and open along its entire course. Subsequent tests in August, 1923, during a spontaneous relapse, gave the same results: persistent spasm was found

(Fig. 76), then the subject relaxed for about 20 minutes, whereupon the barium paste was again given, but this time was seen to pass normally through the esophagus, which evidently was no longer spastic.

At this point I am indebted to Professor Carlson for suggesting that the Roentgen-ray observations be confirmed with kymograph tracings. A rubber bag about 3.8 centimeters long was swallowed, attached to a fine rubber tube connected with a manometer. This device, well known to physiologists, was a U-shaped glass tube placed vertically and partly filled with water. One end contained a light glass float, the bottom of which rested in the water, while the top carried a writing-point which made a curve on the smoked paper as the float rose and fell with the water. To the other end of the U-tube was at-

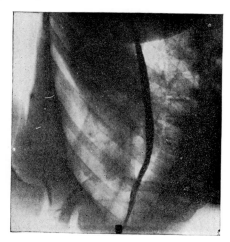

Fig. 76.—(H. A.) Spasm of the esophagus, confirmed under the fluoroscope. A film alone may not establish the presence of spasm, since it is necessary to show that the contraction is persistent. Later, when another portion of barium was given after about 20 minutes of general relaxation, the esophagus appeared wide and relaxed under the fluoroscope, emptying in normal manner practically at once.

tached the tube from the bag in the esophagus. A thread 22.5 centimeters long was attached to the proximal end of the bag and to a front tooth, to prevent the bag slipping too far down. In this way the bag lay in about the middle of the esophagus. When the bag was moderately inflated, a record of contraction and relaxation of the esophagus

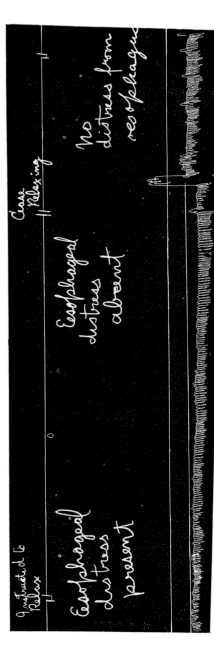

Fig. 77.—(H. A.) Progressive relaxation of the esophagus is shown by the general fall of the curve measured from the base-line. The swallowing reflex is absent here. Up-and-down strokes represent respiration. The total time during record shown is about 25 minutes.

could be obtained, for contraction of the esophagus drove a little air out of the bag via the rubber tube into the U-tube, raising the column of water and moving the writing-point on the drum. During the period of observation the subject lay upon the couch, making no bodily movement, for this would confuse the tracing. I found that leaks of the air system occurred in the usual laboratory apparatus even if only new rubber was used and the connections were sealed with wax; but these were finally completely avoided by ligating the rubber at each connection with surgical knots and covering each junction completely with collodion. Such precautions are required if results are to be trustworthy, for a leak, like relaxation, produces a fall of the writing-point. Tests of the sealed system over a 24-hour period finally disclosed practically no leak. Many periods of practice proved necessary before the subject learned to relax under these conditions, for the inflated balloon in the esophagus added to his distress. In each instance control records were taken in order to establish that the contraction or spasm persisted before the signal to relax was given. A sample of the latter records is shown in Figure 77. They confirm the Roentgen-ray observations that the esophagus in some measure is subject to voluntary relaxation.

Fluoroscopic studies had revealed that the lowermost portion of the esophagus acted like the upper portions during relaxation. To test this, the balloon was introduced with a thread about 36 centimeters long. A greater distance would have placed the balloon in the cardia, as shown by the change in the tracing (Carlson and Luckhardt, 1914). Tracings (Figs. 78 and 79) disclose that the smooth muscle of the esophagus responded to voluntary relaxation.

Study of the records reveals various items of interest. The esophagus responds in its contractions to divers stimuli

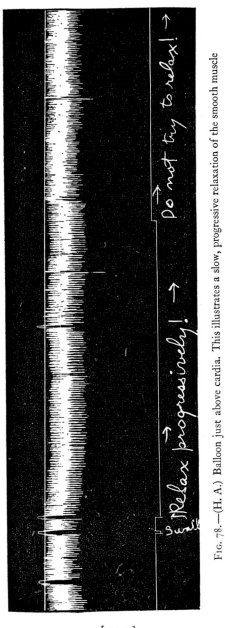

Fig. 78.—(H. A.) Balloon just above cardia. This illustrates a slow, progressive relaxation of the smooth muscle

or disturbances, such as noises, the entrance of a visitor, flies alighting on the skin or even thought-processes. When the disturbance awoke fear, the contraction was increased. Marked spasm seemed identical with fear to this subject. As the esophagus relaxed, the distress vanished correspondingly. He readily gained the ability to judge when the esophagus was tense. Comparison of the subjective reports with the tracings showed noteworthy correspondence. (This har-

Fig. 79.—(H. A.) Balloon just above cardia. The record illustrates almost immediate relaxation of the smooth muscle following the instruction to relax.

monizes with Carlson's observation [1913] that his subjects were able from their sensations to judge the presence of hunger contractions.) Sometimes the attempt to relax was fairly successful practically at once, as shown by an abrupt fall of the curve; but more often the fall was slow and progressive. Not seldom the instruction to relax is followed by an unsuccessful attempt: the subject may fail to relax his skeletal muscles to a marked degree, as seen by gross observation, and may report his failure later. It may be necessary repeatedly to give instructions to relax and even to point out to the subject what muscles appear tense. When the esoph-

[373]

agus once became well relaxed, the subject failed to be able by an effort to make it tense again. We did not try to find out whether such ability could be cultivated with practice. After practice at relaxation has made the matter habitual, a certain degree of relaxation may take place *nolens volens* when the individual takes his accustomed position of rest, even if the instruction not to relax be given. In subsequent months, as relaxation became habitual and distress less fre-- quent, we found it increasingly difficult to excite spasm for experimental purposes by inflation of the balloon.

Owing to accumulation of saliva due to the tube in the mouth, this subject generally swallowed preliminary to carrying out the instruction to relax; therefore in later studies he was directed to swallow before the instruction to relax was given. When he seemed most relaxed, no swallow- ing took place even for periods of half an hour. This con- trasts with the unrelaxed normal esophagus, where re- peated swallowing is characteristic (Daniélopolu, Simici and Dimitriu, 1924).

We return to the clinical notes:

6–12–1923. Period 61. Has neglected to practice for a week. Reports an apparent rise in general nervous tension and a corresponding increase in abdominal pain. (This relapse was brief.) It is his experience that he can relax pain away most successfully when sitting with eyes closed.

7–6–1923. Period 70. States that pain has become much diminished and possibly of about half the daily duration of the original. Instructed today to keep eyes open, permitting them to move idly, but to relax the abdominal region. He fails to do this successfully.

7–10–1923. Period 71. Succeeds in differential relaxation with eyes open. Has practiced diligently at home. Pain diminishes accordingly dur- ing the period of relaxation. Remarks that an extra bowel movement char- acteristically follows a spell of "fear-pain" and brings relief.

7–24–1923. Period 75. Affirms that he is now able, with eyes open, to relax to the point of disappearance of "fear-pain." Objective signs noticed of this increased ability to relax.

[374]

8–2–1923. Period 77. Finds that the greatest factor in causing "fear-pain" is association with other sex. Often now is so engaged. Instructed to read and observe what takes place. Fails spontaneously to report tensions during visual imagery and also in the tongue and lips during inner speech. He believes that the latter are relaxed. He is instructed to read but to keep the tongue and lips perfectly relaxed. Discovering that this is impossible, he corrects his former report and states that slight tensions occurred there during reading.

9–25–1923. Enters pale, with troubled eyes, forehead a little drawn, limbs apparently held somewhat rigidly. Has been so of late, probably due to plans for gastric analysis. Instructed to keep himself more relaxed. He states that this requires constant attention, not yet having become automatic. He relaxes speedily, whereupon the eyes and forehead lose their troubled expression. He adds that the pain now disappears at the moment of general relaxation.

10–9–1923. Graphic records have revealed that each time he was instructed to relax he began with a slight movement. He ascribes this to an uncomfortable position. Instructed to relax without such initial tension, to let go directly, regardless of discomfort or other good reasons for not relaxing.

10–16–1923. Reports that he is making progress: His feet do not fidget as previously. Sleep seems to be more relaxed in character, since now he awakes without pain in contrast to 2 months ago.

10–23–1923. Characteristic appearance now different from months previous: The eyelids droop somewhat, in place of being wide open. Bodily movements deliberate, and the lines of mouth are at ease, with a quiet rather than the previous anxious expression. Reports that he no longer requires "constant attention" to keep relaxed during the day.

11–8–1923. Appears with complaint of distress following a disagreeable argument with his instructor. As he relaxes during the course of the period, reports that distress subsides.

11–15–1923. Observations of primary and secondary tensions during reading.

2–7–1924. Periods now bi-weekly. Has had a recurrence of epigastric distress for 4 days, with constipation. For 3 months has taken mineral oil. Soon later discontinued.

3–13–1924. Working hard as examination time approaches. Gets tense studying calculus. Pain is then increased.

6–24–1924. Reports that he has not felt so well in years. Practice at mathematics: Reports tensions of the eyes, forehead, arms, etc. Instructed to relax, but not to the point where thinking ceases.

7–22–1924. Having neglected to practice, he has had a relapse of pain, present almost all day for a few days.

8–14–1924. Has been tense from entertaining out-of-town guests, with some return of pain. Instructed to eat slowly and to masticate thoroughly.

9–8–1924. States that he has been generally active but with less pain.

11–1–1924. Explains that he fails to relax at times, especially the eyes, because he is not always willing to sacrifice the pleasure of being active.

1–24–1925. Periods now one a week. Has engaged in chemical work, standing many hours each day and has again become overwrought. Has again eaten quickly at meals. Some distress has returned.

1–31–1925. Little distress this week. Tests shows absent knee-jerk during relaxation.

4–8–1925. Little distress present except at time of nervousness.

8–12–1925. Has been almost free from pain as a rule. No longer necessary to come regularly for treatments.

Since the last date, fluoroscopic tests as a rule no longer reveal spasm. From 1925 to 1927 there was occasional return of pain, although slight in comparison with the original. The patient explains this as due to his own neglect at times to keep himself well relaxed. In 1927 he had a relapse, with marked pain for about 2 weeks attended with some nervous hypertensive symptoms. Upon resuming treatment for purposes of review, it was found that his eyes relaxed well but inner speech apparently continued with slight tongue and lip movements. Treatment was directed toward correcting this and for a time he was directed to relax at home each evening, curtailing his activities to this extent. (Occasional relatively mild relapses may have to be met in cases of this type.) From 1929 to 1931, when last seen, he has been practically free from symptoms. On the whole, his general nervous and mental development has been away from his earlier "neurasthenic habitus" and he has become well balanced emotionally. As a rule his general appearance is evidently relaxed (1929). He has become an active business-man (1937). Recurrence of slight but persistent pain at times has been accompanied by characteristic failure to relax residual tension completely, particularly in the abdominal muscles, according to electrical measurements. A present review of treatment appears to be removing this residue of pain.

[376]

Case 2 (I. Q.)

Case 2 was a woman of forty-nine, examined 12–15–1923. Her complaint was of "choking sensations" all day long for the previous 6 weeks, most marked after meals. Since the age of thirty-one, she had been subject to epileptic spells. The Wassermann test was negative; other findings are omitted for brevity. Under the fluoroscope, spasm was severe. A spoonful of paste failed to leave the esophagus completely in 25 minutes (see Fig. 80). Then she was requested to relax so far as possible while continuing to stand. She had no previous training, but her head and limbs drooped unusually well for an unpracticed person. After 20 minutes more, the esophagus was empty. At the same time the

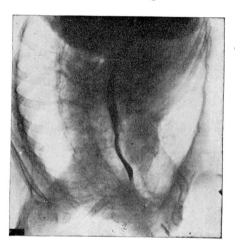

Fig. 80.—(I. Q.) Spasm of the esophagus. Barium persistent for 25 minutes. After a period of general relaxation, the esophagus appeared relaxed under the fluoroscope and emptied in normal time when barium was again given.

"choking sensations" diminished. No advance hint of this result had been given to her.

Tracings were taken during the period of severe spasm. First a control test was made for an hour to see whether the spasm disappeared spontaneously because of the supine position alone. After it had persisted, the instruction to relax was given. As shown in Figure 81, there was a remarkable change in the contractions when the subject relaxed. She reported that during general relaxation, the "choking sensa-

[377]

FIG. 81. —(I. Q.) Relaxation of the esophagus is shown by the fall of the curve corresponding with instructions

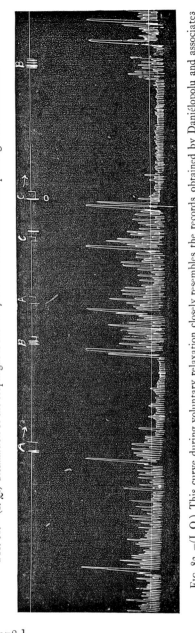

FIG. 82. —(I. Q.) This curve during voluntary relaxation closely resembles the records obtained by Daniélopolu and associates for the effects of calcium chloride and other depressants. Breaks in upper line indicate: *A*, signal to relax; *B*, to cease relaxing; *C*, prolonged instructions to relax.

[378]

tions" disappeared. No previous intimation to this effect had been made to her. As she practiced relaxation from day to day during the following month, marked spasm seemed to become more difficult to evoke upon inflation of the balloon, and it was also absent upon the fluoroscopic examination. Records (Fig. 82) made after the spasm was no longer in evidence suggest that a progressive decline in contraction may follow the instruction to relax, which quite resembles the action of calcium chloride, atropine and other depressant chemicals on the esophagus, as shown in the tracings of Daniélopolu and associates. These writers used small doses, and the effects apparently were shorter in duration than ours. There is similar diminution of swallowing and of height of respiratory contraction, as well as a general decline of the curve as compared with the base-line.

A month later, after she had learned something of habitual relaxation, there was no evidence of spasm under the Roentgen ray, and when last seen, about 2 years later, there had been no further subjective complaints. This result, which followed after only four periods of instruction, contrasts with the many periods required by the preceding case. Doubtless the chronicity of the latter accounts in part at least for the greatly longer period of treatment required. To be sure the technic gained during this longer period of training greatly exceeded the other in many particulars. But the contrast stands, and illustrates the marked range of variability that may be required for the treatment of a similar set of symptoms in different patients.

Case 3 (S. L.)

Case 3 was a married woman of thirty-nine who had been under treatment for nervous hypertension with moderate phobias. For 2 days before the tests, she had complained of almost continual distress as from a "lump" or "ball in the throat." At 1:40 P.M. she took two teaspoonsful of the barium mixture. Persistent spasm apparently involved the entire esophagus. At 1:53 P.M. more barium was given, and part of this remained for 25 minutes. The pylorus opened after 15 minutes, while the cardiac portion of the stomach also appeared spastic. At 2:35 she began to relax and at 3:15 reported that she was free from the feeling "as from a stiff tube." At this time spasm no longer appeared under the fluoroscope, excepting a little, above the cardia. The organ now emptied in 3–4 minutes with the same paste as previously used. Cultivation of an increased measure of habitual relaxation later resulted in practical disappearance of symptoms of spasm.

In summary, these studies indicate that the esophagus responds under emotion by contraction. It is plausible to consider this contraction not merely as a result but as part of the physical occurrence of the emotion. After training in relaxing skeletal muscles, three individuals apparently succeeded at voluntary relaxation of the esophagus. Of course their attempt to relax met with failure on some occasions; but where positive results were attained, the subject reported absence of emotion or distress during the moments of relaxation which corresponded with the tracings (hidden from his view). A close relationship between spasm or relaxation and habit is suggested, since after habitual relaxation had been cultivated, the spasm generally failed to appear. When the subject ceased to relax, while the balloon remained *in situ*, the esophagus sometimes became tense again, but frequently it remained relaxed. The supine position seemed to favor relaxation, but usually did not of itself effect this result. Evidently the lower smooth-muscle portion of the esophagus behaved like the middle portion during relaxation.

MUCOUS COLITIS

If the tonus of the esophageal musculature may vary with that of the skeletal system, it would seem simplest to assume that esophageal relaxation proceeds because of the diminution of spinal impulses from skeletal muscles as these relax.

THE SPASTIC COLON

Closely related to the spastic esophagus both in occurrence and probably also in etiology, is the spastic colon. Mucous colitis is not always sharply demarcated from spastic colon, since the latter condition, if carefully watched, in many instances also reveals occasional passage of mucus. Furthermore, although in chronic mucous colitis the colon may be in places toneless and dilated, yet this is found to be most characteristic of long-standing pathology, and the impression is gained that the atony has resulted from a chronically spastic condition. A tendency toward spasm can be detected at all stages of mucous colitis.

The diagnosis rests as a rule upon the history, perhaps the palpation of a tender, firm colon, the finding of considerable mucus, sometimes with ribbon-like or small ball-formed stools together with the evidence from proctoscopic and Roentgen-ray examination. Here the barium enema may suggest the presence of spasticity, but I have preferred to use the picture 24 hours after the usual barium meal as a more normal test. No cathartic is permitted for at least 24 hours previous to the meal. Under the fluoroscope and before taking a film, palpation of the colon as well as testing for adhesions should be omitted, since this tends to alter the natural picture. I have taken as Roentgen-ray evidence of colonic spasticity the presence of an irregularly narrowed colon and the marked deepening of haustra (cf. Carman, 1921; Heagey, 1924; Case, 1921; Mills, 1922).

It is of interest, when the barium is first given by mouth,

to time with a stop-watch the spontaneous opening of the pylorus, while the radiologist should not manually express the stomach content.

Correct diagnosis is particularly important because of the danger of mistaking these conditions for chronic cholecystitis, chronic appendicitis and other diseases that require surgery. Woodyatt (1927) cites such a case in which operation was almost performed after a severe colonic spasm was revealed by the Roentgen ray. Fortunately, later films proved negative and the operation was averted. But when tenderness is located near the gall-bladder or the appendix or both, as often occurs in this common malady, who can say how many futile operations are performed for this condition each year? My own experience, as previously said, suggests that there are many; and Graham in his recent surgical address at the opening of the Medical School of the University of Chicago issued the same warning.

Seven cases of mucous colitis or spastic colon will be considered. Two of these, living in another city, departed after 2 months of treatment contrary to medical orders, considering themselves recovered. At the present time, 16 months later, they send a favorable report. It is doubtful that a permanent result in long-standing colonic affection can be so soon secured: the patient who so early leaves the guidance of the physician is prone, when alone, to break the strict rules necessary to sustain the improvement that has been gained. Treatment of these colonic conditions generally has required more than a year. The patient continues to practice daily after being discharged by the physician. One patient reported improvement after 6 months; but personal difficulties prevented regular continuation of treatment and no later roentgenograms could be taken; therefore this case will not be considered further. In judging results, reliance has been placed not on the subjective reports of patients but on the objective findings. Mucus has virtually disappeared or its occurrence has become infrequent and in smaller amounts; the stool has lost its ribbon-like character and assumes a normal contour; the colon no longer appears firm and tender upon palpation; bowel action becomes regular; the Roentgen ray finally reveals a colon of normal or nearly normal markings. The clinical observations on the colon will need to be followed by corresponding laboratory investigations, and we should not consider the subject closed until then.

Case 1 (E. T.)

E. T., a woman of fifty-eight, of a notable Irish family, complained in January, 1922, of symptoms of mucous colitis for 30 years. Her attacks formerly had come on about six times a year, but during recent years were on the increase until of late they appeared two or three times a month. They were increasing in severity, for pain had been marked during the last 4 years all over the abdomen, sometimes shooting down the thighs, sometimes passing with a burning sensation up under the sternum. These attacks generally lasted about 2 days continuously, including the nights, relieved only by the electric pad and slightly by baking soda. Cramping abdominal pain occurred with bowel movement, and epigastric distress appeared usually within 5–10 minutes after eating or drinking. This pain generally disappeared spontaneously an hour or two after eating. In addition to mucus she asserted that her stools had seemed to contain blood and pus. She emphasized her general weakness and inability to engage in normal activities. Other complaints were constipation, frequent urination, and pain in the chest, sometimes with shortness of breath. Her husband was living and well. One son, aged thirty-one, was mentally subnormal. She had had one stillbirth at 7 months and another premature delivery at 7 months, with death soon after birth. A boy had died at 5 months, apparently from stomach trouble. Her past history was negative except hemorrhoidectomy 25 years ago and two operations on the uterus.

Physical examination revealed a short, stocky woman with good color, normal pulse and temperature, but with marked signs of suffering when seen during a spell. The heart and lungs were normal, as well as the other

regions not here mentioned. Blood-pressure was 188 systolic and 100 diastolic. The entire colon was palpably firm, and there was slight abdominal
resistance. The Roentgen ray revealed a spastic colon (Fig. 83) with an

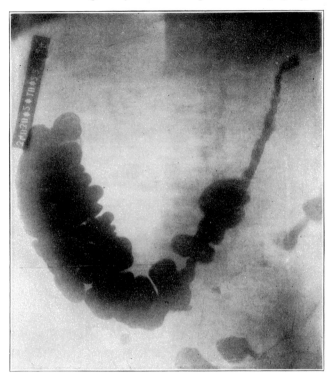

FIG. 83.—(E. T.) Original examination, February 9, 1922, showed a highly
spastic colon, with almost stricture-like appearance near splenic flexure; this filled
out with a barium enema. The diagnosis was "mucous colitis."

area of very marked constriction near the splenic flexure. Tenderness was
severe at the gall-bladder area and near McBurney's point, but marked
also at the hepatic flexure and duodenum and generally throughout the
colon. Rather extreme hypermotility was shown by the barium leaving
the descending colon in 5 hours. The duodenum showed stasis and limited
motility, and the cecum also was not perfectly free. There were several
fairly large hemorrhoids, and the proctoscope revealed some redness of the
mucosa on the right side beginning about 4 cm. from the anal orifice. The

vagina was narrowed, following perinneorrhaphy; and the cervix had been partially amputated. Considerable mucus was found in the stools, which were yellow or greenish. Hemoglobin was 80. Blood otherwise negative, including Wassermann. An Ewald meal gave no evidence of noteworthy findings, and likewise stool culture. Aspiration of the duodenum showed the presence of strings of mucus, pus and squamous epithelial cells in the fraction presumed to be from the gall-bladder. Ferments were normally present. A cystoscopic examination made by a consultant showed a chronic trigonal hyperemia, which was treated with applications of silver nitrate.

When treatment was begun with this patient, it did not occur to me to try progressive relaxation, and for about 3 months the more or less usual therapeutic measures were applied: Diet was carefully limited; liquid petrolatum was given by mouth and olive oil by retention enemas. Much daily rest in bed was prescribed in the ordinary way. But these and other accessory measures did not prove successful. Tincture of belladona, benzyl benzoate, even diathermy for pain were tried and discarded. The failure of these measures, from which much had been hoped, incidentally gave evidence that the patient's condition did not respond to "suggestion." I considered the removal of the appendix and gall-bladder, but a surgical consultant, Dr. E. Willys Andrews, seemed doubtful, and Dr. Edmund Andrews suggested that progressive relaxation be tried.

Treatment with progressive relaxation was accordingly begun in May, 1922. The study was of added interest because other measures, including diet and rest, had been carefully employed without satisfactory effect, thus furnishing a control set of conditions. This patient was not very clever at learning to recognize muscular contractions, which is part of the present method. Among the striking features of her postural tonus were a persistent severe frown and a frequently wrinkled forehead: much practice was devoted to this region (see Fig. 84). As she gradually became relaxed in these and other parts, the hospital record of stool examinations showed a change: May 1–6, there were shreds of mucus; May 11–12 small shreds; May 10–23, no mucus; then a reappearance, May 25. Thereafter mucus was largely absent. The general symptoms and signs gradually abated, excepting a relapse during the last 2 weeks of July. August 15, 1922, while still in the hospital, the gall-bladder region had become entirely free from tenderness, the ileocecal region showed but slight tenderness, and the transverse colon but little. Pain on defecation was much diminished, and often absent. Attacks of pain were now infrequent and no longer of former severity. The stools had become normal as a rule. She was discharged from the hospital, but continued daily practice at home, with practically no further direction from the physician.

At home she continued without further medical aid to improve at relaxing her muscles, just as an individual will improve with practice after instruction at the piano or dancing or other physical feat. After about 6 months more, tenderness had disappeared from the abdomen, and the colon no longer was firm upon palpation. She went to California and nursed her husband who had become ill, drove a motor car and performed other functions with impunity for the first time in many years. A Roent-

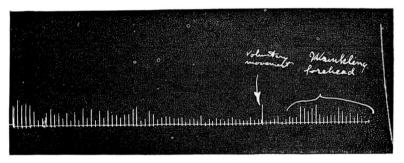

Fig. 84.—(M.S.) Knee-jerk tracing of a normal subject, indicating that wrinkling can have an augmenting (reinforcing) influence on the knee-jerk.

gen-ray film (Fig. 85), very kindly made for me in 1925 by Dr. A. B. Smith at La Jolla, reveals a colon no longer extremely spastic. According to her report received in 1927, her general condition has been highly satisfactory for 5 years, and any relapse has been occasional, mild and brief. In 1930 frowning seems very greatly diminished. She has had no recurrences, is on full diet and appears very well.

Case 2 (E. R.)

E. R., a married woman of thirty-four, complained in June, 1923, of spells of diarrhea, often with mucus, for the preceding 14 years. Occasionally there was tenderness in the right side or umbilical pain. These spells were weakening, interfering with her activities and her general efficiency often for periods of weeks. In childhood, her tonsils had been partially removed, and she had passed through pneumonia and the usual other disorders. At twelve she had an enlarged thyroid, which was treated with

[386]

Roentgen rays until it subsided. Her mother also had colitis for several
years.

The present illness began at the age of twenty, with much distress and
diarrhea after meals. This seemed especially marked after nervous diffi-

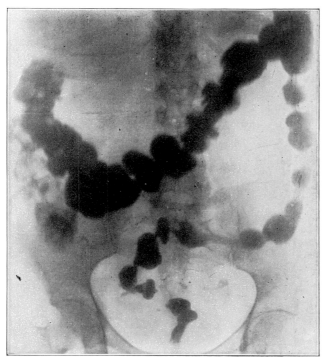

FIG. 85. —(E. T.) May 12, 1925, the colon much improved but still somewhat
spastic on 24-hour observation. The constricted band near the splenic flexure was
now filled out. Later observation (48 hours) showed stasis.

culties over love affairs. About a year later a badly infected appendix was
removed. Distress and diarrhea continued severely during the following
few months, with moderate improvement thereafter. In 1914, there was
occasional vomiting after or before meals. She was married in 1917. Spells
of diarrhea seemed to come on after getting chilled, after eating an ordinary meal, after anxiety about her baby or after worry about some visit
which she was about to pay. In 1919, following the delivery of her first

child, her hemoglobin "sank to a very low figure." In 1921, during the birth of her second child, her perineum was lacerated, impairing control of the anal sphincter.

Examination disclosed a fairly well-nourished woman. Blood-pressure, 116 systolic; 80 diastolic. The eyes, nasal passages and tongue were nor-

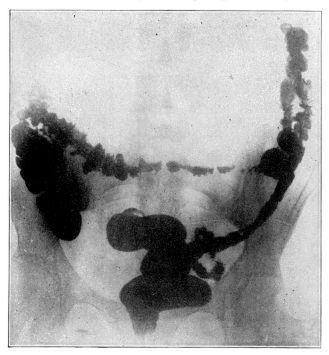

FIG. 86.—(E. R.) July 18, 1923, original examination of spastic colon in mucous colitis.

mal, but the tonsil remnants occasionally showed a little dried streak of pus on the left. Her thyroid was fairly firm but not enlarged. The heart was normal, and the lungs showed no signs of active process. At the time of first examination the colon was not tender or palpable. A midline scar from the pubis to about 2 cm. above the umbilicus dated from a Caesarean section. The deep reflexes were of increased liveliness. Proctoscopic examination revealed a little generalized redness of the mucosa, beginning about 8 cm. from the anus, with a little mucus but no inflammation of the cryptic region. Under the Roentgen ray, the colon appeared highly spastic (Fig. 86). The urine was normal. Culture of the feces disclosed *Bacillus coli* and *Staphylococcus albus*. No tubercle bacilli and no ova or parasites were found. Her basal metabolic test and roentgenograms of teeth and Wassermann test of later date proved negative.

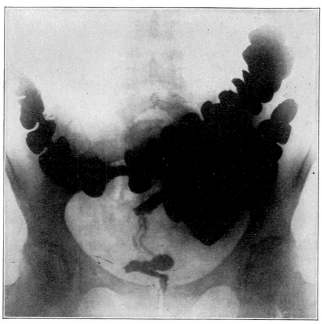

Fig. 87.—(E. R.) August 22, 1924, diminished spasticity of colon after about 1 year of treatment.

Treatment was begun with relaxation and restriction of diet to milk, eggs, stewed fruits, cereals, custards and jelly. Observations disclosed that spells of diarrhea occurred after what for her appeared to be dietary indiscretions, such as eating onions, after coryza, but most often after general overstrain, such as occurred when her husband or children fell ill. It took about 45 days before she began to make apparent progress in relaxation. Diarrhea became markedly diminished at about this time. A recurrence followed illness of her husband. Previous to treatment such spells had lasted 1 week to a month, with a minimum of 3–4 days. During the first year of treatment the spells not alone became less frequent but

[389]

were generally stopped after intensive relaxation during many hours for 1–3 days. Roentgenograms taken in August, 1924 showed a colon not far from normal in appearance (Fig. 87). The pylorus did not open until after 10 minutes. At this time she reported very marked improvement, with spells about one-fifth as often as formerly and lasting about one-fifth as long. The great prolongation of treatment seemed to be due to failure

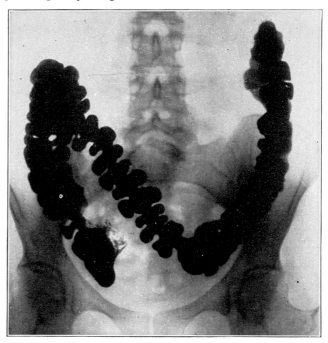

Fig. 88.—(E. R.) February 16, 1926, improved condition of colon 2 years later

of the patient to follow directions about diminishing her domestic and social activities. Instructions had to be repeated many times before they were followed. She continued to practice at home, occasionally calling for treatment. About January, 1925, spells had become infrequent.

Roentgenograms on February 16, 1926, revealed a practically normal colon (Fig. 88), but a little pylorospasm has persisted.

Since that time, there have been occasional spells of colitis, but of markedly less severity and duration than her former experiences. Chronic infection was discovered in the tonsils and removal was performed, with no apparent effect on the colonic condition. There was a lack of co-operation owing to "artistic temperament," and a consequent lack of skill at relaxation, which seems to account for the variations in her case during 1927. But gradually her co-operation increased, so that during 1928 she became fairly relaxed habitually and secured a greater stability of colonic improvement. In 1930 she reports that she has been well, which persisted until death in 1934 from breast cancer.

It is well known that the effect of general "nervous upset" may be to cause a spell of diarrhea. A familiar example is the frequent defecation of soldiers before a battle. In harmony with this is my frequent clinical experience with patients who are nervously hypertense: spells of diarrhea or constipation with ribbon-like stools tend to be present with or to follow periods when they are excited, and tend to disappear when they become relaxed. The Roentgen ray upon original examination generally reveals a spastic colon as the apparent cause of these tendencies to intestinal disorder; but care must be taken to fluoroscope the patient at a time when symptoms are present rather than during an interval of improvement, else the signs of spasticity and hypermotility may be missed. It would appear from clinical experience that intestinal upset with diarrhea or constipation due to nervous hypertension is extremely frequent and widespread. Our view is that nervous hypertension is frequently responsible for symptoms of intestinal spasticity or hypermotility.

Any present-day treatment of alimentary spasm will of course begin with a thorough search and removal, if possible, of focal infections. There is a tendency to relapse for a brief period if the patient is put under a severe nervous strain or at the onset of marked infection, such as severe coryza. In the present studies of chronic mucous colitis, other measures besides progressive relaxation were so far as possible excluded in order to isolate the effects of this method alone. However, this could not always be perfectly accomplished.

PROGRESSIVE RELAXATION

It seemed best, in order to favor the results, to advise the patient to masticate food thoroughly and to avoid such foods as seemed to the patient and the physician to favor the arousal of colonic spells. At times, when there was constipation, petrolatum or petrolagar was used for a period; and in two cases when the proctoscope showed an irritated mucosa, olive-oil retention enemas were employed. However, such accessory measures have been in wide use heretofore, and in my experience and that of others generally fail in themselves to

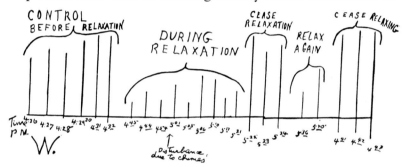

Fig. 89.—(S. W.) Test of ability to relax, as shown by diminution of knee reflex during the treatment of mucous colitis. This record shows fair but not extreme relaxation (which is marked by complete disappearance of the knee-jerk).

effect a satisfactory result in mucous colitis; therefore it may be assumed that they play a minor rôle in the present studies. The effect of progressive relaxation was most clearly isolated with the first patient studied, for here months of treatment in bed at home and in the hospital with diet and other measures, including diathermy, had failed to produce the desired result within that time, showing that the patient did not respond to ordinary rest or to suggestion. It was only when the patient later learned to relax that the tenderness disappeared and the objective indications of spasticity waned. To test the patient's technical progress, the knee-jerk may prove of service (Fig. 89).

In addition to the seven cases of severe chronic colitis or irritable colon discussed above (1929), fourteen cases are now reported (1937), including eleven men and three women. Varying in age from twenty-five to fifty-seven, ten were in the thirties. Depressive tendencies appeared in three.

In seven patients who completed treatment, the reported previous duration of colonic symptoms was respectively 4, 4, 7, 11, 18, 20 and 34 years. In two others whose treatment is almost complete, the previous duration was respectively 9 months and 2 years, preceded by a spell in earlier years.

In the fourteen cases, treatment was limited, sooner or later, approximately to relaxation alone. As explained to the patient, success or failure to attain results could eventually be attributed to relaxation with greater probability. Accordingly full diet was substituted for previous special diets as soon as possible, and medicines were gradually discontinued.

In nine individuals who completed treatment, at least approximately, all became and have continued to be practically free from symptoms according to their latest reports. Relapses, if any, have been mild and transitory. Two summarized case-reports follow.

Spells of diarrhea, of abdominal pain, particularly in the right lower quadrant, and occasional flatus and belching of air were the chief complaints of a practicing physician, age thirty-six, full-time professor in a medical school. These symptoms had been present more or less for 18 years and were particularly aggravated following spells of overwork and nervous strain, notwithstanding dietetic management. In a thorough course, he learned to relax very well, as shown by electrical as well as by clinical observations. Four years after the close of treatment he states that he has been practically free from symptoms, can fall asleep in a few minutes at will, while relapses following overexertion do not occur.

Generalized abdominal pains, four or five loose movements each day, distention by "gas," excessive perspiration, and hemorrhoids had been present in a professor in a scientific department, age thirty-one, ever since he could recall, notwithstanding periods of restricted diet and alkalis under competent management. During treatment by relaxation alone, the colonic symptoms gradually disappeared, and X-ray findings became negative. Hemorrhoidectomy, previously postponed, was performed toward the close of treatment because of slight bleeding. Five years after treatment, he continues much improved, practically free from relapse.

Six patients failed to finish treatment, four for financial reasons, and one because of pregnancy. In these five, marked lessening of symptoms has been persistent for several years. The remaining patient quit treatment during a relapse marked by melancholy.

The results have seemed to justify the exclusion of methods of treatment other than relaxation. Commonly, there has been an amelioration of symptoms beginning within one or two months or a little more. During treatment, symptoms sometimes recur temporarily, often leading to discouragement on the part of the patient. Progress toward recovery, as in most other chronic medical disorders, generally is marked by ups and downs rather than by a straight line of uninterrupted advance. Where relapse follows improvement, there is always a danger that the patient may lose courage and become unwilling to finish treatment. A similar unwillingness sometimes arises during a more or less prolonged period when the patient is free from symptoms and believes that he has had enough treatment. The consequences following incomplete treatment from either of these motives may be relapse later on and failure to achieve full and persistent recovery.

CHAPTER XX
PROGRESSIVE RELAXATION IN DIVERS MEDI-
CAL CONDITIONS—ILLUSTRATIVE CASES[1]

THE purpose and principle of neuromuscular treatment is to reduce the reflex irritation and excitation of the organism. Where this can be done, in the absence of organic nervous lesions, it is reasonable to expect diminution of all symptoms due to excessive or inco-ordinate nervous action. In this connection it is of interest to recall that the symptoms displayed by an individual during organic disease do not depend wholly upon the character of the pathology but are modified according to his nervous and mental reactions. For example, the highly nervous, emotional individual with peptic ulcer is likely to complain every time that he belches or has a sour taste or a sense of fulness and oppression or other discomfort. On the other hand, other patients with little nervous or emotional reaction obviously ignore such symptoms, and the surgeon may be suddenly called to find a ruptured viscus with evident chronic pathology. To seek to explain this on the ground of a supposedly diminished sense of pain alone is obviously inadequate, for in any event we have the absence not merely of a history of pain to account for but also of other symptoms as well, including fulness, epigastric discomfort, belching and nausea. Absence of subjective history of such symptoms apparently points to absence of that excess attention to self which is likely to characterize the nervous and emotional temperament. Accordingly the instance of ruptured viscus without preceding history illustrates that the attention of the patient deter-

[1] Certain of these cases have been previously reported (Jacobson, 1920, 1921).

mines what symptoms shall be reported to the physician. Many other illustrations of the same point can readily be recalled.

As everyone will admit, the patient is benefited if his symptoms due to internal disorder serve to act as signals of something wrong, leading him to consult a physician and to receive proper treatment. But when suitable measures are being taken to relieve a medical or surgical condition, any additional means to diminish symptoms may prove helpful. In many instances this will not mean cure of the underlying condition; for instance, relief produced in the presence of organic heart disease of course will not alter the fundamental pathology. But in the practice of medicine several factors often combine to produce a favorable result, and accessory measures are not to be neglected.

It is often important, as appears in chapter iii, to secure nervous and emotional rest in the course of acute medical conditions. Drugs may prove ineffectual even when aided by suggestion, as illustrated in the first study below. The physician may prefer to use a physiologic method. Abstracts of records follow:

Case 1

Acute endocarditis in a boy of twenty with symptoms of nervous and emotional excitability. At the age of six, acute tonsillitis had been followed by rheumatic fever and mitral endocarditis; something of the same sort had recurred at the age of twelve. The cardiac condition was compensated, but the doctor restrained the patient's athletic activities. His present illness began about June 27, 1921, with symptoms of coryza. Digitalis was prescribed by the first physician who attended him, and this inadvertantly had been prolonged by the assistant: for a condition of heart-block, with an irregular pulse of 39, had developed. When seen on July 11, 1921, a moderate grade of intermittent fever was still present. My treatment consisted principally in discontinuing the digitalis, in prescribing salicylates, and in securing adequate nervous rest, proper diet and elimination.

The patient was apparently in a much-weakened condition, and it seemed of consequence to quiet him. Morphine (gr. ¼, one night; gr. ⅙ with atropine, twice the second night) failed to secure the desired result and was followed by depression the next day. Veronal (gr. x) with sodium bromide (gr. xx) gave a fair night's rest, but seemed later to have an exciting reaction. A larger dose (gr. xv) failed the following night. Adelin likewise was unsuccessful. Suggestions given to the patient at the time of administration that he would rest well were of no avail.

Accordingly he was taught something of relaxation upon several successive occasions. A comfortable position was permitted, following his habit to lie upon the right side. He gently flexed the left arm while the physician pointed out the biceps group. "This is making the muscle tense. Do you feel that muscle?" After an affirmative reply, he was requested to "let the arm go and to continue to relax it until it became perfectly quiet— as limp as a rag." Because of his extreme weakness, he was informed of the location of tenseness and not required to find it for himself, and for the same reason, contrary to usual principles, an element of suggestion was added, telling him that as he relaxed, he would become quiet and go to sleep. A boy of this age is likely to learn any physical act of skill relatively quickly, and the patient evidently acquired something from the instructions, for sleep was fairly well resumed with little further aid from drugs; and as he practiced at relaxation from time to time during the day, he was no longer extremely restless, complained less, and seemed to gain a better spirit. It seems possible that the rest was a favorable element in his subsequent recovery from acute symptoms. After about 4 weeks more in bed and several months of convalescence, he was able to return to school and to resume about the same type of limited activities as previously.

Instances have been brought to my attention where physicians following the present method, as described in previously published articles, have misunderstood the purpose of having the patient contract his muscles. They have assumed that contraction was performed as an exercise to aid relaxation. This, to be sure, is not the purpose. Contraction is performed as a rule in order to acquaint the patient with the experience of tenseness—*in order that he may know what not to do*. It is therefore not well to have him contract during practice when alone: he should relax from the outset and relax only. An exception to this rule is when, in the physician's presence, the patient contracts a part progressively (e.g., stiffens a limb) and then gradually reverses this process, ceasing little by little to contract. This is done to give him the physical experience of "going in the negative direction," in order that he may be led to continue this experience until residual tension disappears. But this exercise is only occasionally, if ever, to be repeated by the patient practicing alone. Accordingly in the case cited above, this exercise was not so repeated, for the patient was instructed when alone simply "to relax further and further."

Case 2

An acute spell of cardiac asthma in a slightly obese man aged forty-six. Chronic bronchitis had been present for years, and there had been several spells of bronchopneumonia. The Roentgen ray revealed an increased transverse diameter of the heart, especially on the right, while the aorta appeared to be dilated. Spells of dyspnea often came on toward evening, and tended to keep him restlessly awake all night unless relieved by hypodermic administration of some form of digitalis or by fairly strong doses of caffeine. According to his wife's observations and impressions, nervousness played an important part in making these spells severe.

Upon examination at about 10:00 P.M. during one such spell in 1920, he appeared pale, perspiring, dyspneic, with marked rhonchi, an anxious expression, while the eyes, in medical parlance, seemed "toxic." The aim of nervous treatment will be to point out to the patient the voluntary element in these symptoms and to guide him to rid himself of them. He was directed to go to bed and lie down while the physician sat close by. He fidgeted, breathed violently, cleared his throat often and complained of substernal distress. His co-operation and insight were secured by calling his attention to his restlessness, to frequent movements of the limbs and to the immoderate movements of his chest; he was informed that the aim must be to bring quiet and ease to these parts. Now upon direction he closed his eyes and flexed his left forearm, while the physician offered passive resistance. He was asked, "Do you note the activity here?" while the physician pointed to the biceps. (When a patient fails at first attempt to recognize the feeling of tenseness in a part, contraction is repeated with passive resistance, until he reports success.) Next the instruction was given, "Now let your left arm go until that part (the biceps group) is just as limp as a rag! Just do nothing with the arm! Let it go further and further every minute!" After a little while of quiet the patient was requested to extend the left forearm against resistance by the physician, noting the tenseness in the triceps group; then he relaxed both extensor and flexor groups for some minutes. Proceeding in this way the right and left flexor and extensor groups of the upper arms and forearms received practice; next we passed directly to the chest because it was heaving violently. He was directed to take an exceptionally long breath, noting the feeling of tenseness or activity spread over the chest wall as he did so. Then he relaxed the chest so far as he was able. Later he was requested to cough several times, noting tenseness the while in the throat and tongue. In the same way he cleared the throat, noting tenseness. After this, he was instructed colloquially, "Now let the tongue and throat relax! Do not bother to clear the throat! If there is a little distress, let it go! Do not bother to cough! Just let the breathing take care of itself! Make no exertion at all! Don't bother about anything at all!" In this manner the patient gradually abandoned his excessive efforts to clear his throat and to cough up secretions; the deep, rapid breathing, which was apparently in part a voluntary and excessive effort to obtain oxygen, gradually subsided; the anxious expression gave way to calm. He finally fell asleep for about 10 minutes, and appeared rested when he awoke. He rested well that night without bromides and even without cardiac stimulation. Only three treatments in relaxation were given in all, yet evidently this patient learned something, for 14 years later he reports that he still relaxes when he has need to quiet himself. There is a recent (1935) history of allergic asthma.

Case 3

Chronic insomnia and nervous hypertension. The patient was a fretful, unmarried woman of about fifty-five years, convalescing after the removal of her gall-bladder, which followed a first operation leaving a fistula for about 1 year. She was undernourished, weak, tense, and irritable and complained of insomnia of more than 20 years' duration. She retired at about 11:00 or 12:00 P.M., going to sleep readily, but awakened at about 2:00 A.M., then heard the clock strike for hours. In the morning she felt fatigued. Some years ago she seemed to have suffered from *globus hystericus*, with difficulty in swallowing. The family history is marked with chronic organic disorders, nervousness and insomnia. One sister died with diabetes and cerebral hemorrhage; another with chronic nephritis, exophthalmic goiter, glycosuria, hypertension; a brother also had nephritis with hypertension. Physical examination revealed two operation scars near the gall-bladder, where there was very marked tenderness on pressure. When she raised herself erect her face was contorted with pain from this region. On her face lines of apprehension were marked. Between the eyebrows there was a persistent furrow, while the forehead was wrinkled and the nasolabial sulci were deep. There were signs of moderate nervous excitement, such as changing facial expression, occasional deep sighs and frequent shifts of position of the limbs, head and trunk.

Only eleven treatments were given, each lasting over an hour, from June 11 to July 7, 1919. The original plan, at her request, was to give an abridged course of only five treatments. In these five periods, accordingly, a rapid but complete survey of the entire body was made. At the end of this time her sleep was considerably improved, being sound until 5:00 A.M., when she awoke, later usually going again to sleep. She decided not to limit herself to so few treatments and the course was therefore prolonged to eleven, when she left for her vacation.

The first period was devoted to relaxation of the arms. The second to the arms and lower limbs. The third to the limbs and trunk. The fourth to what had gone before, along with the shoulders, neck and forehead. The fifth to the foregoing, along with the eyes. The sixth to the same. The seventh took in the lips, tongue and throat. The eighth was devoted to a review along with special practice on the right lower limb, which felt tense to the patient. The ninth was given over to a general review with special practice in noting incipient movements of the eyes, tongue and lips during thought-processes. At the tenth period the attempt was begun to show her how to be differentially relaxed during daytime activities. She sat in a chair, relaxing her limbs, trunk and neck. At the eleventh period this was repeated, including the face, eyes and tongue so that she sat with diminishing mental activity, almost dozing away.

It was carefully explained to her that not ordinary relaxation was meant, not lying on the couch, apparently still and quiet, for a person might lie for hours relaxed in the ordinary sense and yet remain nervous and sleepless. Rather, the goal to be sought was progressive relaxation from moment to moment—a continual letting go of activities in every part of the body.

The abridged method used for this patient was as follows: She noted sensations of contraction while flexing both arms at the same time; then was instructed to relax until no such sensations remained. For the most part she was successful in perceiving and localizing muscular sensations, but at first she was able to detect these only during the first few seconds after gross movement. After considerable practice, according to her reports, she became able to note the dwindling sensations for many minutes. The second step was extension of both arms with passive resistance, followed by relaxation to the end that no sensations should be noticeable. Next her attention was cultivated toward contraction of the flexors and extensors of the forearms during closing of the fist, as well as during movements backward and forward of the hands against resistance made by the physician. A similar bilateral procedure was carried out on the lower limbs. She came to recognize activity in the abdominal muscles by means of panting movements as well as bending forward of the trunk. Likewise, she readily learned to recognize sensations throughout the chest wall in active breathing. Since she sighed frequently and breathed irregularly, special practice was needed at this point. Instructions were, "Just let your breathing go, don't bother to move your chest! It will take care of itself. Let the chest muscles go farther and farther every minute!" Progress was rather slow here, but after a few periods the sighs ceased to be a disturbing element. Thereafter the usual practice was given the back, neck and forehead. She was markedly successful in noting activity of the eyes and speech apparatus during imaginary and thought-processes.

This patient met with several striking difficulties:

(1) She perplexedly asked, "How can I relax my arms or other parts as far as you wish me to?" This question is not seldom put to the physician. The answer should be simple, for the patient does not require a scientific explanation of the physiology, but simply wishes to know what to do. In 1919 I frequently made reply, "Just cease moving the arm—simply do nothing with it; relax it further and further from minute to minute!" This patient was encouraged to practice and was informed that she need not expect to be able to relax any part fully until she became able to relax the rest of her body fairly well. Of late years I have modified the instructions so that this perplexity is not aroused in the patient. Dur-

ing the very first or second period, after progressive stiffening of the muscles of the right or left arm (chap. v), he is directed not to contract quite so much, then again not so much, and so on until the arm appears fairly relaxed. Thereupon the instruction is worded, "Whatever you have been doing (or not doing) up to this point, *that* continue on and on past the point where the arm seems to you fully relaxed." This instruction should be frequently repeated.

(2) The patient continued, as is usual with insomniacs, to make occasional slight shifting movements as she lay on her couch. She was informed that such movements seemed to be among the most important causes of her sleeplessness. It was explained to her that her facial expression suggested that such movements revealed certain mental traits. They showed that she became quickly dissatisfied with any position that her body assumed after a little time, then shifted a little in order to seek more comfort, then repeated this a little later, vainly seeking a position of perfect comfort, and so keeping herself awake by these little movements of dissatisfaction. Her face expressed this dissatisfaction. She replied that such movements were made unconsciously. During this conversation she made such movements and these were pointed out to her, and she was shown and admitted that they resulted in dissatisfaction from the particular position she had assumed. "When you feel like moving some particular member, relax it instead until the inclination to move disappears." As was evident during the fifth period, she misunderstood this direction. After lying almost motionless intermittently during the hour, she vehemently complained that the period had made her nervous—that she became uncomfortable in lying so quietly. The physician informed her that this indicated that she had not been following directions: she had been holding herself tensely motionless. This tenseness had been apparent at times to the observer, and it made itself apparent to the patient in the form of what she called "discomfort and nervousness." She had not been requested to hold herself quiet; it would be much better to yield to a temptation to move than to resist with effort; no effort at all was to be used. She was simply to let go, to relax, to do nothing, to cease activity, to use no effort. She failed to grasp this difference between voluntary inhibition and relaxation until about the sixth or seventh period, after which her improvement was rapid. The patient noted a slight trembling all over the body, as she expressed it, at times of nervousness. When trying to relax, she would often note greatest difficulty with her right leg.

An approach to relaxing her nervous condition was made also from the mental side. She had previously been treated principally for gall-bladder troubles. It was now pointed out to her that she should take daily annoyances less seriously. It was also made clear that a determination to do this should be often repeated. She evidently made a sincere beginning at carrying out the instruction.

She was advised to engage in general relaxation for half an hour or more in the late morning and late afternoon. During the day she was to keep herself in as relaxed a condition as her occupation permitted.

On July 7, 1919, she reported that she had been sleeping better but tended to awake at five o'clock in the morning. Quite unexpectedly she fell asleep during relaxation several times during the afternoon. Apparently, also, she fell asleep several times in the presence of the physician. A year after the last period of treatment, she reported that while she did not sleep so well as during the months following her treatment yet her rest was considerably improved and her general nervousness diminished. Obviously eleven treatments are too few to overcome entirely an insomnia of so many years' standing. In such cases it would seem better to give many more treatments. However, in this postoperative case, where acute conditions added excitement to chronic irritability, the use of progressive relaxation seemed to fulfil the aim of bringing quiet to the nervous system. Favorable results persisted for about fourteen years. She died of carcinoma.

Case 4

"*Compulsion neurosis*" in a married woman, aged thirty-one. An excavated ulcer near the middle of the right sternocleidomastoid muscle, about 3×1.5 cm. in size, was caused by picking at it with her fingers. A few small scars on the face indicated the sites of previous self-mutilations. Local treatment proved ineffective because she failed to control herself. Attracted to a lover, yet longing for her husband from whom she was perhaps to part, she was wretched with indecision. Since those circumstantial causes of disturbance could not be removed, she was shown how to relax but was placed in the hospital in order to have hot boric-acid dressings frequently applied to her wound. Her attention was called to the physical manifestations of her excitement and worry, including the nervous movements of her hands toward her neck. She did not succeed in acquiring relaxation of the eyes sufficient for marked mental repose, but after a month's treatment she felt that her equilibrium was restored and she was able to return to her work and domestic problems with no further medical assistance for many months. When the fingers no longer molested the ulcer, it healed rapidly. Five years later, when last seen, there had been no recurrence.

It is not generally possible to predict what number of treatments will be required in any particular case. This holds for the following instance, which will illustrate the treatment of a condition of chronic nervousness with an element of phobia. Here only twelve treatments were needed, yet usually a much more protracted training would be anticipated. The yielding of deep-seated habits can be explained in the light of diligent daily practice by the patient himself. On the whole, during the following 9 years clinical observations indicated a progressive decrease of nervous symptoms with increasing vigor of personality.

Case 5

Nervous hypertension with mild phobias. A merchant, aged thirty-eight, whose occupation obliged him to travel, complained in the summer of 1918 of nervousness in two forms: (1) fear of travel and of difficulties in business matters; (2) restlessness, quivering and inability to sit without fidgeting. Once he had shown a trace of glycosuria. Typhoid fever in 1898 was the only noteworthy previous disease. He stated that he drank moderately with occasional slight excess, and smoked from six to eight cigars a day. His father was well at the age of seventy-seven. His paternal aunt had *tic douloureux*, but otherwise his father's family history was largely negative. His mother later died with diabetes. A sister of hers had Graves' disease with glycosuria. Their family were all highly nervous with a marked tendency to hypochondria, a disposition which the patient evidently shared. Another maternal aunt had been almost an invalid with a nurse in attendance for years, owing to vague nervous difficulties.

In 1910 he had suffered a nervous shock from an automobile accident. Business matters in 1911 had taken him to Europe, and for the first time he felt fear to travel. In 1913 a telegram from a relative stated, "Father died today." Believing it was his own father, he fainted; but later he discovered that another man was meant. This experience seemed to increase his perturbation. Whenever he took a cold, he feared pneumonia. His fear to travel was greatest in winter, particularly in regard to being "stalled" by snow; this actually took place in 1918. His fears seemed to him unreasonable, for why should he worry about such events? On the day of examination he had worried that there would be no diner so that food would be lacking if the train should be snow-bound. His fears lasted but a few minutes each time, but "they were accompanied by twitching all over the body." Subsequently they seemed to him folly, like a bad dream, and he felt ashamed. When he traveled in company, the feeling was absent.

General examination revealed a tall, somewhat undernourished man, with thin abdominal walls and general muscular weakness. The pharynx was slightly injected. His pulse was 84, regular, strong while sitting. Blood-pressure 124–70. Reflexes were normally present, with increase of those of the arms. The urine showed a specific gravity of 1.030, but was otherwise negative.

12–28–1918. Period 1. Practice at relaxing both arms and left leg. Slept while relaxing. Discussion of the application of relaxation to self-control.

12–30–1918. Period 2. Practice at relaxing limbs, trunk, neck and brow. Relaxes very well. Goes to sleep three times during practice but is interrupted for new instructions. Reports that he has applied what he learned during the first period as follows: During a meal he suddenly realized that he had made a blunder in a business matter. Instead of getting excited as usual, he relaxed. The only feeling of excitement he had was in the epigastrium, and this soon passed away. After he had become quieted he realized how he had done it.

12–31–1918. Period 3. General instruction, "Just let everything go in a perfectly easy manner! Try to relax further and further every moment!" Practice given to muscles of arms, legs, back, abdomen and neck. After about 10 minutes he falls asleep, but awakes with a start a few minutes later. He arose unusually early this morning. Brief practice on noting tensions in brow and about eyes, looking in each direction. Tensions noted in face during smiling, in tongue during protrusion and retraction, in lips upon puckering and smiling, in the tongue and lips upon counting to 10, in the tongue and throat upon swallowing. Brief practice on imagery, including visual, auditory, verbal, and kinesthetic (flexing the index finger). Diminishing tensions practiced with the latter. Practice at relaxing foregoing types of tensions. Appears to relax very well. Reports at first that he continues to pay attention to sounds. Discovers for himself that when this occurs he has a visual image of the source of sound. Thereupon begins to relax all tensions connected with visual and verbal imagery. Reports that for what seemed to him 3 seconds he did not have any visual image or any thought-process. It became clear that he was now supplied with a method of relaxing his body and mind.

1-3-1919. Period 4. Practice at general relaxation; later at relaxation while sitting and while standing. Special drill to relax rapidly, after making his muscles generally tense. Explanation that this negative process is to be carried out at moments of excitement. Instruction to carry out relaxation during social and business hours.

1-4-1919. Period 5. Sitting posture again employed. Reports relaxing when he tended to become angry at the impertinence of an employee.

1-8-1919. Period 6. Reports that periods of tossing about during sleeping-hours have ceased for the present. Has successfully overcome a tendency to "flare up" over business affairs. His attention is called to needless gestures or motions of the limbs. He has recently overcome a tendency to play with pencils and objects. Reports that when he was nervously uncomfortable before a fluoroscopic examination, he relaxed and felt better.

1-8-1919. Period 7. Reports successful relaxation of business worries. Finds that he "stops to reason" if he is relaxed.

1-9-1919. Period 8. Practice at noting sensations of tenseness while thinking of business matters. States that complete relaxation takes his mind off a disturbing matter. Reports that when tense he tends to take things irritably, but when relaxed his "attitude is easy." Instructed to take sufficient sleep, to eat slowly, and to exercise daily.

1-10-1919. Period 9. Further detailed practice. Observation and relaxation of reflections about business. Reports that when he tends to become irritated over business matters he relaxes and the emotions ebb away.

5-23-1919. Reports that he is now able to relax to the point of sleep, if he desires, in about 5 minutes, both lying and sitting. Where formerly he used to tremble all over, there is now no more excitement over business affairs. He no longer has the same need to relax away disturbance, since it does not so much tend to develop. On a recent trip there was no fear of traveling.

1-17-1920. Period 10. Reports that he no longer has fear of slight business difficulties, and was free from fear of travel for about 1 year after the relaxing treatments; that he has been in good nervous condition until the last 2 weeks, when he has had a "cold" and failed to sleep well one night. Then perturbation recurred, perhaps because of the present epidemic of influenza which has tended to make him fearful. Desires to review relaxation.

Practice given with both arms and legs simultaneously. Then with abdomen, back and chest. Readily perceives sensations of contraction. Winking of the eyelids present for a brief period, then relaxes to sleep.

Periods 11 and 12. General review.

Following the preceding periods, the patient reported that he soon lost the remnants of his fear of traveling. During the following three years there occurred several spells of kidney colic, leading to the discovery of an enlarged right kidney with dilated pelvis. Later he developed an epiplocele. Certain other minor details of medical history may be here omitted. From 1918 to 1927 he improved in weight and strength. He showed considerably less of his former habit to worry, greatly diminished hypochondria, and abandoned his former practice of being a frequent visitor at medical offices. In 1927 he had a mild brief relapse, perhaps owing to neglect of practice during the past year. He seemed to be solving his nervous problem by himself when, after several transient spells of somewhat increased blood-pressure (which subsided upon complete rest), renal colic recurred, probably due to a small calculus, soon followed by grippe. These various strains led to a moderate recurrence of nervous hypertension, which apparently diminished after a further brief period of treatment, repeated in 1933. Health continues good in 1937 with blood pressure normal.

In applying the "rest-cure" of Wier-Mitchell, the "neurasthenic" is often removed from his business and social affairs and thrust into solitude. In consequence, he may suffer such financial loss or ruin as may justly give him cause for anxiety during the remainder of his life. In contrast therewith, the plan of the present method is to let the individual remain at his affairs, deriving not only the practical benefits therefrom but also the therapeutic advantages of regular occupation. Where business cares have seemed to produce the exhaustion, they may be diminished. But where some trying situation has produced anxiety, it is often better morale for the patient to face the difficulty, learning to adjust in its presence, than to run away from it with a feeling of defeat. In recounting the following illustration, in which the patient continued at her daily work during the period of treatment, attention may be drawn to some of the finer points of technic in securing mental relaxation:

Case 6

Nervous hypertension, with fatigability (so-called "neurasthenia") in a single woman of thirty-five. Her complaint, November 14, 1925, was of anemia, chronic constipation, "nervous tension" with inability to sit quietly, slight dizzy spells during excitement, and slight discharge from the nose. She was engaged in secretarial work and was known to be highly efficient. In childhood she had varicella, measles, pertussis, typhoid (at thirteen) and mild scarlet fever (at nineteen). Menstrual history was negative. She did not smoke or drink, and slept very well. Her father died after cerebral hemorrhage. One sister was well, while another sister and a brother were of "nervous disposition." About April, 1924 she gradually became severely fatigued, with pallor and loss of 10 pounds. A physician found marked secondary anemia and gave bi-weekly injections probably of iron and arsenic. After European travel for 3 months she again felt well. Since then her chief complaint has been a tendency to fatigue very readily. She did not consider herself a nervous individual but stated that her knees became "fidgety" in the evening. On the average, one bowel movement occurred every other day. Once she noticed mucus. She had a very frequent but slight mucous discharge from the nose and nasopharynx.

Examination disclosed a somewhat pale woman, with a regular strong pulse of 72 while sitting. A little acrid discharge was found in one middle meatus. Blood-pressure was 134–64. Her heart was palpable at the apex in the fifth interspace with no thrill. Its borders were about normal in location, but there was a distinct blow at the apex faintly transmitted to the left of the sternum. On 11–14–1925 frequent extra systole was heard, almost regularly every third beat at first, but about 15 minutes later only occasionally. Such extra beats were absent on 11–23–1925. Physical examination was otherwise negative.

The blood-count was 4,620,000 erythrocytes per cu. mm., 6,700 leucocytes per cu. mm., hemoglobin 60, color index 0.65. Erythrocytes were otherwise negative. Differential count showed 40 per cent small mononuclear, 5 per cent large mononuclear, 3 per cent indentate nucleus lymphocytes, 52 per cent polymorphonuclear neutrophiles. Urine analyses and Wassermann tests proved negative. Her blood had a normal sugar content and 40 mg. of non-protein nitrogen per 100 cc. on general diet. The electrocardiogram showed sinus arhythmia, average rate 62 per min., slight notching of P wave, T of slightly increased amplitude. Roentgen films of the chest, sinuses and teeth showed no noteworthy abnormality. The basal metabolic rate was within normal limits.

In treatment, spleen marrow tablets were prescribed for the anemia, a throat-wash for local purposes, and thorough mastication without haste and with plenty of fruit was advised for the constipation. The results in this case therefore are not to be ascribed exclusively to neuromuscular treatment, but the evidence indicates that the improvement in her nervous reactions was chiefly due to this source.

11–14–1925. Period 1. Tri-weekly periods during following months. Usual practice with right biceps-brachial group.

1–8–1926. Period 12. The arms and most of the groups of the lower limbs have received practice. To date she has shown fair ability at spontaneous localization of tensions. As she lies, the eyelids flutter or wink frequently, while the eyeballs can be seen to move frequently under the closed lids.

1–11–1926. Patient reports diligent practice at home, with marked diminution of dizziness and fatigue. Period devoted to left thigh. The eyelids show less movement than previously.

1–19–1926. A little morning dizziness has reappeared. Also slight blood in stools, which are hard.

2–2–1926. Recognizes tensions mostly +.1S. Reports that it now takes her about 1 hour to get relaxed and that she meets difficult situations with less fatigue than formerly.

2–24–1926. Notes tenseness in frowning. States that she has enjoyed the practice from the outset. Reports spontaneously that eyes seem to relax with the forehead and brow. A little stiffness in the neck muscles found upon passive movement.

2–26–1926. Her eyelids no longer flutter or wink while attempting to relax. (The patient who overcomes this difficulty has gone far.) After each period the eyes now have a typical extremely relaxed appearance. Reports that she has twice fallen asleep during practice at home. Sleeps during present period. This patient readily distinguishes between tenseness and strain, has shown an increasing ability to observe, and gives objective signs of marked progress.

3–1–1926. Practice with visual imagery. Tenseness in eye muscles reported $+1S$, without hint from physician.

3–22–1926. States that she has been "trying hard" to relax. Instructed not to do this. Knee-jerk not diminished as yet during relaxation.

3–24–1926. Complains that "her mind wanders" at times during period. The instruction is given not to concern herself with this. She is not to try hard to relax, but a little directive attention is to be used at the outset, and this is then to be relaxed along with other tensions.

3–31–1926. Differential relaxation begun. After relaxing on a couch, she changed to the sitting posture on a chair, remaining as relaxed as possible while shifting. Practice with right biceps and right triceps.

4–9–1926. This patient can readily distinguish tenseness when due to marked contractions of muscles. Therefore to cultivate a finer sense, the contractions now used are slight; for instance, the leg is moved about an inch or less in flexing.

4–12–1926. During observation of thought-processes the distinction is made clear between muscular contraction, the sensation therefrom, and the meaning. She understands this clearly. It is noticed that she has been sitting more stiffly than formerly, and she explains that she fears to develop a habit of faulty posture. Explanation is offered that while learning to relax, certain faulty habits may be neglected for a time and their correction postponed. Upon inquiry, she states that the tensions which carry for her the meaning or determination to maintain posture are in the back. She is informed that there are two ways to do away with such tensions: (1) on the meaning level, by explaining to her the reasons why she holds stiffly and persuading her to change; (2) on the tensions level, by showing her directly how to relax the muscles.

4–28–1926. Her color is better, eyes brighter, and she looks generally well. She reports that she now realizes that she formerly tried too hard to relax. Yesterday for the first time, after an hour of practice, she felt well rested.

5–14–1926. Feeling better. Rests an hour before dinner each day. Seldom falls asleep, but relaxes more quickly than formerly.

5–21–1926. Constipation considerably diminished. Still eats too quickly. Knee-jerk still present during relaxation.

6–4–1926. Periods now bi-weekly.

6–30–1926. Knee-jerk present during relaxation. Patient reports that she finds herself thinking about the coming test and that this seems to interfere with success.

7–2–1926. Knee-jerk now diminished during relaxation. Patient leaves for vacation.

8–19–26. Patient now appears less restless than formerly. She states that her outlook on life is different. She feels rested, her difficulties appear less, and she enjoys work in place of dreading the day's fatigue. Generally she relaxes well and quickly, but occasionally she has given it up because it has seemed to make her more nervous. (This report indicates faulty technic due to trying too hard which needs to be carefully corrected.) She no longer is subject to dizziness, except a few slight spells recently upon swimming before breakfast. Constipation has been on the whole diminished and hemoglobin has risen to 83. While she is not absolutely free from "nervous tension" and restlessness, she is no longer troubled to any noteworthy extent.

3–1–1928. The foregoing general condition was maintained for more than one year, while nervous symptoms virtually disappeared. During a routine examination in February, 1928, when she reported general absence of symptoms, anemia was discovered again, with hemoglobin at 53, along with considerable cardiac arhythmia (right ventricular extra-systole). Since the patient's nervous condition has remained excellent, in spite of unusual recent trials and strains, we are able to rule out the original increase of hemoglobin as the cause of her nervous improvement. Obviously her anemia, due to non-nervous causes, needs to be traced to its source and treated appropriately; but her nervous symptoms, which interest us here, have subsided with habitual relaxation.

CHAPTER XXI
THE THERAPEUTIC USE OF RELAXATION

THERE is general agreement among writers, as was seen in chapter i, on the widespread importance and many applications of rest in the practice of medicine. I therefore sought to inquire by laboratory methods whether I was really getting a more intensive form of rest by means of progressive relaxation than is secured by the ordinary instructions of the doctor. Various evidence to this effect was disclosed. Furthermore, the review of tonus, of muscular innervation and of pertinent investigations in nervous physiology in earlier chapters showed why, from a physiological standpoint, we should expect extreme muscular relaxation to result in reduction of reflexes, and therefore in an intensive form of rest; and likewise our review of psychological investigations made clear why this should automatically bring with it a reduction of mental activity.

Who Can Be Relaxed?

(THE INFLUENCE OF AGE, PERSONAL TYPE, OCCUPATION, VARIOUS DISEASE CONDITIONS AND OTHER FACTORS)

In practice I found that relaxation could be taught to willing patients much the same as golfing, skating, pianoforte, or driving a motor car are accessible to the average person. Degrees of skill acquired will of course differ greatly according to the character and degree of co-operation of the patient, the distractions produced by physical or emotional conditions or by circumstances, and the efficacy of the teaching. Much depends upon the amount of time allotted for practice, and the physician needs to devote considerable energy and patience to the difficult, distracted patient.

I found that singers, dancers and athletes may learn very rapidly owing to a certain prior familiarity with muscular tension and relaxation. As would be expected, healthy individuals (the subjects were university students and doctors) learn more quickly than do patients with distress. During neurosis, distraction on the part of the patient tends to prolong the learning period. Children may require special methods of training. The youngest in my experience was an intelligent child of eight and one-half years, who learned very readily. I succeeded, not perfectly but very markedly, in relaxing a girl of thirteen years who was of subnormal intelligence (I.Q. 53–57, three tests, Terman modification of Binet-Simon scale), indicating that a high intelligence quotient is not required for the present method. My oldest patient was sixty-eight; she made noteworthy progress, but the results were of course far from perfect. Old age, therefore, is not an insuperable barrier.

It has been said that willingness is the cardinal requirement for the method of relaxation; yet this statement has to be modified. For as a rule, even the healthy adult would rather be up and doing than lying down to relax; in this he is like the child who does not wish to go to bed at night. Unwillingness to give up activities is increased during fretfulness and distress. Thus in disease, where relaxation is most needed, the conditions for inducing it become increasingly difficult. This is particularly true, for instance in toxic goiter, where the individual feels the drive to be doing yet shows evident need for relaxation. It therefore becomes the physician's task to overcome this unwillingness by argument, persuasion and every other means in his power; just as it is the business of the surgeon to overcome the opposition of his patients to needed operations. Fortunately, as the individual progressively relaxes, unwillingness gives way to the desire for rest and for the resulting comfort.

I have investigated whether methods of relaxation can be applied to certain forms of insanity, notably the manic-depressive variety. Here willingness is at times notably absent, and therefore the physician can only hope for rather than expect success. He must seek to induce relaxation by argument, persuasion, example and proper instructions to attendants. He must keep in mind that the disease is characterized by spontaneous recessions and therefore use great caution in attributing observed improvements to his methods. With these considerations in mind, our results, although unsuccessful in some cases, nevertheless suggest that the method can in principle be applied, and that the effort is justified.

The method to produce extreme relaxation is not in principle limited to chronic conditions where we wish to inculcate habits of rest. Often in medical practice a brief period of intensive rest is required for acute conditions of distress or exhaustion, and it may be desired to apply physiologic methods in place of bromides, morphine or other drug sedatives. The physician here encourages the patient to relax extremely, shows him briefly in a period or two how to do this and performs it himself for a few minutes by way of illustration for imitation. Treatment of the patient in acute distress (provided that pain is not too severe at the moment) is therefore open by this route, and the aim will be transitory alleviation of symptoms with consequent favoring of recovery, when possible.

The Christian Scientist, the theosophist, and in general the cultist are difficult subjects. These are the suggestible and dependent types, giving the physician an added task to make them rely upon themselves. They have many bizarre ideas which interfere with their learning to relax, and much time may be lost with them in argument. Patients inclined to credulity and excessive faith in the physician also generally fail to observe for themselves and take longer to learn to relax than do average individuals. One cultist reported that she had hypnotized herself daily since childhood in order to go to sleep. This proved to be a difficulty when treatment by relaxation was begun, for she tended toward a hypnoidal condition. The attempt to relax her was therefore abandoned.

Since suggestibility is to a certain extent an impediment in being taught to relax, the hysterical individual sometimes offers marked difficulties. The physician must give directions precisely and make sure that no hypnoidal condition and no increased suggestibility results. Following these precautions, the hysterical subject, I believe, can generally be relaxed.

To What Disorders Does Relaxation Apply?

A consideration of the widespread application of rest to disease conditions (chap. i) leads us to anticipate widespread opportunity for the present method of intensive rest. In some disorders relaxation may appear as the principal or only method of treatment, while in others it may be adjuvant to other measures of surgery or medicine or hygiene. The physician will recall that the applications of methods of diet to the practice of medicine are manifold, requiring a large volume to describe them. Writers who recognize the importance of diet do not thereby offer it as a panacea, and there is likewise no such intent or purpose in presenting the present method of intensive rest.

An interesting experience is to observe the signs of excitement or distress diminish as the patient relaxes. It is difficult and not quite satisfactory, at the present stage of work, to present results statistically. If we consider the influence of intensive relaxation on the entire course of the malady, we lack an accurate quantitative measure of the effects. An added difficulty is that the maladies of individuals so far treated by the present method vary greatly and are not readily classed together for purposes of statistical calculations. Furthermore, it is difficult for one investigator alone to collect a large series of cases of one particular malady because of the generally long course of the condition and its treatment. Again, how shall we classify the results of treatment of nervousness, when favorable, but existing as a complication of incurable organic disease? Where the malady can be classed under two headings, which heading should be used?

Despite these limitations, the following tables may have some significance. Therapeutic agents other than relaxation were practically or entirely excluded in a number of cases selected for treatment and the results are presented in Table VIII. We present separately (Table IX) the figures for certain other cases where some other form of treatment was used in addition to relaxation; for example, special diet in mucous colitis, iron or spleen-marrow where anemia accompanied nervous hypertension, or surgery for the removal of suspected focal infections. But such measures were generally directed toward the correction of complications and not toward the functional disorder under study; except in a few cases, where some form of psychological treatment supplemented the relaxation. As will be seen, the results have warranted the expenditure of time and patience, and encourage the hope that others will be induced to apply the same methods. "Objective" results in Graves' disease include records of pulse, temperature, basal metabolism. In some instances the pa-

tients failed to remain under treatment for an extended period, but they are included in the tables when the duration
has been long enough to seem significant even if the work was

TABLE VIII

DISEASE	No. of Cases	Quit	Objective Result (Improvement)					Patient's Report (Improvement)			
			None	Slight	Marked	Very Marked	Doubtful	None	Slight	Marked	Very Marked
Nervous hypertension....	31	6	0	6	14	11	0	0	2	11	18
Acute insomnia with nervousness..............	6	0	0	1	1	3	1	0	0	3	3
Anxiety neurosis........	2	0	0	0	0	2	0	0	0	0	2
Cardiac neurosis........	1	0	0	0	0	1	0	0	0	0	1
Chronic insomnia.......	12	7	0	5	2	3	2	0	2	3	7
Compulsion neurosis.....	1	0	0	0	0	1	0	0	0	0	1
Convulsive tic..........	3	1	0	0	1	2	0	0	0	1	2
Cyclothymic exaltation...	2	1	0	0	1	0	1	0	0	0	2
Cyclothymic depression..	7	2	0	0	2	5	0	1	0	1	5
Esophageal spasm.......	3	0	0	0	0	3	0	0	0	0	3
Graves' disease..........	3	0	0	1	2	0	0	0	1	1	1
Hypochondria..........	2	0	0	0	0	2	0	0	0	0	2
Mucous colitis..........	2	1	0	0	0	2	0	0	0	1	1
Spastic paresis..........	1	0	0	0	0	1	0	0	0	0	1
Stuttering, stammering...	3	3	0	0	3	0	0	0	0	3	0
Unclassified psychosis....	1	0	0	0	1	0	0	0	0	1	0
Cardiac asthma.........	1	0	0	0	1	0	0	0	0	1	0

Figures indicate results at end of treatment in cases where relaxation was the
sole therapeutic measure. As a rule, each patient appears but once in the table,
being placed under one heading, e.g., under "Mucous colitis" or "Nervous hypertension," even if his disease symptoms might properly be included under both
headings. A considerable number of patients with symptoms classed under "Nervous hypertension" might equally well be represented under "Hypochondria" or
"Insomnia" but it is believed that singleness of representation of each patient in
the table will give a fairer idea of results.

incomplete. It has seemed best to avoid the use of the term
"cure," partly in order to avoid the difficulty of distinguishing between this and "very marked improvement" and partly in order to err on the side of claiming too little rather than
too much. In many instances, however, where the patient

has persisted at re-education, the consequences have seemed quite thoroughgoing and have persisted for years.

Treatment limited approximately to relaxation alone was applied from 1929 to 1938 in 105 additional patients, as shown in Table X. The patients accepted this limitation in order to learn what results might be due to relaxation and in

TABLE IX

DISEASE	No. of Cases	Quit	Objective Resul (Improvement)					Patient's Report (Improvement)			
			None	Slight	Marked	Very Marked	Doubtful	None	Slight	Marked	Very Marked
Nervous hypertension...	8	1	0	1	4	3	0	0	0	4	4
Anxiety neurosis........	1	0	0	0	0	1	0	0	0	0	1
Chronic insomnia.......	1	1	0	0	0	0	1	1	0	0	0
Cyclothymic exaltation...	1	0	1	0	0	0	0	0	0	1	0
Cyclothymic depression..	2	1	0	0	1	0	1	0	0	1	1
Graves' disease.........	3	1	0	0	1	2	0	0	0	1	2
Mucous colitis..........	6	2	0	0	2	4	0	0	0	0	6
Unclassified colitis.......	1	0	0	0	0	1	0	0	0	0	1
Unclassified psychosis....	1	0	0	1	0	0	0	0	0	1	0
Cardiac asthma.........	1	0	0	1	0	0	0	0	1	0	0
Bronchial asthma........	1	0	0	0	0	1	0	0	0	0	1

Figures indicate results of treatment in cases where relaxation was the sole measure used for treatment of nervousness but where other measures, such as diet in colitis, or iron in anemia, or the surgical removal of infectious foci, were employed for the treatment of concomitant disease conditions.

order to eliminate sedative or other medicines, if taken previously. These medicines were eliminated as soon as possible, even if they seemed indicated. In several instances, minor operations were postponed until toward or after the close of treatment. In colitis, as previously stated, laxatives, belladonna or other agents for chronic pain, as well as restrictions on diet, were gradually abolished.

Psychotherapeutic measures, including assurance to the patient of his ultimate recovery, were likewise, so far as possible, omitted. Deviations from this rule occurred in

THERAPEUTIC USE

TABLE X

Disorder	No. of Cases	Treatment Incomplete	Physician's Report (Improvement)					Patient's Report (Improvement)				
			None	Slight	Marked	Very Marked	Doubtful	None	Slight	Marked	Very Marked	Doubtful
Nervous hypertension	82	35	o	7	33	38	4	1	4	27	49	1
Fatigue states.......	17	4	o	o	5	11	1	o	o	3	14	o
Insomnia...........	34	18	o	6	12	15	1	o	4	13	17	o
Anxiety states.......	5	2	o	2	o	3	o	o	2	o	3	o
Nervous depression..	7	2	o	2	1	4	o	o	2	1	4	o
Hypochondria.......	4	o	o	o	o	4	o	o	o	o	4	o
Phobias...........	5	o	o	o	1	4	o	o	o	1	4	o
Chronic colitis.......	25	11	o	2	6	16	1	o	2	5	18	o
Esophagus spasm....	5	1	o	1	o	4	o	o	1	o	4	o
Sigmoiditis..........	1	o	o	o	o	1	o	o	o	1	o	o
Arterial hypertension.	13	7	2	2	2	5	2	2	1	3	5	2
Nephritic hypertension.............	1	o	o	1	o	o	o	o	o	o	1	o
Tic................	1	o	o	o	1	o	o	o	o	1	o	o
Facial spasm........	1	1	1	o	o	o	o	1	o	o	o	o
Tension headaches...	4	1	o	o	1	3	o	o	o	1	3	o
Occular headaches...	1	o	o	o	o	1	o	o	o	o	1	o
Facial neuralgia.....	1	1	1	o	o	o	o	1	o	o	o	o
Functional tachycardia...............	1	o	o	o	1	o	o	o	o	o	1	o
Coronary disease....	1	o	o	o	1	o	o	o	o	1	o	o
Tremor.............	1	o	o	o	o	1	o	o	o	o	1	o
Dsymenorrhea......	1	o	o	o	o	1	o	o	o	o	1	o
Cyclothymic depression..............	14	3	2	o	2	9	1	2	2	o	8	2
Cyclothymic exaltation..............	1	1	o	o	1	o	o	o	o	o	1	o
Mild alcoholism.....	1	1	o	o	o	1	o	o	o	o	1	o
Neurosis of bladder..	1	1	o	o	1	o	o	o	o	1	o	o
Parkinson's disease...	1	o	o	o	o	o	1	1	o	o	o	o
Stammering, stuttering..............	2	1	o	o	1	1	o	o	o	1	1	o

Figures for 105 additional patients (1929–37) treated by methods of relaxation, omitting other measures (see text).

[419]

cyclothymic depression, particularly when discouragement was so severe as to threaten suicide. In that event, any measure of psychotherapy or medicine was added, provided that it seemed to avert the danger, after which methods were again restricted to relaxation alone.

To complete the data for this table, 66 questionnaires were sent out in November, 1937, to which 38 replies were received. In the remaining 28 instances, data are based upon the condition of the patient as ascertained previously. In some instances the results refer to the condition at the termination of treatment; but in most, they include also a subsequent period up to 6 years, provided that other treatment did not follow.

The present tables are based upon interpretations of complex data secured by various objective clinical and laboratory measures, but also upon the patient's observations. It is difficult, if not sometimes impossible, to estimate therapeutic results accurately. Since the results include the patients whose treatment was not complete, they are in this respect understated.

INDICATIONS FOR TREATMENT BY RELAXATION

The purpose of relaxation is to do away with certain activities that place an undue tax upon the organism. Since relaxation is the intensive form of rest, we may assume that it will be indicated as a rule where rest is indicated, a topic which was discussed in the early chapters. It is well for the physician to bear in mind that patients who are obviously restless and emotional are not the only ones who are in need of intensive rest. He may equally well investigate whether his cases of peptic ulcer, organic heart disease or divers other conditions might not show subsidence of symptoms if in addition to other treatment their reflexes were to be quieted by relaxation. Pending many such investigations by others, we cannot here make final statements of indications for the present method. However, a tentative program may be suggested:

1. *Acute neuromuscular hypertension* (commonly called "nervousness" or "emotional disturbance").—This may occur during the course of a large variety of diseases (chap. iii) and may be treated with sedative drugs, physiologic relaxation, or both.

2. *Chronic neuromuscular hypertension.*—This includes what is commonly called "neurasthenia," but appears also in all the "functional neuroses," such as phobias, tics, habit spasm, insomnia, stammering and stuttering, emotional unbalance, and heightened reflexes without organic derangement. The psychic phases of neuromuscular hypertension are highly variable: they include worry, anxiety, hypochondria, inability to concentrate because of apparent restlessness, compulsions and the many and divers symptoms sometimes termed psychasthenia. In certain conditions, the aim of treatment will be thoroughgoing: to remove by relaxation the groundwork of the neuromuscular disposition to neurosis; to do away with those static and other continual tensions which close clinical observation of the neurotic individual always discloses. In other conditions, the aim will be less ambitious, yet still remain of consequence: palliative, in the sense of removing a set of symptoms, where the underlying malady cannot be successfully reached.

3. *States of fatigue and exhaustion,* alone or in complication with disease. Prophylaxis by means of relaxation is also to be thought of here.

4. *States of debility.*—This would include convalescence from infectious and exhaustive diseases of various types.

The present method appears appropriate in cases of retarded recovery where an element of rest might aid.

5. *Sundry preoperative and postoperative conditions.*— Writers on surgery are beginning to emphasize the importance of rest in preparation for the nervous strain of operation and the subsequent distress. The present method, in my experience, has appeared to reduce the patient's excitability and has seemed therefore to contribute to postoperative comfort. In several instances also, the effects of treating the "chronic complainer" subsequent to operation have seemed helpful.

6. *Toxic goiter.*—Plummer's extensive experience has led him to regard early operation as the most important and indispensable of treatments for the protection of the heart. Even the most successful operative statistics, however, do not claim 100 per cent of cures. In this disease there is room for a method of rest in conjunction with operation as well as where operation is not performed. Indeed a certain reviewer has given rest the place of first importance (Read, 1924).

7. *Disturbances of sleep.*—These topics will be discussed in a later work. Treatment is commonly effective.

8. *Alimentary spasm,* including mucous colitis, colonic spasm, cardiospasm, and other esophagospasm.

9. *Peptic ulcer,* ascribed to nervous tendencies.

10. *Chronic pulmonary tuberculosis.*—Increasing popularity of rest in the treatment of this disease suggests that intensive relaxation may prove of service. The tendency might be to favor the healing of diseased tissue, in effect like pneumothorax. No reference is here intended to any process of immunity.

11. *Organic and functional heart disorders,* requiring rest.

12. *Vascular hypertension.*—This matter has been under investigation since 1921. In essential hypertension there is increased irritability in arterial and arteriolar musculature

and eventual hypertrophy (Lange, 1933; Chase, 1937). Both systolic and diastolic pressure tend to fall as residual tension in skeletal muscles nears zero, as measured by action-potentials (in press). Treatment seems promising even in difficult cases. One patient, for example, with diastolic pressure above 130 at least 6 years previously, now shows marked reduction with normal levels at times. High systolic pressures may yield more readily to prolonged treatment.

Since the importance of prevention in present-day medicine is generally recognized, it is evident that in disorders where the efficacy of relaxation has been established, the method of relaxation should be employed for prophylaxis as well as for treatment.

THE RELATION OF RELAXATION TO VARIOUS OTHER METHODS

The most common means to quiet the nervous system in the current practice of medicine is the use of bromides and other drug sedatives. They produce their results by toxic action on nerve centers or nerve endings, usually thus involving some degree of muscular relaxation.

The present method apparently has three points of advantage over drug sedatives for most chronic conditions: (1) Prolonged use of bromides and other sedatives has not generally proved successful in the treatment of most chronic nervous disorders. While recognizing this, the practicing physician, as is well known, often prescribes them because he does not know what else to do and therefore writes a prescription rather than let the patient go away dissatisfied with nothing at all. The use of relaxation is therefore in my experience to be preferred in most instances. (2) That sedative drugs tend to have deleterious effects is often admitted and has been particularly discussed by E. L. Hunt (1921). The physiologic method obviates such objectionable possibilities. (3) As habituation develops, a sedative tends to lose in effect and larger doses are required. In contrast thereto, habituation increases the efficacy of relaxation.

Our conclusions, therefore, are in the line of current opinion that sedative drugs on the whole have been too widely used in the past. They are of course useful at times, as when in a severe acute malady, complicated by insomnia or other nervous symptoms, a quick and marked depressant action is desired. Even in acute conditions, however, relaxation may often be used alone or jointly with sedatives, if desired. For it is well known to pharmacologists and clinicians that a quiet state favors the action of a sedative and that a larger dose may be required if there is marked activity.

Next to drug sedatives in order of popularity in current medical practice is the employment of several types of measures which may be conveniently considered together. They are largely empirical procedures. Frequently the physician advises a trip to the seashore or other change of scene. If the patient fails to be able to leave the situation that seems to excite him, some effort may be made to ameliorate his position. Again, visits to the theater or the concert or the like may be urged for purposes of amusement, or warm milk may be provided to promote sleep at night. Sometimes hydrotherapy is employed. Evidently the effort is to modify or reduce an undesired mental or emotional content by distraction or inhibition. The foregoing types of treatment, while sometimes helpful in an accessory manner, are notoriously for the most part makeshifts if used as the chief treatment of severe or persistent nervous disorder; they are not founded on the rational analysis and understanding of symptoms and should be relegated to a minor position in the practice of systematic medicine.

Frequently the physician, confronted with a patient who

suffers not alone nervously but from other sources as well, strives in every way to treat these other sources under the assumption that they are the cause of the nervous condition. In some instances this may be true; but it is doubtful if it is characteristically true, as often seems to be believed. Modern specialization perhaps overencourages this tendency, as when the rhinologist seeks in the deflected septum, the throat surgeon in the infected tonsil, the oculist in the astigmatism, the gastroenterologist in the constipation, the general surgeon in the appendix to discover the etiology of the so-called "nervous breakdown." Some truth may lie in the view that each of the preceding sources may, at times at least, add to a sum total of nervous irritation. But "snap diagnoses" on this matter are scarcely to be encouraged; and attention to local pathology, if really present, may improve the patient in this respect yet fail to penetrate to the basis of his nervous disorder. Decision on such matters to be scientific must rest upon carefully controlled investigations. But medical offices are literally overrun with patients who have had septums and astigmatisms corrected, tonsils and appendices removed and often various other operations in addition, but yet continue to suffer from nervous complaints. This is striking evidence that such measures as the foregoing commonly fail to be effective in the therapy of the nervous or neurotic patient.

Of late a few neurologists have strongly advocated what they name "occupational therapy" in the treatment of nervous disorder. If the sufferer has no occupation, he is provided with one; if he already has one, it is changed for another: from this, cure is supposed to follow. Doubtless there are instances where this approach is appropriate. No one will deny that suitable occupation is important for everyone, including the healthy as well as the neurotic. But common experience attests that *many a patient engaged in a highly satisfactory occupation nevertheless suffers from nervous disorder*, and that change of occupation often may prove deleterious as well as practically inadvisable. The clerk advised to take an out-of-door position may in some respects of health profit by the change while in others he may lose. As a rule the neurologist is in no position to regulate the hourly work of the sufferer in such a way as comports with his varying needs. It seems reasonable to conclude that so-called "occupational therapy," commonly founded on no detailed physiological analysis of the patient and relatively uncontrollable in its effects, is, in itself, as a rule superficial and ineffective in meeting serious nervous disorder.

As previously said, progressive relaxation stands in contrast with suggestive procedures: Suggestion often brings its effects with well-known rapidity. Hysterical paralyses and anesthesias may disappear during or after a single session of treatment. Such apparent miracles do not occur with progressive relaxation, for re-education is generally a slow process. On the other hand, as is well known, suggestion often has merely a superficial effect, influencing a particular set of symptoms but failing to remove the basic tendencies and habits which are responsible for the disorder. Nervous re-education in the form of progressive relaxation, although slower in its working, seeks to remove the roots of the disorder and with co-operation, in my experience, appears to succeed.

The physician confronted with a severe psychoneurosis may be glad to use any measures that may be available. This may be true of depressed or maniac states. In this event suggestion may be combined with relaxing methods. It is well known that cyclothymic states are not effectively handled with suggestion alone. However, in other maladies suggestion is to be avoided where feasible, for it throws responsibility on the methods of the physician in place of having the patient do the work himself.

Measures of persuasion, like that of Dubois, are often helpful in psychotherapy. Here the physician deals with the "meanings" of the mental processes of the patient (see chapter xi for discussion of "meanings") and by arguments seeks to select among and modify those processes in order to bring about better adjustment to enrivonment. If the physician succeeds in convincing the patient of the logical absurdity of his fears or other emotions, relaxation may be automatically induced. Such devices also may be combined, if desired, with relaxing methods. The objection to the former is their lack of systematic character. Their physiology is unknown and unmeasured. On the other hand, progressive relaxation has an anatomic basis and can be expressed in terms of physiologic principles. When persuasion proves effective, the result would seem to be due in large measure to relaxation of the disturbance. But the means used to produce relaxation is indirect, the results are therefore likely to be uncertain, and residual tension is not as a rule habitually done away with.

The present method has certain features in contrast with what is commonly known as the "psychoanalysis" of Freud, Rank and others:

1. Relaxation is a physiological method, and in line with the rest of the practice of medicine can be tested by laboratory procedures. As related in earlier chapters, it has been so tested on various points. Modifications thus can be made from time to time as results warrant. There are no symbols, no mysticism and no recondite philosophy. To be sure, several investigators (Jung, Smith and others) have ingeniously applied one type of laboratory test to psychoanalysis. But the test has been made only at one point of a vast field, and

[427]

much remains that is largely speculative and difficult to determine with scientific measures.

2. Analysts assume that the source of nervous disturbances is to be sought in the "unconscious," and that if the proper elements are made conscious, a cure can be effected. According to Freud, to analyze a condition back to the infantile experience is therapy that removes the fundamental cause of the disorder. If, for purposes of discussion, we waive the objections of critics to psychoanalysis and assume that some of the theories may eventually be verified, it still seems doubtful from a physiological standpoint that this method really can remove the cause of serious nervous disturbance completely. For observations of patients who have been analyzed indicate that the nervous individual who suffers from "suppressions" still remains the nervous individual after these difficulties have been removed; still retains the tendency to neurosis, which I have sought to express in terms of nervous hypertension. The tendency to neurosis in such patients both "cured" and uncured is revealed by the admission of analysts that normal persons have the same type of "suppressions" as the neurotic but have a better balance against them: Their theory evidently leads us to look beyond the "suppressed" content for the full explanation of the disorder. It seems reasonable to recall that it is generally agreed that psychic events, conscious and unconscious, have their physiological correlates. This leads me to suggest that the tendency to neurosis, which as we have seen is tacitly admitted by analysts, goes back on the physiological side to what I have termed "neuromuscular hypertension." If so, the need is discerned of supplementing analytic measures with progressive relaxation in the hope of doing away with the disposition of neuromuscular hypertension.

3. Psychoanalysis centers about the motives and desires

that determine conduct, which are said to be generally un-

that determine conduct, which are said to be generally unconscious and to require a special method to disclose them.
Obviously it would be difficult to arrive at strictly scientific
conclusions when dealing with conscious desires, much less
with unconscious ones; and with any particular patient there
is likely to be considerable uncertainty about the authentic
character of motives so disclosed. But waiving these objections, it still remains questionable whether this is a necessary procedure. I have found that it is by no means always necessary to inquire in fine detail into the possible mo-
tivation of disorder or disturbance; for in any event the present result is of neuromuscular character and can therefore
be met on a physiological basis by the foregoing methods.

4. The method of Freud, practiced thoroughly as he rec-
ommends, requires so much of the physician's time that it is
not available for the average patient. Relaxation requires
much less time, can be practiced with several patients simultaneously, with intermittent attention to other matters, and
is therefore generally employable.

5. Analysts commonly agree that their methods are relatively ineffectual with patients aged about fifty-five or more.
This age limit, as previously noted, does not apply to neuromuscular methods.

6. Relaxation evidently applies to a larger field of disorders, since various conditions in internal medicine which
are not psychogenic come within its scope.

A few words may be added concerning what might be
called the "eclectic psychotherapy" which has of late arisen
in nervous therapeutics, descending partly from Freudian
viewpoints and their offshoots and partly from various departments of earlier-day psychotherapy. The personality of
the patient is "analyzed," and he is aided to adjust himself
to his environment. Much here depends upon the personal-

ity of the neurologist and upon the particular theories he adopts. It is impossible to generalize as to the effects produced upon patients by this manner of treatment, but it is obvious that such treatment is an artistic rather than a highly systematic and scientific procedure.

A survey of the foregoing procedures suggests that while each has a place, they are for the most part indirect and uncertain in producing their results. But a more important criticism is the frequent absence of definite analysis in physiological terms: efforts are made to overcome the emotional state of the patient with little clear-cut realization on either side that the aim is to lead him to relax. Accordingly the symptoms and signs of nervous hypertension receive scanty recognition, much less study by laboratory methods; and a sea of speculative neurology results from the absence of a direct and clear-cut approach to the abnormal physiology which underlies the patient's problem. Simplicity and directness of approach are therefore not the only arguments in favor of a physiological method: of no less importance is the fact that its procedures can one by one be put to the tests of experimental investigation, permitting the scientific correction of errors as well as the proper positive development of knowledge in this field.

CONCLUSION

Many more observations doubtless would be required for each therapeutic condition in order to warrant final conclusions. But in the foregoing chapters the general significance of the neuromuscular factor as a cause and element of the symptomatology of various disease processes has seemed evident. This factor has appeared to be in large measure the exaggeration of normal physiological phenomena, including tonus, muscular contraction and emotional experience, so that the present point of view has seemed to throw light upon the physiology and psychology of normal behavior, and consequently also of pathological behavior, opening a wide field for investigation. I have offered evidence that proprioceptive sensations from muscular contraction are important elements in the "stream of consciousness"; that as these diminish with advancing relaxation, not alone kinesthetic but also visual and auditory images become fewer, until for recurrent brief periods, at least, mental activity is so to speak shut off. Corresponding to the excess of voluntary and reflex muscular activities occurring in nervous hypertension, it was submitted that external observation of the individual reveals various evidences of phasic and static contractions of skeletal and visceral musculature. These muscular states are of clinical no less than of laboratory interest. Thus a trifold correspondence (which is at large an identity) is found to exist between the muscular states, the proprioceptive sensory impulses and the conscious processes of the individual, apparently yielding to us a particularly intimate and detailed insight into matters concerning the relation of "the mind and the body."

Assuming that these interpretations are correct, I have sought to show that physiologic methods of reducing reflex activity are not fixed in type, but can be made plastic and flexible to the needs of the moment and therefore possibly can, like dietetics in general medicine, serve a useful purpose under various medical and surgical conditions, for both prophylaxis and treatment. It has been sufficiently emphasized that in the general practice of medicine and surgery, neuromuscular methods may be used along with diet, drugs, operation, vaccines, hydrotherapy, electrotherapy and other therapeutic measures.

[431]

BIBLIOGRAPHY
1690–1929

ACH, N. (1905): *Über d. Willenstätigkeit u. d. Denken* (Göttingen), p. 21.

ACKLAND, W. R. (1925): "Focal Sepsis of Dental Origin Causing Mental Impairment," *Brit. Med. Jour.*, II, 12.

ADRIAN, E. D. (1912): "On the Conduction of Subnormal Disturbances in Normal Nerve," *Jour. of Physiol.*, XLV, 389.

——— (1913): "Wedenskij Inhibition in Relation to the 'All or None' Principle in Nerve," *ibid.*, XLVI, 384–412.

——— (1914): "The 'All or None' Principle in Nerve," *ibid.*, XLVII, 460–74.

——— (1924): "Some Recent Work on Inhibition," *Brain*, XLVII, 399–416.

——— (1926): "Sympathetic Innervation of Striated Muscle," *ibid.*, XLIX, 135.

ADRIAN, E. D., and K. LUCAS (1912): "On the Summation of Propagated Disturbances in Nerve and Muscle," *Jour. of Physiol.*, XLIV, 69.

ADRIAN, E. D., and Y. ZOTTERMAN (1926): "The Impulses Produced by Sensory Nerve-Endings. Part II. The Response of a Single End-Organ," *ibid.*, LXI, 151–71.

AEBLY, J. (1910): *Zur Analyse der physikal. Vorbedingungen des psychogalv. Reflexes mit exsomatischer Stromquelle* (Zurich).

AGDUHR, E. (1919a): "Über die plurisegmentelle Innervation der einzelnen quergestreiften Muskelfasern," *Anat. Anz.*, LII, 273–91.

——— (1919b): "Are the Muscle Fibers of the Extremities Also Innervated Sympathetically," *Proc. Konin. Akad. Wet. Amsterdam*, XXI, 1231–37.

ALLERS, R., and J. BORAK (1920): "Zur Frage des Muskelsinnes," *Wiener Med. Wochenschr.*, LXX, 1166–68.

ALVAREZ, W. C. (1922): *The Mechanics of the Digestive Tract* (New York), p. 10.

AMSLER, C. (1923): "Beiträge zur Pharmakologie des Gehirns," *Arch. f. exper. Path. u. Pharm.*, XCVII, 1–14.

ARONSOHN, E. (1902): "Über den Ort der Wärmebildung in dem durch Gehirnstich erzeugten Fieber," *Virch. Archiv*, CLXIX, 501.

BIBLIOGRAPHY

ASHER, L. (1926): "Zur Frage der sympathetischen Innervation der willkürlichen Muskulatur," *Schweizerische medicinische Wochenschrift*, VII, 537.

ASHER, L., and M. FLACK (1910): "Nachweis der Wirkung eines inneren Sekretes der Schilddrüse und die Bildung desselben unter dem Einfluss der Nerven," *Zentralbl. f. Physiol.* (Leipzig u. Wien), XXIV, 211–13.

BAEYER, H. (1903): "Das Sauerstoffbedürfnis des Nerven," *Zeitschrift f. allg. Physiol.*, II, 169.

BAIN, A. (1855): *Senses and Intellect* (London; 4th ed., New York, 1894).

——— (1859): *The Emotions and the Will* (London).

BALLIF, L.; J. F. FULTON; and E. G. T. LIDDELL (1925): "Observations on Spinal and Decerebrate Knee-Jerks with Special Reference to Their Inhibition by Single Break-Shocks," *Proc. Roy. Soc.*, Ser. B, XCVIII, 589–607.

BARENNE. *See* Dusser de Barenne.

BASTIAN, H. (1890): "On the Symptomatology of Total Transverse Lesions of the Spinal Cord, with Special Reference to the Condition of Various Reflexes," *Medico-Chirurgical Trans.*, XXIII, 151–217.

BAYLISS, W. M., and E. H. STARLING (1899): "The Movements and Innervation of the Small Intestine," *Jour. of Physiol.*, XXIV, 99.

BAZETT, H. C., and W. G. PENFIELD (1922): "A Study of the Sherrington Decerebrate Animal in the Chronic as Well as in the Acute Condition," *Brain*, XLV, 185.

BEARD, G. (1869): "Neurasthenia," *Boston Med. and Surg. Jour.*, LXXX, 217.

——— (1880): *A Practical Treatise on Nervous Exhaustion* (New York).

BEATTIE, F., and T. H. MILROY (1925): "The Rôle of Phosphates in Carbohydrate Metabolism in Skeletal Muscle," *Jour. of Physiol.*, LX, 379–401.

BECHTEREW, W. VON (1908): *Functionen der Nervencentra* (Jena), p. 43.

BECK, A., and G. BIKELES (1912): "Versuche über die gegenseitige funktionelle Beeinflussung von Gross- und Kleinhirn," *Pflüger's Arch.*, CXLIII, 283.

BEEVOR, C. E. (1904): *The Croonian Lectures on Muscular Movements and Their Representation in the Central Nervous System* (London).

BELL, C. (1806): *The Anatomy and Philosophy of Expression* (London).

——— (1830): *The Nervous System of the Human Body* (London).

BENTLEY, I. M. (1927): "Environment and Context," *Amer. Jour. Psychol.*, XXXIX, 54–61.

BERGER, O. (1879): "Über Sehnenreflexe," *Centralblatt f. Nervenheilkunde*, II, 73.

BERITOFF, J. S. (1913): "Zur Kenntnis der Erregungsrhythmik des Nerven- und Muskelsystems," *Zeitschrift f. Biol.*, LXII, 125–201.

———— (1914): "Die tonische Innervation der Skelettmuskulatur und der Sympathicus," *Folia Neurobiol.*, VIII, 421.

BERNHEIM, H. (1896): *Die Suggestion und ihre Heilwirkung*, translated by S. Freud (Leipzig).

BERNIS, W. J., and E. A. SPIEGEL (1925): "Die Zentren der statischen Innervation und ihre Beeinflussung durch Klein- und Grosshirn," *Arb. a. d. Neurol. Inst. Wien. Univer.*, XXVII, 197.

BETHE, A. (1911): "Die Dauerverkürzung der Muskeln," *Pflüger's Arch.* CXLII, 291.

BICKEL, A. (1897): "Über den Einfluss der sensibilen Nerven und de. Labyrinthe auf die Bewegungen der Thiere," *Arch. f. d. ges. Physiol.* LXVII, 299.

BICKEL, H. (1916): *Die wechselseitigen Beziehungen zwischen psychischen Geschein und Blutkreislauf* (Leipzig).

BILLINGS, F.; G. COLEMAN; and W. HIBBS (1922): "Chronic Infectious Arthritis," *Jour. Amer. Med. Assoc.*, LXXVIII, 1097.

BINET, A. (1886): *La psychologie du raisonnement* (Paris), p. 25.

———— (1903): *L'étude expérimentale de l'intelligence* (Paris).

BINET, L. (1918): "Recherches sur le tremblement" (thèse de Paris).

BING, R. (1906): "Experimentelles zur Physiologie der Tactus Spino-Cerebellares," *Arch. f. Anat. u. Physiol.*, *Physiol. Abt.*, p. 250.

BINSWANGER, O. (1896): *Die Pathologie und Therapie der Neurasthenie* (Wien: Gustav Fischer).

BLATZ, W. E. (1925): "The Cardiac, Respiratory, and Electrical Phenomena Involved in Fear," *Jour. Exp. Psychol.*, VIII, 109.

BOEHM, G. (1913): "Über den Einfluss des Nervus Sympathicus und anderer autonomer Nerven auf die Bewegungen des Dickdarmes," *Arch. f. exper. Path. u. Pharmakol.*, LXXII, 5.

BOEHM, R., and F. A. HOFFMANN (1878): "Beiträge zur Kenntniss des Kohlehydratstoffwechsels," *Archiv für experimentelle Path. u. Phar.*, VIII, 271–308.

BOEKE, J. (1913): "Die doppelte (motorische und sympathetische) efferente Innervation der quergestreiften Muskelfasern," *Anat. Anz.*, XLIV, 343–56.

BIBLIOGRAPHY

BOEKE, J., and J. G. DUSSER DE BARENNE (1919): "The Sympathetic Innervation of the Cross-striated Muscle Fibres of Vertebrates," *Proc. Konin. Akad. v. Weten.* (Amsterdam), XXI, 1227–30.

BOER, S. DE (1913*a*): "Die quergestreiften Muskeln erhalten ihre tonische Innervation mittels der Verbindungsäste des Sympathicus (thoraceles autonomes System)," *Folia-Neurobiolog.*, VII, 378–85.

———— (1913*b*): "Über den Skelettmuskeltonus, 2te Mitteilung. Die tonische Innervation der quergestreiften Muskeln bei Warmblütern," *ibid.*, 837–40.

BOIS-REYMOND, E. DU (1877): *Gesammelte Abhandlungen zur allgemeinen Muskel- und Nervenphysik* (Leipzig), II, 5–36.

BORING, E. G. (1920): "The Control of Attitude in Psychophysical Experiments," *Psychol. Rev.*, XXVII, 440–52.

BORUTTAU, H., and F. FRÖHLICH (1904): "Elektropathologische Untersuchungen," *Arch. f. d. ges. Physiol.*, CV, 444.

———— (1909): "Beiträge zur Analyse der Reflexfunction des Rückenmarks mit besonderer Berücksichtigung von Tonus, Bahnung und Hemmung," *Ztschrift f. allgem. Physiol.*, IX, 71.

BOSTROEM, A. (1925): *Abderhalden's Handbuch der biologischen Arbeitsmethoden*, Abt. 6, Teil C, Heft 5, "Hypnose," p. 291.

BOTEZAT, E. (1906): "Die Nervenendapparate in den Mundteilen der Vögel und die einheitliche Endigungsweise der Peripheren Nerven bei den Wirbeltieren," *Zeitschrft. f. wiss. Zool.*, LXXXIV, 228.

BOTTAZZI, P. (1897): "The Oscillations of the Auricular Tonus in the Batrachian Heart with a Theory on the Function of Sarcoplasm in Muscular Tissues," *Jour. of Physiol.*, XXI, 1.

BOWDITCH, H. (1871): Über die Eigenthümlichkeiten der Reizbarkeit, welche die Muskelfasern des Herzens zeigen," *Ber. d. Verhand. Königl. Sächs. Ges. d. Wissen.*, XXIII, 652–89.

BOWDITCH, H., and J. W. WARREN (1890): "The Knee-Jerk and Its Physiological Modifications," *Jour. of Physiol.* XI, 25.

BREMER, F. (1922*a*): "Contribution à l'étude de la physiologie du cervelet. La fonction inhibitrice du paléo-cerebellum," *Arch. Inter. Physiol.*, XIX, 189–226.

———— (1922*b*): "Contribution à l'étude de la physiologie du cervelet," *Comptes rend. de la Soc. de Biol.*, LXXXVI, p. 955.

BREMER, L. (1882): "Über die Endigungen der markhaltigen und marklosen Nerven im quergestreiften Muskel.," *Arch. f. micr. Anatomie*, XXI, 164.

BRONDGEEST, P. Q. (1860): "Über den Tonus der willekeurige Spieren" (dissertation Utrecht). Abstract in *Arch. f. Anat. u. Physiol.*, 1860, pp. 703–4.

BROOKS, H. (1921): In Tice's *Practice of Medicine*, Vol. VII (*Medicine*), chap. x, p. 379.

BROWN, G. (1913): "On Postural and Non-Postural Activities of the Mid-Brain," *Proc. Roy. Soc.*, Ser. B, LXXXVII, 145.

――― (1915): "Studies in the Physiology of Nervous System," *Quar. Jour. of Exper. Physiol.*, IX, 81, 116–17, 140.

BROWN, L., and W. W. TUTTLE (1926): "The Phenomenon of Treppe in Intact Human Skeletal Muscle," *Amer. Jour. Physiol.*, LXXVII, 483.

BRUNSWICK, D. (1924): "The Effects of Emotional Stimuli on the Gastro-Intestinal Tone," *Jour. Comp. Psych.*, IV, 19–80, 225–88.

BRYAN, W. R., and N. HARTER (1897): "Studies in the Physiology and Psychology of the Telegraphic Language," *Phys. Rev.*, IV, 27.

BUBNOFF, N., and R. HEIDENHAIN (1881): "Über Erregungs- und Hemmungsvorgänge innerhalb der motorischen Hirncentren," *Pflüger's Arch.*, XXVI, 137.

BUCHANAN, F. (1908): "The Electrical Response of Muscle to Voluntary, Reflex, and Artificial Stimulation," *Quar. Jour. Exper. Physiol.*, I, 211–42.

BUGBEE, E., and A. E. SIMOND (1926): "The Increase of Voluntary Activity of Ovariectomized Albino Rats Caused by Injections of Ovarian Follicular Hormone," *Endocrinology*, X, 349.

BURTT, H. E., and W. W. TUTTLE (1925): "The Patellar Tendon Reflex and Affective Tone," *Amer. Jour. Psychol.*, XXXVI, 553.

CAJAL, R. S. (1888): "Terminaciones nerviosas en los husos musculares de la rana," *Rivista trim. Histol. Norm. y Patol.*, Vol. I, Fasc. 1.

CALL, A. P. (1902): *Power through Repose* (Boston), pp. 16 ff.

CAMPBELL, A. W. (1905): *Histological Studies on the Location of Cerebral Function* (London: Cambridge University Press).

CAMPBELL, W. R. (1925): *Insulin* (Baltimore), p. 105.

CANNON, W. B. (1902): "The Movements of the Intestines Studied by Means of Röntgen Rays," *Amer. Jour. Physiol.*, VI, 251–77.

――― (1915): *Bodily Changes in Pain, Hunger, Fear, and Rage* (New York and London).

BIBLIOGRAPHY

Cannon, W. B. (1919): "Studies on the Conditions of Activity in Endocrine Glands," *Amer. Jour. Physiol.*, L, 399.

Carey, E. B. (1921): "Studies of the Anatomy and Muscular Action of
the Small Intestine," *International Jour. of Gastro-Enterol.*, Vol. I.

Carlson, A. J. (1913): "Contributions to the Physiology of the Stomach.
II. The Relation between the Contractions of the Empty Stomach and
the Sensation of Hunger. IV. The Influence of the Contractions of
the Empty Stomach in Man on the Vasomotor Center, on the Rate of
the Heart Beat and on the Reflex Excitability of the Spinal Cord,"
Amer. Jour. Physiol., XXXI, 175, 318.

Carlson, A. J.; T. E. Boyd; and J. F. Pearcy (1922*a*): "Studies on the
Visceral Sensory Nervous System," *Arch. Int. Med.*, XXX, 409–33.

———— (1922*b*): "Studies on the Visceral Sensory Nervous System. XIII.
The Innervation of the Cardia and the Lower End of the Esophagus
in Mammals," *Amer. Jour. Physiol.*, LXI, 14–41.

Carlson, A. J., and A. Luckhardt (1914): "The Condition of the Oe-
sophagus during the Periods of Gastric Hunger Contractions," *ibid.*,
XXXII, 129.

Carman, R. D. (1921): *Roentgen Diagnosis of Diseases of Alimentary Canal*
(2d ed., Philadelphia), pp. 529 ff., 595.

Case, J. T. (1921): "Surgical Physiology and Pathology of the Colon
from the X-Ray Standpoint," *New York State Jour. Med.*, XXI, 158.

Church, A., and F. Peterson (1923): *Nervous and Mental Diseases* (9th
ed., Philadelphia and London).

Cobb, S. (1918*a*): "A Note on the Supposed Relation of the Sympathetic
Nerves to Decerebrate Rigidity, Muscle Tone, and the Tendon Reflexes," *Amer. Jour. Physiol.*, XLVI, 478–82.

———— (1918*b*): "Electromyographic Studies of Clonus," *Johns Hopkins
Hosp. Bull.*, XXIX, 247–54.

———— (1925): "Review on the Tonus of Skeletal Muscles," *Phys. Reviews*, V, 518.

Cobb, S.; A. A. Bailey; and P. R. Holtz (1917): "On the Genesis and
Inhibition of Extensor Rigidity," *ibid.*, XLIV, pp. 239–58.

Cobb, S., and A. Forbes (1923): "Electromyographic Studies of Muscular Fatigue in Man," *Amer. Jour. of Physiol.*, LXV, 234–51.

Coman, F. D. (1926): "Observations on the Relation of the Sympathetic
Nervous System to Skeletal Muscle Tonus." *Johns Hopkins Hosp.
Bull.*, XXXVIII, 163.

[437]

COOMBS, H. C., and J. TULGAN (1925): "Some Phenomena of Decerebrate Rigidity," *Amer. Jour. Physiol.*, LXXIV, 314.

CONKLIN, V., and F. L. DIMMICK (1925): "An Experimental Study of Fear," *Amer. Jour. Psychol.*, XXXVI, 96.

CROCQ, M. (1901): "Physiologie et pathologie du tonus musculaire, des réflexes et de la contracture," *Gazette des Hôpitaux*, LXXIV, 850.

———— (1902): *Abstract u. Jahresbericht für Neurol. u. Psychiatrie*, p. 82.

CURSCHMANN-ROSTOCK, H. (1924): "Neurasthenie," *Münchner Med. Wochnschr.*, LXXI, 407–9.

DANIÉLOPOLU, D., and A. CARNIOL (1922): "Recherches sur le tonus des muscles voluntaires," *Rev. Neurol.*, XXIX, pt. 2, 1186–1203.

DANIÉLOPOLU, D.; D. SIMICI; and C. DIMITRIU (1924): "Recherches sur la motilité de l'oesophage chez l'homme," *Jour. de Physiol. et de Path.*, XXII, 596.

DARROW, C. (1927): "Sensory, Secretory and Electrical Changes in the Skin following Bodily Excitation," *Jour. Exp. Psychol.*, X, 197–226.

DARWIN, C. (1890): *The Expression of the Emotions in Man and Animals* (2d ed., London).

DAVIS, H.; A. FORBES; D. BRUNSWICK; and A. McHOPKINS (1926): "Studies of the Nerve Impulse," *Amer. Jour. of Physiol.*, LXXVI, 448–71.

DAVIS, L., and A. KANAVEL (1926): "The Effect of Sympathectomy on Spastic Paralysis of the Extremities," *Jour. Amer. Med. Assoc.*, LXXXVI, 1890.

DEICKE, E. (1922): "Die Beziehungen des vegetativen Nervensystems zum Tonus der Skelettmuskulatur," *Pflüger's Arch.*, CXCIV, 473.

DELSARTE, F. (1894): *Delsarte System of Expression*, by G. Stebbins (New York).

DERCUM, F. X. (1917): *Rest, Suggestion and Other Therapeutic Measures in Nervous and Mental Diseases* (2d ed., Philadelphia).

DERRIEN, E., and H. PIÉRON (1923): "Réaction glycémique émotionnelle," *Jour. de Psych.*, XX, 533.

DESCARTES, R. (1650): *Méditation 6; traité des passions* (Amsterdam).

———— (1664): *Traité de l'homme et de la formation du foetus* (Paris).

DEWEY, J. (1910): *How We Think* (Boston).

DITTLER, R., and H. GÜNTHER (1914): "Über die Aktionsströme menschlicher Muskeln, bei natürlicher Innervation, nach Untersuchungen an gesunden und kranken Menschen," *Arch. f. d. ges. Physiol.*, CLV, 251.

BIBLIOGRAPHY

DODGE, R. (1910): "A Systematic Exploration of a Normal Knee-Jerk, Its Technique, the Form of the Muscle Contraction, Its Amplitude, Its Latent Time, and Its Theory," *Zeits. f. allg. Physiol.*, XII, 1–58.

DODGE, R., and C. M. LOUTTIT (1926): "Modification of the Pattern of the Guinea Pig's Reflex Response to Noise," *Jour. Comp. Psych.*, VI, 267.

DOWNING, A. C.; R. W. GERARD; and A. V. HILL (1926): "The Heat Production of Nerve," *Proc. Roy. Soc.*, Ser. B., C, 223–51.

DRESER, H. (1887): "Ein Vorlesungsversuch, betreffend die Säurebildung bei der Muskelthätigkeit," *Centrlblt. f. Physiol.*, I, 195.

DRYSDALE, H. H. (1924): "Neglected Factors in the Prevention of Apoplexy," *Jour. Amer. Med. Assoc.*, LXXXIII, Part I, 104.

DUBOIS, P. (1909): *The Psychic Treatment of Nervous Disorders* (N. Y. and London).

DUCHENNE, G. B. (1855): *De l'électrisation localisée* (Paris), p. 767.

DUMAS, G. (1900): *La tristesse et la joie* (Paris).

——— (1906): *Le sourire* (Paris).

——— (1923): *Traité de psychologie* (Paris), p. 1.

DUMAS, G., and L. LAVASTINE (1913): *Recherches sur les variations du liquide céphalo-rachidien dans leur rapport avec les émotions* (Paris).

DUNBAR, W. P. (1905): "Aetiologie und specifische therapie des Heufiebers," *Berlin Klin. Wchnschr*, XLII, 942.

DUSSER DE BARENNE, J. G. (1910): "Die Strychninwirkung auf das Zentralnervensystem. I. Die Wirkung des Strychnins auf die Reflextätigkeit der Intervertebralganglia," *Folia Neuro-biol.*, IV, 467–74; "II. Mitt. Zur Wirkung des Strychnins bei lokaler Applikation auf das Rückenmark," *ibid.*, V, 42–58; "III. Mitt.," *ibid.*, pp. 342–59.

——— (1911): "Die electromotorischen Erscheinungen im Muskel bei der reziproken Innervation der quergestreiften Skelettmuskuletur" (prelim. note), *Zentralbl. f. Physiol.*, XXV, 334–36.

——— (1917): "Über die Innervation und den Tonus der quergestreiften Muskeln," *Pflüger's Archiv*, CLXVI, 145.

——— (1920): "Recherches expérimentales sur les fonctions du système nerveux central, faites en particulier sur deux chats dont le néopallium avait été enlevé," *Arch. Neerland. Physiol.*, IV, 31.

——— (1922): "Sur l'excitation artificielle du cervelet," *ibid.*, VII, 112–15.

DUSSER DE BARENNE, J. G., and G. C. E. BURGER (1924): "A Method for the Graphic Registration of Oxygen Consumption and Carbon

Dioxide Output: The Respiratory Exchange in Decerebrate Rigidity," *Jour. Physiol.*, LIX, 17–29.

EBBINGHAUS, H. (1885): *Über das Gedächtnis* (Leipzig).

—— (1897): *Grundzüge der Psychologie* (Leipzig: Veit und Comp.), I, 646.

EINTHOVEN, W. (1918): "Sur les phénomènes électriques du tonus musculaire," *Arch. Neerland.*, II, 489–99.

EINTHOVEN, W., and W. A. JOLLY (1908): "The Form and Magnitude of the Electrical Response of the Eye to Stimulation by Light at Various Intensities," *Quar. Jour. of Exper. Physiol.*, I, 373.

ELLIOTT, T. R., and E. BARCLAY-SMITH (1904): "Antiperistalsis and Other Muscular Activities of the Colon," *Jour. Physiol.*, XXXI, 272–304.

EMBDEN, G. (1924): "Untersuchungen über der Verlauf den Phosphorsäure- und Milchsäurebildung bei der Muskeltätigkeit," *Klin. Wchnschr.*, III, 1393–96.

—— (1925): "Chemismus der Muskelkontraktion und Chemie der Muskulatur," *Bethe's Handbuch Norm. u. Path. Physiol.*, VIII, 369–475.

EMBDEN, G., and C. HAYMANN (1924): "Über die Bedeutung von Ionen für die Muskelfunktion. IV. Über fermentative Lactacidogensynthese unter dem Einfluss von Ionen. I. Mitteilung," *Zeitschr. f. Physiol. Chem.*, CXXXVII, 154–75.

EMBDEN, G., and H. LANGE (1923a): "Muskelatmung und Sarkoplasma," *ibid.*, CXXV, 258–83.

—— (1923b): "Der Eintritt von Chlorionen in den arbeitenden Muskel," *ibid.*, CXXX, 350–73.

EMBDEN, G., and E. LEHNARTZ (1924): "Über die Bedeutung von Ionen für die Muskelfunktion. I. Die Wirkung verschiedener Anionen auf den Lactacidogenwechsel in Froschmuskelbrei," *ibid.*, CXXXIV, 243–75.

EPPINGER, H., and L. HESS (1915): *Vagotonia*, translated by W. M. Kraus and S. E. Jelliffe (New York).

ERB, W. (1875): "Über Sehnenreflexe bei Gesunden und bei Rückenmarkskranken," *Arch. f. Psychiatrie*, V, 792.

ERLANGER, J.; G. H. BISHOP; and H. S. GASSER (1926): "Experimental Analysis of the Simple Action Potential Value in Nerve by the Cathode Ray Oscillograph," *Amer. Jour. of Physiol.*, LXXVIII, 537–73.

BIBLIOGRAPHY

ERLANGER, J., and H. S. GASSER (1922): "A Study of the Action Currents of the Nerve with the Cathode Ray Oscillograph," *Amer. Jour. Physiol.*, LXII, 496.

EVANS, C. L. (1923): "Studies on the Physiology of Plain Muscle," *Jour. of Physiol.*, LVIII, 22–32.

—— (1926): *Recent Advances in Physiology* (Phil.).

EVANS, C. L., and S. W. F. UNDERHILL (1923): "Studies on the Physiology of Plain Muscle," *Jour. of Physiol.*, LVIII, 1–22.

EWALD, J. R. (1892): *Physiol. Untersuchungen u. d. Endorgan des Nervus Octavus* (Wiesbaden).

EXNER, S. (1882): "Zur Kenntniss von der Wechselwirkung von der Erregungen im Centralnervensystem, *Pflüger's Arch.*, XXVIII, 487.

—— (1877): "In welcher Weise tritt die negative Schwankung durch das Spinalganglion ?" *Arch. f. Physiol.*, pp. 567–70.

—— (1894): *Entwurf z. einer physiologischen Erklärung der psychischen Erscheinungen* (Wien).

FERNBERGER, S. (1921): "Experimental Study of the Stimulus Error," *Jour. of Exper. Psychol.*, IV, 63–76.

FÉRÉ, C. (1888): "Note sur des modifications de la résistance électrique dans l'influence des excitations sensorielles et des émotions," *Compte Rend. de la Soc. de Biol.*, XL, 217.

—— (1900): *Sensation et mouvement* (Paris).

FERNALD, M. R. (1912): "The Diagnosis of Mental Imagery," *Psychol. Rev.*, Mon. Supp. No. 14, p. 128.

FICK, A. (1903–6): *Gesam. Schriften* (Würzburg: O. Stabel), III, 109.

FILLIÉ, H. (1908): "Studien über die Erstickung u. Erholung der Nerven in Flüssigkeiten," *Zeitschrft. f. allg. Physiol.*, VIII, 492.

FLETCHER, W. M., and F. G. HOPKINS (1907): "Lactic Acid in Amphibian Muscle," *Jour. of Physiol.*, XXXV, 247.

FLOURENS, J. P. M. (1842): *Recherches expérimentales sur les propriétés et les fonctions du système nerveux dans les animaux vertébrés* (Paris).

FOIX, C. (1924): "Sur le tonus et les contractures," *Rev. Neurologique*, XXXI, Part II, 1.

FORBES, A. *See* Cobb and Forbes.

FORBES, A.; C. J. CAMPBELL; and H. B. WILLIAMS (1924): "Electrical Records of Afferent Nerve Impulses from Muscular Receptors," *Amer. Jour. Physiol.*, LXIX, 283–303.

FORBES, A., and M. CATTELL (1924): "Electrical Studies in Mammalian Reflexes. IV. The Crossed Extension Reflex," *ibid.*, LXX, 140–73.

FORBES, A., and S. COBB (1926): "Physiology of Sympathetic Nervous System in Relation to Certain Surgical Problems," *Jour. Amer. Med. Assoc.*, LXXXVI, 1884–86.

FORBES, A.; S. COBB; and C. J. CATTELL (1923): "Electrical Studies in Mammalian Reflexes. III. Immediate Changes in the Flexion Reflex after Spinal Transection," *Amer. Jour. Physiol.*, LXV, 30.

FORBES, A., and W. C. RAPPLEYE (1917): "The Effect of Temperature Changes on Rhythm in the Human Electromyogram," *ibid.*, XLII, 228–55.

FORBES, A.; L. H. RAY; and F. R. GRIFFITH, JR. (1923): "The Nature of the Delay in the Response to the Second of Two Stimuli in Nerve and in Nerve-Muscle Preparation," *ibid.*, LXVI, 553–677.

FORBES, A.; L. R. WHITAKER; and J. F. FULTON (1927): "The Effect of Reflex Excitation on the Response of a Muscle to Stimulation through Its Motor Nerve," *ibid.*, LXXXII, 693–716.

FOSTER, N. B. (1927): "Psychic Factors in the Course of Cardiac Disease," *Jour. Amer. Med. Assn.*, LXXXIX, 1017.

FRANZ, S. I. (1923): *Nervous and Mental Re-education* (New York).

FREUD, S. (1916): *The Interpretation of Dreams* (3d ed., New York and London).

FREUND, H., and R. STRASMANN (1912): "Zur Kenntnis des nervösen Mechanismus der Wärmeregulation," *Arch. f. Exper. Path. u. Phar.*, LXIX, 12.

FREUSBERG, A. (1874): "Reflexbewegungen beim Hunde," *Pflüger's Arch.*, IX, 358.

FRITSCH, G., and E. HITZIG (1870): "Über die elektrische Erregbarkeit des Grosshirns," *Archiv. f. Anat. und Physiol. u. wissenschaftliche Med.*, p. 300.

FRÖHLICH, A., and H. H. MEYER (1912): "Untersuchung über die Aktionsströme anhaltend verkürzter Muskeln," *Zbl. Physiol.*, XXVI, 269–77.

FRÖHLICH, F. W. (1905): "Über die scheinbare Steigerung der Leistungsfähigkeit des quergestreiften Muskels im Beginn der Ermüdung (Muskeltreppe), der Kohlensäurewirkung und der Wirkung anderer Narkotika (Aether, Alkohol)," *Zeitschrft. f. allg. Physiol.*, V, 288–316.

BIBLIOGRAPHY

FRÖHLICH, F. W. (1909): "Beiträge zur Analyse der Reflexfunction des Rückenmarks mit besonderer Berücksichtigung von Tonus, Bahnung und Hemmung," *Zts. f. allgemeine Physiol.*, Vol. IX.

FULTON, J. F. (1925): "Fatigue and Plurisegmental Innervation of Individual Muscle Fibers," *Proc. Roy. Soc.*, Ser. B, XCVIII, 493.

———— (1925): "On the Summation of Contractions in Skeletal Muscle," *Amer. Jour. Physiol.*, LXXV, 211.

———— (1926a): *Muscular Contraction* (Baltimore).

———— (1926b): "The Influence of Temperature upon Muscular Activity," *Quar. Jour. Exper. Physiol.*, Vol. XVI.

FULTON, J. F., and E. G. T. LIDDELL (1925): "Electrical Responses of Extensor Muscles During Postural (Myotatic) Contraction," *Proc. Roy. Soc.*, Ser. B, XCVIII, 577–89.

FULTON, J. F., and J. PI-SUÑER (1928): "A Note Concerning the Probable Function of Various Afferent End-Organs in Skeletal Muscle," *Amer. Jour. Physiol.*, LXXXIII, 554.

FURUSAWA, K. (1925): "Muscular Exercises, Lactic Acid, and the Supply and Utilization of Oxygen. Parts IX, X," *Proc. Roy. Soc.*, Ser. B, XCVIII, 65–76, 287–89.

GALTON, F. (1883): *Inquiries into Human Faculty and Its Development* (New York).

GERARD, R. W. (1927a): "The Two Phases of the Heat Production of Nerve," *Jour. of Physiol.*, LXII, 349–63.

———— (1927b): "Studies on the Nerve Metabolism, I. The Influence of Oxygen Lack on Heat Production and Action Current," *ibid.*, LXIII, 280–98.

———— (1927c): "Studies on Nerve Metabolism, II. Respiration in Oxygen and Nitrogen," *Amer. Jour. Physiol.*, LXXXII, 381–404.

GERARD, R. W.; A. V. HILL; and Y. ZOTTERMAN (1927): "The Effect of Frequency of Stimulation on the Heat Production of Nerve," *Jour. of Physiol.*, LXIII, 130–43.

GIACOMINI, E. (1898): "Sui Fusi Neuro-Muscolari dei Sauropsidi," *Atti. Accad. Fisocrit.*, IX, 215.

GILDEMEISTER, M. (1913): "Über die im tierischen Körper bei elektrischer Durchströmung entstehenden Gegenkräfte," *Pflüger's Arch.*, CXLIX, 389–401.

———— (1915): "Der sogenannte psychogalvanische Reflex und seine physikalisch-chemische Deutung," *ibid.*, CLXII, 489–506.

[443]

GILDEMEISTER, M. (1919): "Über elektrischen Widerstand, Kapazität und Polarisation der Haut," *ibid.*, CLXXVI, 84–105.

GLEY, E. (1903): *Études de psychologie physiologique et pathologique* (Paris).

GOLDSCHEIDER, A. (1889): "Über den Muskelsinn und die Theorie der Ataxie," *Zeitschrft. f. klin. Med.*, XV, 109 ff.

—— (1898): *Die Bedeutung der Reize für Pathologie und Therapie im Lichte der Neuronlehre* (Leipzig), p. 16.

GOLLA, F. L. (1921): "Croonian Lectures on the Objective Study of Neurosis," *Lancet*, Part II, pp. 265 ff.

GOLLA, F. L., and J. HETTWER (1922): "The Influence of Various Conditions on the Time Relations of Tendon Reflexes in the Human Subject," *Proc. Roy. Soc.*, Ser. B, XCIV, 92–98.

—— (1924): "A Study of the Electromyogram of Voluntary Movement," *Brain*, XLVII, 57–69. Discussion of paper by Gordon Holmes, Cobb, etc., *ibid.*, pp. 101–2.

GOLTZ, F. (1892): "Der Hund ohne Grosshirn," *Pflüger's Arch.*, LI, 570–614.

GOLTZ, F., and A. FREUSBERG (1874): "Über die Funktionen des Lendenmarks des Hundes," *ibid.*, VIII, 460.

GOTCH, F. (1902): "The Submaximal Electrical Response of Nerve to a Single Stimulus," *Jour. of Physiol.*, XXVIII, 395.

GOTCH, F., and G. J. BURCH (1899): "The Electrical Response of Nerve to Two Stimuli," *ibid.*, XXIV, 410–26.

GOTTLIEB, R. (1891): "Calorimetrische Untersuchungen über die Wirkungsweise, des Chinins und Antipyrins," *Arch. f. Exp. Path.*, XXXVIII, 167.

GRABOWER (1902): "Über Nervenendigungen im menschlichen Muskel," *Arch. f. Micr. Anat.*, erstes Heft, LX, 1.

GREGOR, A., and S. LOEWE (1912): "Zur Kenntnis der physikalischen Bedingungen des psychogalvanischen Reflexphänomens," *Ztschrft. f. d. ges. Neurol. u. Psychiat.*, XII, 411–46.

GRÜNBAUM, A. S. F., and C. S. SHERRINGTON (1902): *Reports of the Thompson, Yates and Johnson Laboratories*, IV, 351.

—— (1903): *ibid.*, V, 55.

HALL, M. (1833): "On the Reflex Function of the Medulla Oblongata and Medulla Spinalis," *Phil. Trans. Roy. Soc.* (London), CXXIII, 635.

BIBLIOGRAPHY

HANSEN, K., and P. HOFFMANN (1922): "Weitere Untersuchungen über die Bedeutung der Eingenreflexe für unsere Bewegungen," *Zeitschr. f. Biol.*, LXXV, 293–304.

HANSEN, K.; P. HOFFMANN; and V. VON WEIZÄCKER (1922): "Der Tonus des quergestreiften Muskels," *ibid.*, LXXV, 121–54.

HARTLEY, D. (1749): *Observations on Man* (London), Part I, chap. 1, par. 2.

HARTLINE, H. K. (1925): "The Electrical Response to Illumination of the Eye in Intact Animals, Including the Human Subject; and in Decerebrate Preparations," *Amer. Jour. Physiol.*, LXXIII, 600.

HARTMANN, F. A.; H. A. McCORDOCK; and M. M. LODER (1923): "Conditions Determining Adrenal Secretion," *ibid.*, LXIV, 1.

HARTREE, W., and A. V. HILL (1921): "The Nature of the Isometric Twitch," *Jour. of Physiol.*, LV, 389.

HEAD, H. (1921): "Release of Function in the Nervous System," *Proc. Roy. Soc.*, Ser. B, XCII, 184.

HEAGEY, F. W. (1924): "Symptoms of Colonic Disturbances," *Jour. Radiol.*, V, 261.

HEIDBREDER, E. (1924): "An Experimental Study of Thinking," *Arch. of Psychol.*, XI, No. 73.

HEIDENHAIN, R. (1883): "Über pseudomotorische Nervenwirkungen," *Arch. f. Physiol. Suppl.*, p. 133.

HERING, H. E. (1898): "Beitrag zur experimentellen Analyse coordinirter Bewegungen," *Pflüger's Archiv*, LXX, 559–623.

———— (1902): "Die intracentralen Hemmungsvorgänge in ihrer Beziehung zur Skelettmuskulatur," *Ergebn. der Physiol.*, XII, 513.

HERZEN, A. H. (1864): *Expériences sur les centres modérateurs de l'action réflexe* (Florence: Bettini; also Turin: Loescher), p. 72.

HILL, A. V. (1911): "The Position Occupied by the Production of Heat in the Chain of Processes Constituting a Muscular Contraction," *Jour. of Physiol.*, XLII, 1.

———— (1912): "The Absence of Temperature Changes during the Transmission of a Nervous Impulse," *ibid.*, XLIII, 433.

———— (1926): *Muscular Activity* (Baltimore), p. 67.

HILL, A. V.; C. N. H. LONG; and H. LUPTON (1924): "Muscular Exercise, Lactic Acid, and the Supply and Utilization of Oxygen. Parts I–III." *Proc. Roy. Soc.*, Ser B, XCVI, 438–75; "Parts IV–VI," *ibid.*, XCVII, 84–138. For "Parts VII————," see Furusawa.)

HINES, H. J. G.; L. N. KATZ; and C. N. H. LONG (1925): "Lactic Acid in Mammalian Cardiac Muscle. Part II, The Rigor Mortis Maximum and the Normal Glycogen Content," *Proc. Roy. Soc.*, Ser. B, XCIX, 20–26.

HINES, M. (1927): "Nerve and Muscle," *Quar. Rev. of Biol.*, II, 149.

HOFFMANN, P. (1914): "Über die doppelte Innervation der Krebsmuskeln," *Ztschrft. f. Biol.*, LXIII, 411–42.

——— (1918): "Über die Beziehungen der Sehnenreflexe zur willkürlichen Bewegung und zum Tonus," *ibid.*, LXVIII, 351–70.

——— (1921): "Lassen sich im quergestreiften Muskel des normalen Erscheinungen nachweisen, die auf innere Sperrung deuten," *ibid.*, LXXIII, 247–62.

——— (1922): *Untersuchungen über die Eigenreflexe (Sehnenreflexe) menschlicher Muskeln* (Berlin).

HOFMANN, F. B. (1904): "Studien über den Tetanus," *Pflüger's Arch.*, CIII, 291–346.

HOLMES, G. (1917): "The Symptoms of Acute Cerebellar Injuries Due to Gunshot Injuries," *Brain*, XL, 461–534.

——— (1922): "The Clinical Symptoms of Cerebellar Disease and Their Interpretation," *Lancet*, CCII, 1177–82, 1231–37; CCIII, 59, 111, 115.

HORSLEY, V., and R. H. CLARKE (1908): "The Structure and Functions of the Cerebellum Examined by a New Method," *Brain*, XXXI, 45–124.

HOWELL, W. H. (1925): *A Text Book of Physiology* (Philadelphia and London), p. 155.

HUBER, G. C., and L. M. A. DeWITT (1897): "A Contribution on the Motor Nerve-Endings and on the Nerve-Endings in the Muscle Spindles," *Jour. Comp. Neurol.*, VII, 169–230.

HUGGETT, A. S., and J. MELLANBY (1925): "The Influence of the Sympathetic, Parasympathetic and Somatic Systems of Nerves on the Tonus of Muscle in the Intact and Decerebrate Cat," *Jour. Physiol.*, LX, 8.

HUME, D. (1739): *Treatise on Human Nature* (London), 3 vols. See especially Bk. I, Part I, par. 4.

——— (1740): *An Enquiry Concerning Human Understanding* (London), Sec. II, p. 14.

HUNT, E. L. (1921): "The Deleterious Effects of the Bromide Treatment in the Diseases of the Nervous System," *Med. Record*, C, 103.

BIBLIOGRAPHY

HUNT, J. R. (1921): "Dyssynergia Cerebellaris Myoclonica. Primary Atrophy of the Dentate System," *Brain*, XLIV, 490.

———— (1927): "Nature and Treatment of Psychic and Emotional Factors in Disease," *Jour. of Amer. Med. Assoc.*, LXXXIX, 1014.

HUNTER, J. I. (1924): "The Influence of the Sympathetic Nervous System in the Genesis of the Rigidity of Striated Muscle in Spastic Paralysis," *Surg., Gyn. and Obstet.*, XXXIX, 721–43.

———— (1925): "The Sympathetic Innervation of Striated Muscles," *Brit. Med. Jour.*, January 31–February 28, Pt. I, pp. 197–201, 251–56, 298–301, 350–53, 398–400.

INGBERT, C. (1903a): "An Enumeration of the Medullated Nerve Fibers in the Dorsal Roots of the Spinal Nerves of Man," *Jour. of Comp. Neurol.*, XIII, 53.

———— (1903b): "On the Density of the Cutaneous Innervation in Man," *ibid.*, p. 209.

ISHIKAWA, H. (1910): "Über die scheinbare Bahnung," *Zeitschrft. f. d. ges. Physiol.*, XI, 150.

JACKSON, H. (1884): "The Croonian Lectures on Evolution and Dissolution of the Nervous System," *Brit. Med. Jour.*, Pt. I, pp. 591–93, 660–63, 703–7.

JACOBSON, E. (1911): "On Meaning and Understanding," *Amer. Jour. Psychol.*, XXII, 553–57.

———— (1911b): "Experiments on the Inhibition of Sensations," *Psychol. Rev.*, XVIII, 24–53.

———— (1912): "Further Experiments on the Inhibition of Sensations," *Amer. Jour. of Psychol.*, XXIII, 345–69.

———— (1920a): "Use of Relaxation in Hypertensive States," *N.Y. Med. Jour.*, CXI, 419.

———— (1920b): "Reduction of Nervous Irritability and Excitement by Progressive Relaxation," *Jour. of Nerv. and Ment. Dis.*, LIII, 282.

———— (1921a): "Treatment of Nervous Irritability and Excitement," *Ill. Med. Jour.*, XXXIX, 243.

———— (1921b): "The Use of Experimental Psychology in the Practice of Medicine," *Jour. Amer. Med. Assoc.*, LXXVII, 342–47.

———— (1924a): "The Physiology of Globus Hystericus," *ibid.*, LXXXIII, 911–13.

———— (1924b): "The Technic of Progressive Relaxation," *Jour. of Nerv. and Ment. Dis.*, LX, 568–78.

JACOBSON, E. (1925a): "Progressive Relaxation," *Amer. Jour. Psychol.*, XXXVI, 73–87.

——— (1925b): "Voluntary Relaxation of the Esophagus," *Amer. Jour. of Physiol.*, LXXII, 387–94.

——— (1926): "Response to a Sudden Unexpected Stimulus," *Jour. of Exp. Psychol.*, IX, 19–25.

——— (1927a): "Spastic Esophagus and Mucous Colitis," *Arch. of Int. Med.*, XXXIX, 433–45.

——— (1927b): "Action Currents from Muscular Contractions during Conscious Processes," *Science*, LXVI, 403.

——— (1928a): "Quantitative Recording of the Knee-Jerk by Angular Measurement," *Amer. Jour. of Physiol.*, LXXXVI, 15–19.

——— (1928b): "Differential Relaxation during Reading, Writing, and Other Activities as Tested by the Knee-Jerk," *ibid.*, pp. 675–93.

JACOBSON, E., and A. J. CARLSON (1925): "The Influence of Relaxation upon the Knee-Jerk," *ibid.*, LXXIII, 324–28.

JAMES, W. (1890): *Principles of Psychology* (New York), Vol. I, p. 158; Vol. II, chap. xxv.

JANET, P. (1905): "Mental Pathology," *Psychol. Rev.*, XII, 101.

JAQUET, A. (1922): "Ueber nervöse und psychische Störungen bei Herzkranken, Pt. I," *Schweizer. Med. Wochenschrft.*, LII, 246.

JENDRÁSSIK, E. (1883): "Beiträge zur Lehre von den Sehnenreflexen," *Deutsch. Archiv f. klin. Med.*, XXXIII, 177.

JOHNSON, C. A. (1927): "Studies on the Knee-Jerk. I. A Simple, Dependable and Portable Knee-Jerk Apparatus for Use on Higher Mammals and Man," *Amer. Jour. Physiol.*, LXXXII, 75–83.

JOHNSON, C. A., and A. J. CARLSON (1927): "Studies on the Knee-Jerk. V. The Effect of Hunger Contractions upon the Knee-Jerk," *ibid.*, LXXXIV, 189.

JOHNSON, C. A., and A. B. LUCKHARDT (1928): "Studies on the Knee-Jerk. III. The Effects of Raised Intrapulmonic Pressure upon the Knee-Jerk, Arterial Blood Pressure and the State of Consciousness," *ibid.*, LXXXIV, 642–52.

JOLLY, W. A. (1911): "On the Time Relations of the Knee-Jerk and Simple Reflexes," *Quar. Jour. Exp. Physiol.*, IV, 67–87.

JUNG, C. (1919): *Studies in Word Association* (New York).

KAFKA, G. (1921): *Die Grundlagen der psychischen Entwicklung* (München), p. 73.

KAHN, M. (1927): "Present Status of Curability of Bronchial Asthma," *Arch. of Int. Med.*, XXXIX, 621.

BIBLIOGRAPHY

KATO, G. (1924): *The Theory of Decrementless Conduction in Narcotized Region of Nerve* (Tokyo: Nankōdō).

KERSCHNER, L. (1888): "Bemerkungen über ein besonderes Muskelsystem im willkürlichen Muskel," *Anat. Anz.*, III, 126.

KING, C. E. (1924): "Studies on Intestinal Inhibitory Reflexes," *Amer. Jour. Physiol.*, LXX, 183–93.

KING, W. T. (1927): "Observations on the Rôle of the Cerebral Cortex in the Control of the Postural Reflex," *ibid.*, LXXX, 311.

KLAUDER, J. V. (1925): "The Cutaneous Neuroses," *Jour. Amer. Med. Assoc.*, LXXXV, 1683.

KLEITMAN, N., and G. CRISLER (1927): "A Quantitative Study of a Salivary Conditioned Reflex," *Amer. Jour. of Physiol.*, LXXIX, 571–614.

KRAFFT-EBING, R. (1895): *Nervosität und neurasthenische Zustände* (Wien).

KRONECKER, H. (1871): "Über die Ermüdung und Erholung der quergestreiften Muskeln," *Berichte der Verhandlungen d. k. Sächsischen Gesell. der Wiss. zu Leipz.*, XXIII, 718.

KRONECKER, H., and S. MELTZER (1883): "Der Schluckmechanismus, seine Erregung und seine Hemmung," *Arch. f. Physiol.*, Suppl. B, p. 355.

KRONER, K. (1907): "Über Bahnung der Patellarreflexe," *Neurol. Centralbl. Leipz.*, XXVI, 700–703.

KÜHNE, W. (1863): "Die Muskelspindeln, ein Beitrag zur Lehre von der Entwickelung der Muskeln und Nervenfasern," *Virchow's Arch.*, XXVIII, 528.

KÜLPE, O. (1895): *Outlines of Psychology*, translated by E. B. Titchener (New York), pp. 187, 446.

KUHLMAN, F. (1906): "Analysis of Memory Consciousness," *Psychol. Rev.*, XIII, 316.

KULCHITSKY, N. (1924): "Nerve-Endings in Muscles," *Jour. Anat.*, LVIII, 152.

KUNDE, M. (1927): "Experimental Hyperthyroidism," *Amer. Jour. Physiol.*, LXXXII, 195–215.

KUNTZ, A., and A. H. KERPER (1924a): "The Sympathetic Innervation of Voluntary Muscles," *Proc. Soc. Exper. Biol. and Med.*, XXII, 23–24; see also, experimental data regarding the "Sympathetic Innervation of Skeletal Muscles and Its Rôle in Muscle Tonus," *Proc. Amer. Assoc., Anat. Rec.*, XXXII, 214.

[449]

KUNTZ, A., and A. H. KERPER (1924*b*): "Experimental Observations on the Functional Significance of the Sympathetic Innervation of Voluntary Muscles," *Proc. Soc. Exper. Biol. and Med.*, XXII, 25–28.

——— (1925*a*): "The Sympathetic Innervation of Voluntary Muscles and Its Effect on Their Contractile Power and Resistance to Fatigue," *Anat. Rec.*, XXIX, 366.

——— (1925*b*): "Posture of the Extremities as Related to the Sympathetic Nervous System," *ibid.*, pp. 366–67.

——— (1926): "An Experimental Study of Tonus in Skeletal Muscles as Related to the Sympathetic Nervous System," *Amer. Jour. Physiol.*, LXXVI, 121–44.

LANDIS, C. (1924): "Studies of Emotional Reactions," *Jour. of Comp. Psychol.*, IV, 447.

——— (1925): "Studies of Emotional Reactions. IV. Metabolic Rate," *ibid.*, V, 188–203.

——— (1926): "Studies of Emotional Reactions. V. Severe Emotional Upset," *ibid.*, VI, 221.

LANDIS, C., and GULLETTE, R. (1925): "Studies of Emotional Reactions. III. Systolic Blood Pressure and Inspiration-Expiration Ratios," *ibid.*, V, 221–24.

LANGE, C. (1887): *Über Gemüthsbewegungen*, translated by Kurella (Leipzig).

LANGELAAN, J. W. (1915): "On Muscle Tonus," *Brain*, XXXVIII, 235.

LANGLEY, J. N., and H. K. ANDERSON (1895): "On the Innervation of the Pelvic and Adjoining Viscera. I. The Lower Portion of the Intestine," *Jour. Physiol.*, XVIII, 67–105.

LANGWORTHY, C. F., and H. C. BARROTT (1920): "Energy Expenditure in Household Tasks," *Amer. Jour. Physiol.*, LII, 400.

LANGWORTHY, O. R. (1926): "Abnormalities of Posture and Progression in the Pigeon following Experimental Lesions of the Brain," *Amer. Jour. Physiol.*, LXXVIII, 34.

——— (1928): "The Control of Postures by the Central Nervous System," *Physiol. Rev.*, VIII, 151–90.

LAPICQUE, L. and M. (1912): "Mesure analytique de l'excitabilité réflexe," *C. R. Soc. Biol.*, LXXII, 871–74.

LAUGHTON, N. B. (1924): "Studies on the Nervous Regulation of Progression in Mammals," *Amer. Jour. Physiol.*, LXX, 358.

LAVATER, J. (1804): *Essays on Physiognomy* (2d ed., London).

BIBLIOGRAPHY

LEE, F. S. (1907a): "The Cause of the Treppe," *Amer. Jour. Physiol.*, XVIII, 267.

——— (1907b): "The Action of Normal Fatigue Substances on Muscle," *ibid.*, XX, 170–79.

LEE, M. A. M., and N. KLEITMAN (1923): "Studies on the Physiology of Sleep," *ibid.*, LXVII, 147.

LEHMANN, A. (1899): *Die körperlichen Äusserungen psychischer Zustände* (Leipzig), Vol. I.

LEWINSKI, L. (1879): "Über den Kraftsinn," *Virchow's Arch.*, LXXVII, 145.

LEYTON, A. S. F., and C. S. SHERRINGTON (1916): "Observations on the Excitable Cortex of the Chimpanzee, Orang-Utan, and Gorilla," *Quar. Jour. Exper. Physiol.*, XI, 135–227.

LIDDELL, E. G. T., and C. S. SHERRINGTON (1924): "Reflexes in Response to Stretch (Myotatic Reflexes)," *Proc. Roy. Soc.*, Ser. B, XCVI, 212–42.

——— (1925a): "Further Observations on Myotatic Reflexes," *ibid.*, XCVII, 267–83.

——— (1925b): "Recruitment and Some Other Features of Reflex Inhibition," *ibid.*, pp. 488–518.

LILJESTRAND, G., and R. MAGNUS (1919): "Über die Wirkung des Novokains auf den normalen und den tetanusstarren Skelettmuskel und über die Entstehung der lokalen Muskelstarre beim Wundstarrkrampf," *Pflüger's Archiv*, CLXXVI, 168–208.

LOCKE, J. (1690): *Essay Concerning Human Understanding* (London).

LODHOLZ, E. (1913): "Über die Gültigkeit des 'Alles-oder-Nichts-Gesetzes' für die markhaltige Nervenfaser," *Ztschrift. f. allg. Physiol.*, XV, 269.

LÖWENFELD, L. (1894): *Pathologie und Therapie der Neurasthenie und Hysterie* (Wiesbaden).

LÖWENTHAL, M., and V. HORSLEY (1897): "On the Relations between the Cerebellar and Other Centers, with Special Reference to the Action of the Antagonistic Muscles," *Proc. Roy. Soc.*, LXI, 21.

LOMBARD, W. P. (1887): "The Variations of the Normal Knee-Jerk, and Their Relation to the Activity of the Central Nervous System," *Amer. Jour. Psychol.*, I, 5.

LOONEY, J. M. (1924): "The Relation between Increased Muscle Tension and Creatin," *Amer. Jour. Physiol.*, LXIX, 638–44.

PROGRESSIVE RELAXATION

Lucas, K. (1912): "Croonian Lecture. The Process of Excitation in Nerve and Muscle," *Proc. Roy. Soc.*, Ser. B, LXXXV, 495–524.

——— (1917): *The Conduction of the Nervous Impulse* (London), chap. xiii.

Luciani, L. (1891): *Il Cervelletto. Nuovi Studi di fisiologia normale e patologica* (Firenze: Monnier).

——— (1915): *Human Physiology*, translated by Welby (London), III, 419–60.

Luckhardt, A. B., and C. A. Johnson (1928): "Studies on the Knee-Jerk. VI. An Additional Note on the Effect of Raised Intrapulmonic Pressure on the Knee-Jerk, Heart Rate and State of Consciousness," *Amer. Jour. of Physiol.*, LXXXIV, 453–60.

Lunt, L. K., and R. A. F. Riggs (1924): "The Ophthalmologist and the Psychoneuroses," *Jour. Amer. Med. Assoc.*, LXXXIII, Part II, 1968.

McCarrison, R. (1919): "The Pathogenesis of Deficiency Disease," *Indian Jour. Med. Research*, VII, 185.

McLester, J. S. (1927): "Psychic and Emotional Factors in Their Relation to Disorders of the Digestive Tract," *Jour. Amer. Med. Assoc.*, LXXXIX, 1019.

Magnus, R., and A. de Kleijn (1912): "Die Abhängigkeit des Tonus der Extremitätenmuskeln von der Kopfstellung," *Pflüger's Arch.*, CXLV, 455 ff.

——— (1915): "Weitere Beobachtungen über Hals- und Labyrinthreflexe auf die Gliedermuskeln des Menschen," *Arch. f. d. ges. Physiol.* (Bonn), CLX, 429–44.

Magnus, R., and G. G. J. Rademaker (1923): "Die Bedeutung des Roten Kernes für die Körperstellung (Vorl. Mitt.)," *Schweiz. Arch. f. Neurol. u. Psych.*, XIII, 408–11.

Maloney, W. J. (1918): *Locomotor Ataxia* (New York and London).

Marañón, G. (1922): *The Emotive Action of Epinephrin* (Libro en Honor de S. Ramón y Cajal), II, 291–310; abstracted by Nonidez in *Arch. of Neuro. and Psych.*, XII (1924), 566.

Marbe, K. (1901): *Experimentelle Untersuchungen über des Urteils* (Leipzig).

Marburg, O. (1904): "Die physiologische Function der Kleinhirnseitenstrangbahn," *Arch. f. Physiol.*, Suppl. No. 457.

Marey, E. (1866): "Études graphiques sur la nature de la contraction musculaire," *Journ. de l'Anatomie et de la Physiologie*, III, 225.

BIBLIOGRAPHY

MARTIN, L. J., and G. E. MÜLLER (1899): *Zur Analyse der Unterschieds-empfindlichkeit.* Experimentelle Beiträge (Leipzig).

MAY, W. P. (1904): "The Innervation of the Sphincters and Musculature of the Stomach," *Jour. of Physiol.*, XXXI, 260–71.

MEAKIN, F. (1903): "Mental Inhibition of Memory Images," *Psychol. Rev.*, Vol. IV, Suppl. No. 17, p. 235.

MENZ, P. (1895): "Die Wirkung akustischer Sinnesreize auf Puls u. Athmung," *Phil. Stud.*, II, 61.

MEUMANN, E., and P. ZONEFF (1901): "Über Begleiterscheinungen psychischer Vorgänge in Athem und Puls," *Phil. Stud.*, XVIII, 3.

MEYNERT, T. (1882): "Über Functionelle Nervenkrankheiten," *Anz. d. k. k. Ges. d. Aerzte in Wien.*, p. 23; "Zum Verständnis der functionellen Nervenkrankheiten," *Wien. Med. Bl.*, pp. 481–517.

MIKULICZ, J. VON (1903): "Beiträge zur Physiologie der Speiseröhre und der Cardia," *Mitt. a. d. Grenzgeb. d. Med. u. Chir.*, XII, 569–601.

MILL, J. S. (1829): *Analysis of the Phenomena of the Human Mind* (London).

MILLER, F. R. (1926): "The Physiology of the Cerebellum," *Physiol. Rev.*, VI, 124.

MILLER, F. R., and F. G. BANTING (1922): "Observations on Cerebellar Stimulations," *Brain*, XLV, 104–12.

MILLER, M. (1926): "Changes in the Response to Electric Shock by Varying Muscular Conditions," *Jour. Exper. Psychol.*, IX, 26.

MILLS, R. W. (1922): "X-Ray Evidence of Abdominal Small Intestinal States," *Amer. Jour. Roentgenology*, IX, 199.

MITCHELL, S. WIER: See Wier-Mitchell, S.

MÖBIUS, P. J. (1894): *Neurolog. Beiträge*, II, p. 82.

MOORE, C. S. (1903): "Control of Memory Images," *Psychol. Rev.*, Vol. IV, Suppl. No. 17, p. 277.

MOOS, E. (1928): "Zur Behandlung des Asthma Bronchial," *Münchener Mediz. Wchnschr.*, XLIII, 1841.

MORA, J. M.; L. E. AMTMAN; and S. J. HOFFMAN (1926): "Effect of Mental and Emotional States on the Leukocyte Count," *Jour. Amer. Med Assoc.*, LXXXVI, 945.

MOSSO, A. (1881): *Über den Kreislauf des Blutes im menschlichen Gehirn* (Leipzig), p. 72.

——— (1888): "Le Leggi della Fatica Studiate nei Muscoli dell'uomo," *Memorie della R. Accademia dei Lincei*, ser. 4, IV, 198. See also, *La Fatigue*, trans. by P. Langlois (Paris), 1896.

Mosso, A. (1904): "Théorie de la tonicité musculaire basée sur la double innervation des muscles striés," *Arch. Ital. de Biol.*, XLI, 183.

Mosso, A., et P. Pellacani (1882): "Sur les fonctions de la vessie," *ibid.*, I, 97–127.

Mueller, E. F. (1926): "Evidence of Nervous Control of Leukocytic Activity by the Involuntary Nervous System," *Arch. of Inter. Med.*, XXXVII, 268.

Mueller, J. (1838): *Elements of Physiology* (London: Taylor and Walton).

Münnich, F. (1915): "Über die Leistungsgeschwindigkeit im motorischen Nerven bei Warmblütern," *Zeitschrft. f. Biol.*, LXVI, 1.

Murray, E. (1906): "Peripheral and Central Factors in Memory Images of Visual Form and Color," *Amer. Jour. Psychol.*, XVII, 227.

Mussen, A. T. (1927): "Demonstration on the Red Nucleus with a Criti cism of the Conclusions of Magnus and De Kleijn," *Brain*, L, 52.

Naunyn, B. (1906): *Diabetes Melitus* (2d ed., Wien), p. 321.

Neuhof, S. (1922): "The Neurotic Element in Organic Cardiovascular Disease," *N.Y. Med. Jour.*, CXV, 80.

Neilson, C. H. (1927): "Emotional and Psychic Factors in Disease," *Jour. Amer. Med. Assoc.*, LXXXIX, 1020.

Nikolaïdes, R., and S. Dontas (1907): *Sitzsungsber. d. Kön. Preuss. Akad. d. Wiss.*, I, 364.

———— (1908): "Zur Frage über hemmende Fasern in den Muskelnerven," *Arch. f. Anat. u. Physiol. (Physiol. Abteilung)*, p. 133.

Nixon, J. A. (1925): "Focal Sepsis as a Factor in Causation of Neurasthenia and Insanity," *Brit. Med. Jour.*, II, 12.

Noïca (1921): "Le rôle de fixité du cervelet dans l'exécution des mouvements volontaires des membres," *Rev. Neurol.*, XXVIII, 164.

Noyes, W. (1892): "On Certain Peculiarities of the Knee-Jerk in Sleep in a Case of Terminal Dementia," *Amer. Jour. Psychol.*, IV, 343.

Oddi, R. (1895): "Le cerveau et la moelle épinière comme centres d'inhibition," *Arch. Ital. de Biol.*, XXIV, 360.

Oechsler (1914): "Über den Einfluss der psychischen Erregung auf die Sekretion der Galle und der Pankreas," *Internationelle Beiträge zur Pathologie und Therapie der Ernährungstörungen*, V, 1.

Olmsted, J. M. D., and H. P. Logan (1925): "Cerebral Lesion and Ex tension Rigidity in Cats," *Amer. Jour. Physiol.*, LXXII, 570.

Oppenheim, H. (1923): *Lehrbuch der Nervenkrankheiten* (Berlin), Vol. II

BIBLIOGRAPHY

ORBELI, L. (1923): "Die sympathetische Innervation der Skelettmuskeln," *Jour. Petrograd Med. Inst.*, VI, 8.

OSLER, W. (1910): "Lumleian Lectures on Angina Pectoris," *Lancet*, Part I, pp. 697–702.

OSLER, SIR W., and T. McCRAE (1926): *Principles and Practice of Medicine* (New York and London).

OSTWALD, W. (1890): "Elektrische Eigenschaften halbdurchlässiger Scheidewände," *Zeitschrft. Physik. Chem.*, VI, 71.

PARNAS, J. (1910): "Energetik glatter Muskeln," *Pflüger's Arch.*, CXXXIV, 441–95.

PAVLOW, J. P. (1902): *The Work of the Digestive Glands* (London).

———— (1923): "Die normale Tätigkeit und allgemeine Constitution der Grosshirnrinde," *Skandinav. Arch. f. Physiol.*, XLIII–XLIV, 32–41.

PEKELHARING, C. A. (1911): "Die Kreatininansscheidung beim Menchen unter dem Einfluss von Muskeltonus," *Zeitschrft. Physiol. Chem.*, LXXV, 207–15.

PEKELHARING, C. A., und C. C. VAN HOOGENHUYZE (1910): "Die Bildung des Kreatins im Muskel beim Tonus und bei der Starre," *ibid.*, LXIV, 262–93.

PEMBERTON, R. (1925): "Influence of Focal Infection and the Pathology of Arthritis," *Jour. Amer. Med. Assoc.*, LXXXV, Part II, 1793.

———— (1926): "The Etiology and Pathology of Arthritis," *ibid.*, LXXXVII, 1256.

PERKY, C. W. (1910): "An Experimental Study of Imagination," *Amer. Jour. Psychol.*, XXI, 422 ff.

PERRONCITO, A. (1901): "Sur la terminaison des nerfs dans les fibers musculaires striées," *Arch. Ital. de Biol.*, XXXVI, 245.

PETERS, W. (1905): "Aufmerksamkeit und Zeitverschiebung in der Auffassung disparater Sinnesreize," *Ztschrft. f. Psychol. u. Physiol. d. Sinnesorgane*, XXXIX, 427.

PETERSON, F. (1925): "Nutrition and Nerves," *Arch. of Neuro. and Psychiat.*, XIV, 435.

PFAHL (1921): "Über die reziproke Innervation," *Pflüger's Arch.*, CLXXXVIII, 298–302.

PIÉRON, H. (1910): "Les variations physiogalvaniques comme phénomène d'expression des émotions," *Rev. de Psychiat.*, XIV, 486.

———— (1913): *Le problème physiologique du sommeil* (Paris).

PIKE, F. H. (1909): "The General Phenomena of Spinal Shock," *Amer. Jour. Physiol.*, XXIV, 125–52.

—— (1912*a*): "The Effect of Repeated Injuries to the Spinal Cord during Spinal Shock," *ibid.*, XXX, 436–50.

—— (1912*b*): "The General Condition of the Spinal Vaso-Motor Paths in Spinal Shock," *Quar. Jour. Exp. Physiol.*, VII, 1–29.

PILLSBURY, W. B. (1908): *Attention* (New York).

PIPER, H. (1907): "Über den willkürlichen Muskeltetanus," *Pflüger's Arch. f. d. ges. Physiol.*, CXIX, 301–38.

—— (1908*a*): "Neue Versuche über den willkürlichen Tetanus der quergestreiften Muskeln," *Zeitschrft. f. Biol.*, L, 393; "Weitere Beiträge zur Kenntnis der willkürlichen Muskelkontraktion," *ibid.*, p. 504.

—— (1908*b*): "Über die Leitungsgeschwindigkeit in den markhaltigen menschlichen Nerven," *Arch. f. d. ges. Physiol.*, CXXIV, 591.

—— (1912): *Elektrophysiologie menschlicher Muskeln* (Berlin), p. 79.

—— (1914): "Die Aktionsströme der menschlichen Unterarmflexoren bei normaler Kontraktion und bei Ermüdung," *Arch. f. Physiol., Physiol. Abt.*, 345–64.

POTTENGER, F. M. (1924): "Physiologic Basis of Rest as Therapeutic Measure in Pulmonary Tuberculosis," *Annals of Clin. Med.*, III, 209.

PRATT, F. H. (1917*a*): "The Excitation of Microscopic Areas. A Non-polarizable Capillary Electrode," *Amer. Jour. Physiol.*, XLIII, 159–68.

—— (1917*b*): "The All-or-None Principle in Graded Response of Skeletal Muscle," *ibid.*, XLIV, 517–42.

—— (1925): "Response of a Muscle Fibre-Group to Apparent Stimulation of a Single Motor Fibre in the Nerve Trunk," *ibid.*, LXXII, 179.

PRIDEAUX, E. (1920): "The Psychogalvanic Reflex; A Review," *Brain*, XLIII, 50–73.

RAHE, J. M.; J. ROGERS; G. G. FAWCETT; and S. P. BEEBE (1914): "The Nerve Control of the Thyroid Gland," *Amer. Jour. Phys.*, XXXIV, 72.

RANKE, J. (1865): *Tetanus* (Leipzig: Engelmann).

RANSON, S. W. (1926): "The Rôle of the Dorsal Roots in Muscle Tonus," *Proc. Soc. Exper. Biol. and Med.*, XXIII, 594–96.

BIBLIOGRAPHY

RANSON, S. W. (1928): "The Rôle of the Dorsal Roots in Muscle Tonus," *Arch. of Neurol. and Psych.*, XIX, 201.

READ, J. M. (1924): "Management of Exophthalmic Goitre," *Jour. Amer. Med. Assoc.*, LXXXIII, 1963.

RECHÈDE, J. V. (1913): *Recherches sur les variations de pression du liquide cephalorachidien dans leur rapport avec les émotions* (Paris).

REDFIELD, A. C., and J. T. EDSALL (1927): "Factors Influencing the Anaërobic Activity of Cardiac Muscle," *Amer. Jour. Physiol. (Proc. Amer. Physiol. Soc.)*, LXXXI, 505–6.

REDFIELD, A. C., and D. N. MEDEARIS (1926): "The Content of Lactic Acid and the Development of Tension in Cardiac Muscle," *Amer. Jour. Physiol.*, LXXVII, 662.

REYNOLDS, E. S. (1923): "Hysteria and Neurasthenia," *Brit. Med. Jour.*, II, 1193–96.

RIBOT, TH. (1879): "Les Mouvements," *Revue Philos. d. la France et d. l'étranger*, VIII, 380.

RICHTER, C. P (1928): "The Electrical Skin Resistance. Diurnal and Daily Variations in Psychopathic and in Normal Persons," *Arch. of Neurol. and Psych.*, XIX, 488.

RIDDOCH, G. (1918): "The Reflex Functions of the Completely Divided Spinal Cord in Man, Compared with Those in Less Severe Lesions," *Brain*, XL, 264–402; see also "The Mass Reflex," *Lancet*, Part II, pp. 839–41.

ROAF, H. E. (1912): "The Influence of Muscular Rigidity on the CO_2 Output of the Decerebrate Cat," *Quar. Jour. Exper. Physiol.*, V, 31–53.

—— (1914): "Time Relation of Acid Production in Muscle during Contraction," *Jour. of Physiol.*, XLVIII, 380.

ROLLET, A. (1896): "Über die Veränderlichkeit des Zuckungsverlaufs quergestreifter Muskeln bei fortgesetzter periodischer Erregung und bei der Erholung nach derselben," *Pflüger's Arch.*, LXIV, 596.

ROMANES, G. J. (1876): "Further Observation on the Locomotor System of Medusae," *Phil. Trans. Roy. Soc.*, CLXVII, 659.

ROSENTHAL, I. (1873): "Studien über Reflexe," *Sitzungsberichte der Königlichen Preussischen Akad. d. Wiss.* (Berlin), p. 104.

ROSSBACH, M. J., and K. HARTENECK (1877): "Muskelversuche an Warmblütern," *Pflüger's Archiv*, XV, 1.

Rossi, G. (1912): "Ricerche sulla eccitabilità della corteccia cerebrale in cani sottoposti ad emiestirpazione cerebellare," *Arch. di fisiol.* (Firenze), X, 257–60, 389.

——— (1920): "A Proposito di Recenti Richerche sui Riflessi Laberintici," *Rassegna delle Science Biol.*, II, 124.

Rossi, G., and G. Simonelli (1922): "Recherches Expèrimentales et Considérations sur la Fonction Cérébelleuse," *Schweizer Arch. f. Neur. u. Psychiat.*, XII, 32 ff.

Royle, N. (1924): "The Treatment of Spastic Paralysis by Sympathetic Ramisection," *Surg., Gynec. and Obs.*, XXXIX, 701.

Ruffini, A. (1893): "Considerazioni critiche sui recenti studi dell' apparato nervoso nei fusi muscolari," *Anat. Anzeiger*, IX, 80.

Ruger, H. A. (1910): "Psychology of Efficiency," *Arch. of Psychol.*(" Columbia University Contributions to Phil., Psychol., and Edu."), XIX, 1.

Russell, J. S. R. (1893): "On Some Circumstances under Which the Normal State of the Knee-Jerk Is Altered," *Proc. Roy. Soc.*, LIII, 430–58.

Samajloff, A. (1908): "Aktionsströme bei summierten Muskelzuckungen," *Arch. f. Anat. u. Physiol.*, Suppl., pp. 1–23.

Schroetter, H. (1925): "Zur Kenntnis des Energieverbrauches bei emotiven Äusserungen des Seelenlebens," *Monatsschr. f. Ohrenheilkunde*, LIX, 82.

Schultze, O. (1899): "Über den Wärmehaushalt des Kaninchens nach dem Wärmestich," *Arch. f. exp. Path.*, XLIII, 193.

Setschenow, J. (1863): *Physiologische Studien über die Hemmungs Mechanismen für die Reflexthätigkeit des Rückenmarks im Gehirne des Frosches* (Berlin: Hirschwald).

Sherrington, C. S. (1892): "Notes on the Arrangement of Motor Fibers in the Lumbo-Sacral Plexus," *Jour. Physiol.*, XIII, 621–772.

——— (1893a): "Further Experimental Note on the Correlation of Action of Antagonistic Muscles," *Proc. Roy. Soc.*, LIII, 407–20.

——— (1893b): "Note on the Knee-Jerk and the Correlation of Action of Antagonistic Muscles," *ibid*, LII, 556.

——— (1894a): "On the Anatomical Constitution of Nerves of Skeletal Muscles; with Remarks on Recurrent Fibres in the Ventral Spinal Nerve-Root," *Jour. Physiol.*, XVII, 211–58.

BIBLIOGRAPHY

SHERRINGTON, C. S. (1894*b*): "Experimental Note on Two Movements of the Eye," *Jour. Physiol.*, XVII, 27.

—— (1897*a*): "On Reciprocal Innervation of Antagonistic Muscles. Third Note," *Proc. Roy. Soc.*, LX, 414–17.

—— (1897*b*): "Further Note on the Sensory Nerves of Muscle," *ibid.*, LXI, 247–49.

—— (1898): "Decerebrate Rigidity, and Reflex Co-ordination of Movements," *Jour. Physiol.*, XXII, 319–32.

—— (1899): "On the Reciprocal Innervation of Antagonistic Muscles. Fifth Note," *Proc. Roy. Soc.*, LXIV, 179–81.

—— (1900*a*): "On the Innervation of Antagonistic Muscles. Sixth Note," *ibid.*, LXVI, 66–67.

—— (1900*b*): "Experiments on the Value of Vascular and Visceral Factors for the Genesis of Emotion," *ibid.*, 390–403.

—— (1900*c*): "The Spinal Cord" in Sharpey-Schafer's *Text-Book of Physiology*, XX, 782–883.

—— (1904*a*): "Correlation of Reflexes and the Principle of the Common Path," *Brit. Assoc. Report*, pp. 728–41.

—— (1904*b*): "On Certain Spinal Reflexes in the Dog," *Proc. Physiol. Soc.*, XXXI (March *Jour. of Physiol.*), xvii.

—— (1905*a*): "Über das Zusammenwirken der Rückenmarksreflexe und das Prinzip der gemeinsamen Strecke," *Ergeb. Physiol.*, IV, 797–850.

—— (1905*b*): "On Reciprocal Innervation of Antagonistic Muscles. Seventh Note," *Proc. Roy. Soc.*, LXXVI, 160–63.

—— (1905*c*): "On Reciprocal Innervation of Antagonistic Muscles. Eighth Note," *ibid.*, pp. 269–97.

—— (1906*a*): "On Innervation of Antagonistic Muscles. Ninth Note. Successive Spinal Induction," *ibid.*, Ser. B, LXXVII, 478–97.

—— (1906*b*): *The Integrative Action of the Nervous System* (New Haven: Yale University Press; London: Humphrey Milford).

—— (1907): "On Reciprocal Innervation of Antagonistic Muscles. Tenth Note," *Proc. Roy. Soc.*, Ser. B, LXXIX, 337–40.

—— (1908*a*): "On Reciprocal Innervation of Antagonistic Muscles. Eleventh Note. Further Observations on Successive Induction," *Folia Neuro-Biologica*, I, 365–83; *Proc. Roy. Soc.*, Ser. B, LXXX, 53–71.

SHERRINGTON, C. S. (1908*b*): "On Reciprocal Innervation of Antagonistic Muscles. Twelfth Note. Proprioceptive Reflexes," *Proc. Roy. Soc.*, Ser. B, LXXX, 552–64.

—— (1908*c*): "On Reciprocal Innervation of Antagonistic Muscles. Thirteenth Note. On the Antagonism between Reflex Inhibition and Reflex Excitation," *ibid.*, pp. 565–78.

—— (1909*a*): "On Reciprocal Innervation of Antagonistic Muscles. Fourteenth Note. On Double Reciprocal Innervation," *ibid.*, Ser. B, LXXXI, 249–68; reprinted in *Folia Neuro-Biol.*, III (1910), 447–96.

—— (1909*b*): "On Plastic Tonus and Proprioceptive Reflexes," *Quar. Jour. Exper. Physiol.*, II, 109–56.

—— (1915): "Postural Activity of Muscle and Nerve," *Brain*, XXXVIII, 191–234.

—— (1924): "Problems of Muscular Receptivity," *Nature*, CXIII, 732, 892–94, 924–32.

SHERRINGTON, C. S., and E. H. HERING (1898): "Antagonistic Muscles and Reciprocal Innervation. Fourth Note," *Proc. Roy. Soc.*, LXII, 183–87.

SHERRINGTON, C. S., and S. C. M. SOWTON (1915): "Observations on Reflex Responses to Single Break-Shocks," *Jour. Physiol.*, XLIX, 331–48.

SHIMBERG, M. (1924): "The Rôle of Kinaesthesis in Meaning," *Amer. Jour. Psychol.*, XXXV, 167.

SIMONELLI, G. (1921*a*): "La Dottrina di Luciani sulla Funzione del Cervelletto," *Arch. Fisiol.*, XIX, 355–89.

—— (1921*b*): "Le Insufficienze dell'Attivita Postural nelle Affezioni del Cervelletto," *Riv. Crit. di. Clin. Med.*, XXII, 267.

SINGER, H. (1925): *Tice's Practice of Medicine* (Maryland), X, 269.

SKAGGS, E. B. (1926): "Changes in Pulse, Breathing and Steadiness under Conditions of Startledness and Excited Expectancy," *Jour. Comp. Physiol.*, VI, 303.

SLAUGHTER, J. W. (1902): "Behavior of Mental Images," *Amer. Jour. Psychol.*, XIII, 548.

SMITH, W. W. (1922): *Measurement of Emotion* (New York and London).

SNYDER, C. D. (1910): "The Latency of Knee-Jerk Response in Man as Measured by the Thread Galvanometer," *Amer. Jour. of Physiol.*, XXVI, 474–82.

SOMMER, R. (1913): "Elektrochemische Therapie," *Klinik für psychische und nervöse Krankheiten* (Halle a. S.: C. Marhold), p. 351.

BIBLIOGRAPHY

SPENCER, H. (1855): *Principles of Psychology* (London).

SPIEGEL, E. A. (1923): *Zur Physiologie und Pathologie des Skelettmuskeltonus* (Berlin).

STEINDL, H. (1924): "Über Darmspasmen," *Wien. Klin. Wchschrft.*, Vol. XXXVII, Part II, Suppl. No. 37, p. 1.

STERNBERG, M., and W. LATZKO (1903): "Studien über einen Hemicephalus," *Deutsch. Zeitschr. f. Nervenk.*, XXIV, 209–73.

STEWART, G. N., and J. M. ROGOFF (1917): "The Alleged Relation of the Epinephrin Secretion of the Adrenals to Certain Experimental Hyperglycemias," *Amer. Jour. Physiol.*, XLIV, 543.

——— (1918): "The Relation of the Adrenals to Piqure Hyperglycemia and to the Glycogen Content of the Liver," *ibid.*, XLVI, 90.

——— (1920a): "Essentials in Measuring Epinephrin Output with Further Observations on Its Relation to the Rate of the Denervated Heart," *ibid.*, LII, 521.

——— (1920b): "Further Observations on the Relation of the Adrenals to Certain Experimental Hyperglycemias (Ether and Asphyxia)," *ibid.*, LI, 366.

——— (1922): "Morphine Hyperglycemia and the Adrenals," *ibid.*, LXII, 93.

——— (1923): "The Average Epinephrin Output in Cats and Dogs," *ibid.*, LXVI, 235.

STIRLING, W. (1875): *Arbeiten aus d. Physiol. Anstalt zu Leipzig* (Leipzig), p. 223.

STRICKER, S. (1880): *Studien über die Sprachvorstellungen* (Wien).

——— (1882): *Studien über die Bewegungsvorstellungen* (Wien).

STRÜMPELL, A. (1878): "Beobachtungen über ausgebreitete Anästhesien und deren Folgen für die willkürliche Bewegung und das Bewusstsein," *Deutsch. Arch. f. klin. Med.*, XXII, 347.

SYMES, W. L., and V. H. VELEY (1910): "The Effect of Some Local Anaesthetics on Nerve," *Proc. Roy. Soc.*, Ser. B, LXXXIII, 421.

TARCHANOFF, J. (1890): "Über die galv. Erscheinungen in der Haut des Menschen bei Reizungen der Sinnesorgane und bei verschiedenen Formen der psychischen Thätigkeit," *Pflüger's Arch.*, XLVI, 46–56.

TASHIRO, S. (1913): "Carbon Dioxide Production from Nerve Fibers when Resting and When Stimulated," *Amer. Jour. of Physiol.* XXXII, 107.

THIELE, F. H. (1905): "On the Efferent Relationship of the Optic Thalamus and Deiter's Nucleus to the Spinal Cord, with Special Reference

[461]

to the Cerebellar Influx of Dr. Hughlings Jackson and the Genesis of Decerebrate Rigidity of Ord and Sherrington," *Jour. Physiol.*, XXXII, 358–84.

THÖRNER, W. (1908): "Die Ermüdung des markhaltigen Nerven," *Ztschrft. f. allg. Physiol.*, VIII, 531.

——— (1909): "Weitere Untersuchungen über die Ermüdung des markhaltigen Nerven," *ibid.*, X, 351.

THOMAS, J. E., and A. KUNTZ (1926): "A Study of Gastro-Intestinal Motility in Relation to the Enteric Nervous System," *Amer. Jour. Physiol.*, LXXVI, 606–26.

THOMPSON, H. CAMPBELL (1926): "The Work of Sir Charles Bell in Relation to Modern Neurology," *Brain*, XLVIII, 449–57.

THORNDYKE, E. (1923): *Educational Psychology*, II, *Psychology of Learning* (New York), pp. 85–115.

THORSON, A. M. (1925): "Relation of Tongue Movements to Internal Speech," *Jour. Exp. Psychol.*, VIII, 1.

TIEGEL (1875): *Berichte der mathem.-phys. Classe der K. S. Gesellschaft der Wissenschaften zu Leipz.*, X, 1.

TIEGS, O. W. (1925): "The Function of Creatin in Muscular Contraction," *Australian Jour. Exper. Biol., and Med. Sci.* II, 1–19.

TILNEY, F., and F. H. PIKE (1925): "Muscular Co-ordination Experimentally Studied in Its Relation to the Cerebellum," *Arch. Neurol. and Psychiat.*, XIII, 289–334.

TITCHENER, E. B. (1909): *Experimental Psychology of the Thought Processes* (New York), p. 145.

TOTTEN, E. (1925): "Oxygen Consumption during Emotional Stimulation," *Comp. Psychol. Monographs*, III, 13.

TOWER, S. S. (1926): "A Study of the Sympathetic Innervation of Skeletal Muscle," *Amer. Jour. Physiol.*, LXXVIII, 462.

TRAVIS, L. E. (1927): "Studies in Stuttering," *Arch. of Neurol. and Psychiat.*, XVIII, 675 ff.

TRZECIESKI, A. VON (1905): "Zur Lehre von den Sehnenreflexen. Coordination der Bewegungen und zweifache Muskelinnervation," *Arch. f. Anat. u. Physiol.*, *Physiol. Abt.*, pp. 306–79.

TSCHIRIEW, S. (1879): "Sur les terminaisons nerveuses dans les muscles striés," *Arch. de Physiol.*, VI, 89–116, 295–329.

TUTTLE, W. W. (1924): "The Effect of Sleep upon the Patellar Tendon Reflex," *Amer. Jour. Physiol.*, LXVIII, 345.

BIBLIOGRAPHY

TUTTLE, W. W., and E. WILLIAM (1925): "The Effect of Autocondensation and High Frequency Electric Current on the Tonus of Skeletal Muscle," *Amer. Jour. Physiol.*, LXXIV, 650–55.

UYENO, K. (1922): "On the Supposed Relation of the Sympathetic to Muscle Tone," *Jour. of Physiol.*, LVI (*Proc. Physiol. Soc.*), xliii.

VERAGUTH, O. (1907): "Das psychogalvanische Reflexphänomen," *Monatschr. f. Psychol. u. Neurol.*, XXI, 387–424.

—— (1908). "Das psychogalvanische Reflexphänomen," *ibid.*, XXIII, 204–40.

VERWORN, M. (1900*a*): "Zur Physiologie der nervösen Hemmungserscheinungen," *Arch. f. Anat. u. Physiol.*, *Physiol. Suppl.*, 105–23.

—— (1900*b*): "Ermüdung, Erschöpfung und Erholung der nervösen Centra des Rückenmarks," *ibid.*, 152–76.

—— (1913): *Irritability* (London), pp. 178 ff., 218–34.

VÉSZI, J. (1910): "Der einfachste Reflexbogen im Rückenmark," *Zeitschr. f. allg. Physiol.*, XI, 168–76.

—— (1911–12): "Untersuchungen über die Ermüdbarkeit des markhaltigen Nerven und über die Gültigkeit des Alles oder nichts- Gesetzes bei demselben," *ibid.*, XIII, 321–36.

—— (1913): "Untersuchungen über die rhythmisch-intermittierenden Entladungen des Strychninrückenmarks," *ibid.*, XV, 245.

VIETS, H. (1920): "Relation of the Form of the Knee-Jerk and Patellar Clonus to Muscle Tonus," *Brain*, XLIII, 269–89.

VIGOUREUX, R. (1893): *Neurasthénie et Arthritisme* (Paris).

VINSON, P. P. (1924): "Diagnosis and Treatment of Cardiospasm," *Jour. Amer. Med. Assoc.*, LXXXII, 859–61.

WACHHOLDER, K. (1923*a*): "Untersuchungen über die Innervation und Koordination der Bewegungen mit Hilfe der Aktionsströme. I. Mitt. Die Aktionsströme menschlicher Muskeln bei willkürlicher Innervation," *Pflügers Arch.*, CXCIX, 595–624; "II. Mitt. Die Koordination der Agonisten und Antagonisten bei den menschlichen Bewegungen," *ibid.*, 625–50.

—— (1923*b*): "Über den Kontraktionszustand der Muskeln der Vorderextremitäten des Frosches während der Umklammerung," *ibid.*, CC, 511–18.

—— (1925*a*): "Beiträge zur Physiologie der willkürlichen Bewegung. I. Mitt. Über Inhalt und Aufgaben einer Physiologie der willkürlichen . . . .," *ibid.*, CCIX, 218–47.

WACHHOLDER, K. (1925*b*): "III. Mitt. Über die Form der Muskeltätig-
keiten . . . . .2. Die Antagonisten," *Pflügers Arch.*, CCIX, 266–85.

WACHHOLDER, K., and H. ALTENBURGER (1925*a*): "Beiträge zur Physiologie der willkürlichen Bewegung. . . . . II. Mitt. Über die Form
der Muskeltätigkeit bei der Ausführung einfacher willkürlicher Ein-
zelbewegungen. 1. Allgemeines. Die Agonisten," *ibid.*, 248–65.

———— (1925*b*): "Beiträge . . . . IV. Mitt. Über die Form der Muskel-
tätigkeit. . . . . 3. Die Synergisten," *ibid.*, pp. 286–300: "V. Mitt.
Vergleich der Tätigkeit verschiedener Faserbündel eines Muskels bei
Willkürinnervation," *ibid.*, CCX, 646–60; "VI. Mitt. Über die Bezie-
hungen der Agonisten und Synergisten und über die Genese der
Synergistentätigkeit," *ibid.*, pp. 661–71.

WAGENER, R. (1925): "Über die Zusammenarbeit der Antagonisten bei
der Willkürbewegung. I. Mitt. Abhängigkeit von mechanischen Bedingungen," *Zeitschr. f. Biol.*, LXXXIII, 59–93; "II. Mitt.," *ibid.*,
pp. 120–44.

WALLER, A. D. (1896): "Observations of Isolated Nerve. Electrical
Changes a Measure of Physico-Chemical Change," *Proc. Roy. Soc.*,
LIX, 308.

WALSHE, F. M. R. (1921): "On Disorders of Movement Resulting from
Loss of Postural Tone, with Special Reference to Cerebellar Ataxy,"
Brain, XLIV, 539–56.

———— (1922): "Decerebrate Rigidity in Animals and Its Recognition in
Man," *Proc. Roy. Soc. of Med.*, Vol. XV, Sec. of "Neurology," Part
II, pp. 41–47.

WANG, G. H. (1923): "The Relation between Spontaneous Activity and
Oestrous Cycle in the White Rat," *Comp. Psychol. Monographs*, II,
1–27.

WARNER, W. P., and J. M. D. OLMSTED (1923): "The Influence of the
Cerebrum and Cerebellum on Extensor Rigidity," *Brain*, XLVI,
189–99.

WASHBURN, M. F. (1916): *Movement and Mental Imagery* (Boston and
New York).

WASTL, H. (1925): "The Effect on Muscle Contraction of Sympathetic
Stimulation and of Various Modifications of Conditions," *Jour. of
Physiol.*, LX, 109.

WATERMANN, O., and F. L. BAUM (1906): "Die Arteriosklerose eine Folge
des psychischen und physischen Traumas," *Neurol. Zentralbl.*, XXV,
1137.

BIBLIOGRAPHY

WATSON, J. B. (1919): *Psychology, from the Standpoint of a Behaviorist* (Philadelphia and London).

WATSON, WM. (1925): "Focal Infection. Nasal, Aural and Other Focal Sepsis as a Source of Neurasthenia," *Brit. Med. Jour.*, II, 9–11.

WATT, H. J. (1905): "Experimentelle Beiträge zu einer Theorie des Denkens," *Arch. f. d. ges. Psychol.*, IV, 289–436.

WEBER, E. (1910): "Physiologische Begleiterscheinungen psychischer Vorgänge," *Lewandowsky's Handbuch der Neurologie*, I, 446.

WECHSLER, D. (1925): "Measurement of Emotional Reaction," *Arch. of Psychol.*, XII, 5–81.

WEDENSKY, N. (1891): "Du rhythme musculaire dans la contraction normale," *Arch. de Physiol.*, III, 58.

WEED, L. H. (1914): "Observations upon Decerebrate Rigidity," *Jour. Physiol.*, XLVIII, 205–27.

WESTPHAL, C. (1875): "Über einige Bewegungs-Erscheinungen an gelähmten Gliedern," *Arch. f. Psychiat.*, V, 803.

WHEELER, R. H. (1923): "Some Problems of Meaning," *Amer. Jour. Psychol.*, XXXIV, 185.

WIER-MITCHELL, S. (1879): *Fat and Blood and How To Make Them* (Philadelphia).

WIER-MITCHELL, S., and M. LEWIS (1886a): "Physiological Studies of the Knee-Jerk and of the Reactions of Muscles under Mechanical and Other Excitants," *Med. News*, XLVIII (February 13–20), 169–73, 198–203.

——— (1886b): "The Tendon-Jerk and Muscle-Jerk in Disease, and Especially in Posterior Sclerosis," *Trans. of Assoc. of Amer. Phys.*, I, 11.

WILSON, S. A. K. (1924a): "Some Problems in Neurology. II. Pathological Laughing and Crying," *Jour. Neurol. and Psychopathol.*, IV, 299, 313–14.

——— (1924b): "The Old Motor System and the New," *Arch. Neurol. and Psychiat.*, XI, 385.

——— (1925): "Disorders of Motility and of Muscle Tone, with Special Reference to Corpus Striatum," *Lancet*, Part II, pp. 1–10, 53–62, 169–78, 215–19, 268–76; see also Editorial, *ibid.*, p. 289.

WILSON, S. A. K., and F. M. R. WALSHE (1914): "The Phenomenon of Tonic Innervation and Its Relation to Motor Apraxia," *Brain*, XXXVII, 199.

WINTER, J. E. (1912): "The Sensation of Movement," *Psychol. Rev.*, XIX, 374.

WOODWORTH, R. (1903): *Le mouvement* (Paris).

—— (1906): "Imageless Thought," *Jour. Philos., Psychol., Sci. Methods*, III, 703 ff.

WOODYATT, R. T. (1927): "Psychic and Emotional Factors in General Diagnosis and Treatment," *Jour. Amer. Med. Assoc.*, LXXXIX, 1013.

WOOLLEY, V. J. (1907): "On an Apparent Muscular Inhibition Produced by Excitation of the Ninth Spinal Nerve of the Frog, with a Note on the Wedensky Inhibition," *Jour. of Physiol.*, XXXVI, 177.

WUNDT, W. (1894): "Über psychische Causalität und das Princip des psychophysischen Parallelismus," *Philos. Studien.*, X, 123.

—— (1904): *Principles of Physiological Psychology* (New York).

—— (1907): *Outlines of Psychology* (translated by C. H. Judd), p. 92.

ZIEGLER, L. H., and B. S. LEVINE (1925): "Influence of Emotional Reactions on Basal Metabolism," *Amer. Jour. Med. Sci.*, CLXIX, 68–76.

ZYLBERLAST-ZAND, N. (1925): "Base anatomique de la rigidité décérebrée, *Rev. Neurol.*, XXXII, 998.

BIBLIOGRAPHY
1929–37

ALVAREZ, W. C., and M. F. BENNETT (1931): "Inquiries into the Structure and Function of the Myenteric Plexus. I. Differences in the Reaction of the Muscle and Nerves of the Bowel to Constant and Interrupted Currents," *Amer. Jour. Physiol.*, XCIX, 179–96.

ANDERSON, O. D., and H. S. LIDDELL (1935): "Observations on Experimental Neurosis in Sheep," *Arch. of Neur. and Psych.*, XXXIV, 330–54.

BACQ, Z. M., and G. L. BROWN (1937): "Pharmacological Experiments on Mammalian Voluntary Muscle, in Relation to the Theory of Chemical Transmission," *Jour. of Physiol.*, LXXXIX, 45–60.

BAETJER, A. M. (1930): "The Relation of the Sympathetic Nervous System to the Contractions and Fatigue of Skeletal Muscle in Mammals," *Amer. Jour. Physiol.*, XCIII, 41–56.

—— (1932): "The Effect of Muscular Fatigue upon Resistance," *Physiol. Rev.*, XII, 453–68.

[466]

BIBLIOGRAPHY

BARBIROLI, M. (1936): "Variationi umorali nell émotività preoperatoria," *Boll. d. Soc. med. chir. di Modena,* XXXVI, 110–13.

BARCROFT, J. (1930): "Some Effects of Emotion on Volume of Spleen," *Jour. of Physiol.,* LXVIII, 375–82.

BARD, P., and C. M. BROOKS (1934): "Localized Cortical Control of Some Postural Reactions in the Cat and Rat," *Proc. Assoc. for Research in Nerv. and Ment. Dis.* (Baltimore), XIII, 151.

BARD, P.; C. BROOKS; and C. N. WOOLSEY (1934): "Decorticate and Decerebrate Rigidities in the Cat," *Amer. Jour. Physiol.,* CIX, 5.

BARD, P., and D. M. RIOCH (1937): "Study of Emotions of Four Cats Deprived of Neocortex and Additional Portions of the Forebrain," *Bull. Johns Hopkins Hosp.,* LX (February), 73–147.

BARRIS, R. W. (1937): "Deficiencies in the Righting Reflexes of Cats Following Bilateral Cortical Lesions," *Amer. Jour. Physiol.,* CXX, 225–32.

BARTLEY, S. H., and E. B. NEWMAN (1931): "Studies on the Dog's Cortex. I. The Sensori-motor Areas," *Amer. Jour. Physiol.,* XCIX, 1–8.

BERGER, H. (1929): "Über das Elektrenkephalogramm des Menschen," *Archiv. f. Psych.,* LXXXVII, 527–70; (1931) XCIV, 1–16; (1932) XCVII, 6–28; (1932) XCVIII, 231–54; (1933) XCIX, 555–74; (1933) C, 301–20; (1933) CI, 452–69; (1934) CII, 538–57; (1935) CIII, 444 54; (1936) CIV, 678–89; (1936) CVI, 163–87.

BROWN, G. L. (1937): "Transmission at Nerve Endings by Acetylcholine," *Physiol. Rev.,* XVII, 485–513.

——— (1937): "Action-Potentials of Normal Mammalian Muscle. Effects of Acetylcholine and Eserine," *Jour. of Physiol.,* LXXXIX, 220–37.

BROWN, G. L.; H. H. DALE; and W. FELDBERG (1936): "Reactions of Normal Mammalian Muscle to Acetylcholine and to Eserine," *Jour. Physiol.,* LXXXVII, 394–424.

CANNON, W. B. (1929): *Bodily Changes in Pain, Hunger and Rage,* 2d ed. (New York: D. Appleton–Century).

CANNON, W. B., and A. ROSENBLUETH (1933): "Studies on Conditions of Activity in Endocrine Organs. XXIX. Sympathin E and Sympathin I," *Amer. Jour. Physiol.,* CIV, 557–74.

CANNON, W. B., and A. ROSENBLUETH (1937): "Transmission of Impulses through a Sympathetic Ganglion," *Amer. Jour. Physiol.,* CXIX, 221–36.

CHASE, W. H. (1937): "Hypertensive Apoplexy and Its Causation," *Archives of Neurol. and Psych.,* XXXVIII, 1176–89.

CROUCH, R. L., and W. H. ELLIOT, JR. (1936): "The Hypothalamus as a Sympathetic Center," *Amer. Jour. Physiol.*, CXV, 245–48.

DALE, H. H., and W. FELDBERG (1934): "Chemical Transmitter of Vagus Effects to Stomach," *Jour. of Physiol.*, LXXXI, 320–34.

——— (1934): "Chemical Transmission of Secretory Impulses to Sweat Glands of Cat," *ibid.*, LXXXII, 121–28.

DARROW, C. W., and A. P. SOLOMON (1934): "Galvanic Skin Reflex and Blood Pressure Reactions in Psychotic States," *Arch. of Neurol. and Psych.*, XXXII, 273–99.

DAVIS, D.; B. STERN; and G. LESNICK (1937): "The Lipid and Cholesterol Content of the Blood of Patients with Angina Pectoris and Arteriosclerosis," *Annals of Int. Med.*, XI, 354–69.

DAVIS, H., and P. A. DAVIS (1932): "Fatigue in Skeletal Muscle in Relation to the Frequency of Stimulation," *Amer. Jour. Physiol.*, CI, 339–56.

DAVIS, R. C. (1936): "American Psychology, 1800–1885," *Psychol. Rev.*, XLIII, 471–93.

DAVISON, C., and I. BIEBER (1934): "The Premotor Area: Its Relation to Spasticity and Flaccidity in Man," *Arch. of Neurol. and Psych.*, XXXII, 963–72.

DELCHEF, J., and G. ROUDIL (1934): "Le traitement des paralysies spasmodiques. Syndromes de Little et encéphalopathies," *Rev. d'orthopédie*, 3ᵉ, serie 21, pp. 434–84.

DIETHELM, O. (1936): "Influence of Emotions on Dextrose Tolerance," *Arch. Neurol. and Psych.*, XXXVI, 342–61.

DILL, D. B. (1936): "The Economy of Muscular Exercise," *Physiol. Rev.*, XVI, 263–91.

DOBREFF, M.; L. PENEFF; and E. WITTKOWER (1936): "Über den Einflus von Gemütsbewegungen auf den Blutcholesteringehalt," *Zeitschr. f. d. Ges. Exper. Med.*, XCVIII, 428–31.

DUFFY, E. (1932): "The Measurement of Muscular Tension as a Technique for the Study of Emotional Tendencies," *Amer. Jour. Psychol.*, XLIV, 146–62.

DUSSER DE BARENNE, J. G., and A. A. WARD, JR. (1937): "Reflex Inhibition of the Knee-Jerk from Intestinal Organs," *Amer. Jour. Physiol.*, CXX, 340–44.

EAGLE, E.; S. W. BRITTON; and R. KLINE (1932): "Influence of Corticoadrenal Extract on Energy Output," *Amer. Jour. Physiol.*, CII, 707–13.

BIBLIOGRAPHY

Eccles, J. C. (1937): "Synaptic and Neuro-muscular Transmission," *Physiol. Rev.*, XVII, 538–55.

Eggleton, P. (1929): "Position of Phosphorus in Chemical Mechanism of Muscular Contraction," *Physiol. Rev.*, IX, 432–61.

——— (1930): "Diffusion of Creatine and Urea through Muscle," *Jour. of Physiol.*, LXX, 294–300.

——— (1935): "The Chemistry of Muscle," *Ann. Rev. Biochem.*, IV, 413–29.

Erlanger, J., and H. S. Gasser (1930): "Action Potential in Fibers of Slow Conduction in Spinal Roots and Somatic Nerves," *Amer. Jour. Physiol.*, XCII, 43–82.

Esser, W. (1936): "Recherches sur l'influence du système nerveux sympathique dans le comportement du muscle strié; les mémoire: Les consequences de la faradisation de la chaine sympathique sur la contraction du gastrocnémien de Grenouille," *Arch. Int. de Physiol.*, XLII, 473–515.

Faust, J. (1936): *Aktive Entspannungs Behandlung* (Leipzig).

Fenn, W. O., and D. M. Cobb (1936): "Electrolyte Changes in Muscle during Activity," *Amer. Jour. Physiol.*, CXV, 345–56.

Fenn, W. O., and P. H. Garvey (1934): "Measurement of Elasticity and Viscosity of Skeletal Muscle in Normal and Pathological Cases: Study of So-called 'Muscle Tonus,'" *Jour. Clin. Invest.*, XIII, 383–97.

Fisher, R. E., and G. T. Cori (1935): "Hexosemonophosphate Changes in Muscle in Relation to Rate of Stimulation and Work Performed," *Amer. Jour. Physiol.*, CXII, 5–14.

Fiske, C. H., and Y. Subbarow (1929): "Phosphocreatine," *Jour. Biol. Chem.*, LXXXI, 629–79.

Freeman, G. L. (1933): "The Facilitative and Inhibitory Effects of Muscular Tension upon Performance," *Amer. Jour. Psych.*, XLV, 17–52.

Freeman, G. L., and C. W. Darrow (1935): "Insensible Perspiration and the Galvanic Skin Reflex," *Amer. Jour. Physiol.*, CXI, 55–63.

Fulton, J. F., and M. A. Kennard (1934): "A Study of Flaccid and Spastic Paralyses Produced by Lesions of the Cerebral Cortex in Primates," *Proc. Assoc. for Research in Nerv. and Ment. Dis.* (Baltimore), XIII, 158–210.

Gelfan, S. (1933): "The Submaximal Responses of the Single Muscle Fibre," *Jour. of Physiol.*, LXXX, 285–95.

GILDEA, E. F.; V. L. MAILHOUSE; and D. P. MORRIS (1935): "Relation-ship between Various Emotional Disturbances and Sugar Content of Blood," *Amer. Jour. Psych.*, XCII, 115–30.

GUTTMAN, S. A.; R. G. HORTON; and D. T. WILBER (1937): "Enhance-ment of Muscle Contraction after Tetanus," *Amer. Jour. Physiol.*, CXIX, 463–72.

HARE, W. K.; H. W. MAGOUN; and S. W. RANSON (1936): "Electrical Stimulation of the Interior of the Cerebellum in the Decerebrate Cat," *Amer. Jour. Physiol.*, CXVII, 261–67.

HARRIS, R. E., and D. J. INGLE (1937): "The Influence of Destruction of the Adrenal Medulla on Emotional Hyperglycemia in Rats," *Amer. Jour. Physiol.*, CXX, 420–22.

HENLEY, E. H. (1935): "Factors Related to Muscular Tension," *Arch. of Psychol.*, CLXXXIII, 1–44.

HERREN, R. Y., and D. B. LINDSLEY (1931): "Central and Peripheral Latencies in Some Tendon Reflexes of the Rat," *Amer. Jour. Physiol.*, XCIX, 1, 167–71.

HERREN, R. Y.; L. E. TRAVIS; and D. B. LINDSLEY (1936): "The Effect of Lesions in the Central Nervous System of the Rat upon Reflex Time," *Jour. Comp. Neurol.*, LXIII, 241–49.

HESS, W. R. (1932): *Beiträge zur Physiologie des Hirnstammes*, I. Teil: "Die Methodik der lokalisierten Reizung und Ausschaltung sub-kortikaler Hirnabschnitte" (Leipzig).

——— (1932): "The Autonomic Nervous System," *Lancet*, CCXXIII, 1259–61.

HILDEN, A. H. (1937): "An Action Current Study of the Conditioned Hand Withdrawal," *Psychol. Monographs*, 217, XLIX, No. 1, pp. 173–204.

HOFF, E. C., and H. D. GREEN (1936): "Cardiovascular Reactions In-duced by Electrical Stimulation of the Cerebral Cortex," *Amer. Jour. Physiol.*, CXVII, 411–22.

HUGHES, J.; G. P. MCCOUCH; and W. B. STEWART (1937): "Cord Poten-tials in Spinal Cat," *Amer. Jour. Physiol.*, CXVIII, 411–21.

HUNT, W. A. (1936): "Studies of the Startle Pattern. II. Bodily Pat-tern," *Jour. of Psychol.*, II, 207–13.

HUNT, W. A., and C. LANDIS (1936): "Studies of the Startle Pattern. I. Introduction," *Jour. of Psychol.*, II, 201–5.

INGLE, D. J. (1936): "Work Capacity of the Adrenalectomized Rat Treated with Cortin," *Amer. Jour. Physiol.*, CXVI, 622–25.

BIBLIOGRAPHY

INGRAM, W. R.; S. W. RANSON; and R. W. BARRIS (1934): "The Red Nucleus: Its Relation to Postural Tonus and Righting Reactions," *Arch. of Neurol. and Psych.*, XXXI, 768–86.

JACOBSEN, C. F. (1935): "Functions of Frontal Association Area in Primates," *Arch. of Neurol. and Psychol.*, XXXIII, 558–69.

JACOBSON, E. (1930): "Electrical Measurements of Neuromuscular States during Mental Activities. I. Imagination of Movement Involving Skeletal Muscle," *Amer. Jour. Physiol.*, XCI, 567–608.

——— (1930): "Electrical Measurements of Neuromuscular States during Mental Activities. II. Imagination and Recollection of Various Muscular Acts," *ibid.*, XCIV, 22–34.

——— (1930): "Electrical Measurements of Neuromuscular States during Mental Activities. III. Visual Imagination and Recollection," *ibid.*, XCV, 694–702.

——— (1930): "Electrical Measurements of Neuromuscular States during Mental Activities. IV. Evidence of Contraction of Specific Muscles during Imagination," *ibid.*, pp. 703–12.

——— (1931): "Electrical Measurements of Neuromuscular States during Mental Activities. V. Variation of Specific Muscles Contracting during Imagination," *ibid.*, XCVI, 115–21.

——— (1931): "Electrical Measurements of Neuromuscular States during Mental Activities. VI. A Note on Mental Activities Concerning an Amputated Limb," *ibid.*, pp. 122–25.

——— (1931): "Electrical Measurements of Neuromuscular States during Mental Activities. VII. Imagination, Recollection and Abstract Thinking Involving the Speech Musculature," *ibid.*, XCVII, 200–209.

——— (1932): "Electrophysiology of Mental Activities," *Amer. Jour. Psych.*, XLIV, 677–94.

——— (1933): "Measurement of the Action-Potentials in the Peripheral Nerves of Man without Anesthetic," *Proc. Soc. for Exper. Biol. and Med.*, XXX, 713–15.

——— (1934): *Electrical Measurement of Activities in Nerve and Muscle* (reprinted from *The Problem of Mental Disorder* [New York: McGraw-Hill Book Co.]), pp. 133–45.

——— (1934): "Electrical Measurements Concerning Muscular Contraction (Tonus) and the Cultivation of Relaxation in Man," *Amer. Jour. Physiol.*, CVII, 230–48.

——— (1934): "Electrical Measurements Concerning Muscular Contraction (Tonus) and the Cultivation of Relaxation in Man—Relaxation-Times of Individuals," *ibid.*, CVIII, 573–80.

———— (1936): "The Course of Relaxation in Muscles of Athletes," *Amer. Jour. Psychol.*, XLVIII, 98–108.

———— (in press): "Variation of Blood Pressure with Skeletal Muscle Tension and Relaxation," *Ann. of Int. Med.*

KABAT, H.; B. J. ANSON; H. W. MAGOUN; and S. W. RANSON (1935): "Stimulation of the Hypothalamus with Special Reference to Its Effect on Gastro-Intestinal Motility," *Amer. Jour. Physiol.*, CXII, 214–26.

KATZ, H. L., and L. B. NICE (1934): "Changes in the Chemical Elements of the Blood of Rabbits during Emotional Excitement," *Amer. Jour. Physiol.*, CVII, 709–16.

KELLER, A. D.; R. S. ROY; and W. P. CHASE (1937): "Extirpation of the Neocerebellar Cortex without Eliciting So-called Cerebellar Signs," *Amer. Jour. Physiol.*, CXVIII, 720–33.

KISS, F. (1932): *Anatomisch-histologische Untersuchungen über das Sympathische Nervensystem.*

KROUT, M. H. (1935): "Autistic Gestures: An Experimental Study in Symbolic Movement," *Psychol. Monographs*, XLVI, 1–126.

KUNTZ, A. (1934): *The Autonomic Nervous System* (Philadelphia).

LABHART, F. (1929): "Fortgesetzte Untersuchungen über den Einfluss des Nervus Sympathicus auf die Ermüdung des quergestreiften Muskels," *Zeitschr. f. Biol.*, LXXXIX, 217–36.

LANDIS, C. (1930): "Psychology and the Psychogalvanic Reflex," *Psychol. Rev.*, XXXVII, 381–98.

LANDIS, C., and H. N. DE WICK (1929): "The Electrical Phenomena of the Skin (Psychogalvanic Reflex)," *Psychol. Bull.*, XXVI, 64–119.

LANDIS, C., and T. W. FORBES (1934): "The Relation of Startle Reactions to the Cardiac Cycle," *Psych. Quart.*, VIII, 235–42.

LANGE, F. (1933): "Hypertension in Relation to Arteriosclerosis," in Cowdry's *Arteriosclerosis* (New York), pp. 501–31.

LAPHAM, R. F. (1937): *Disease and the Man*, discussed in editorial, *Jour. Amer. Med. Assoc.*, CIX, 2144.

LIDDELL, H. S.; W. T. JAMES; and O. D. ANDERSON (1934): "The Comparative Physiology of the Conditioned Motor Reflex Based on Experiments with the Pig, Dog, Sheep, Goat and Rabbit," *Comparative Psych. Mono.*, XI, 1–77.

LINDSLEY, D. B. (1935): "Electrical Activity of Human Motor Units during Voluntary Contractions," *Amer. Jour. Physiol.*, CXIV, 90–99.

BIBLIOGRAPHY

LOHMANN, K. (1933): "Über Phosphorylierung und Dephosphorylierung. Bildung der natürlichen Hexosemonophorsäure aus ihren Komponenten," *Biochemisch. Zeit.*, CCLXII, 137–56.

LUNDSGAARD, E. (1930): "Weitere Untersuchungen über Muskelkontraktionen ohne Milchsäurebildung," *Biochem. Zeitschr.*, CCXXVII, 51–83.

——— (1930): "Untersuchungen über Muskelkontraktionen ohne Milchsäurebildung," *ibid.*, pp. 162–77.

MACY, J. W., and E. V. ALLEN (1933): "A Justification of the Diagnosis of Chronic Nervous Exhaustion," *Proc. Staff Meet. Mayo Clinic*, VIII, 396–98.

MAGOUN, H. W.; D. ATLAS; E. H. INGERSOLL; and S. W. RANSON (1937): "Associated Facial, Vocal and Respiratory Components of Emotional Expression; Experimental Study," *Jour. Neurol. and Psych.*, XVII, 241–55.

MAGOUN, H. W.; W. K. HARE; and S. W. RANSON (1937): "Rôle of the Cerebellum in Postural Contractions," *Arch. Neurol., and Psych.*, XXXVII, 1237–50.

MASSERMAN, J. H. (1937): "Effects of Sodium Amytal and Other Drugs on the Reactivity of the Hypothalamus of the Cat," *Arch. of Neurol. and Psych.*, XXXVII, 617–28.

MASSIONE, R., and G. DONINI (1937): "Sulla presenza di sostanze imidazoliche nelle orine di affaticati," *Clinica med. ital.* (Milan), LXVIII, 35–52.

MAX, L. W. (1933): "An Experimental Study of the Motor Theory of Consciousness," *Psychol. Bull.*, XXX, 714.

——— (1934): "An Experimental Study of the Motor Theory of Consciousness. I. History and Critique," *Jour. Gen. Psychol.*, XI, 112–25.

——— (1935): "An Experimental Study of the Motor Theory of Consciousness. II. Method and Apparatus," *ibid.*, XII, 159–75.

——— (1935): "An Experimental Study of the Motor Theory of Consciousness. III. Action-Current Responses in Deaf-mutes during Sleep, Sensory Stimulation and Dreams," *Jour. Comp. Psychol.*, XIX, 469–86.

——— (1937): "Experimental Study of the Motor Theory of Consciousness. IV. Action-Current Responses in the Deaf during Awakening, Kinaesthetic Imagery and Abstract Thinking," *ibid.*, XXIV, 301–44.

MILROY, T. H. (1931): "The Present Status of Chemistry of Skeletal Muscular Contraction," *Physiol. Rev.*, XI, 515–48.

MORISON, R. S., and D. McK. RIOCH (1937): "The Influence of the Fore-brain on an Autonomic Reflex," *Amer. Jour. Physiol.*, CXX, 257–77.

MÜLLER, L. R. (1931): *Lebensnerven und Lebenstriebe* (Berlin: J. Springer).

NEEDHAM, D. N. (1937): "The Biochemistry of Muscle," *Ann. Rev. Biochem.*, VI, 395–418.

NEWMAN, E. V.; D. B. DILL; H. T. EDWARDS; and F. A. WEBSTER (1937): "The Rate of Lactic Acid Removal in Exercise," *Amer. Jour. Physiol.*, CXVIII, 457–62.

NEWMAN, I. (1936): "Cannon's Theory of Emotion and an Alternative Thalamic Theory," *Jour. of Abn. and Soc. Psychol.*, XXXI, No. 3, 253–59.

NICE, L. B., and D. FISHMAN (1936): "The Specific Gravity of the Blood of Pigeons in the Quiet State and during Emotional Excitement," *Amer. Jour. Physiol.*, CXVII, 111–12.

NICE, L. B., and H. L. KATZ (1934): "Changes in Chemical Elements of Blood of Rabbits during Emotional Excitement," *Amer. Jour. Physiol.*, CVII, 709–16.

——— (1934): "Blood Volume and Hematocrit Determinations in Rabbits before and during Emotional Excitement," *ibid.*, CVIII, 349–54.

——— (1936): "Emotional Leucopenia in Rabbits," *ibid.*, CXVII, 571–75.

NÓ, R. L. DE (1935): "Facilitation of Motoneurones," *Amer. Jour. Physiol.*, CXIII, 505–23.

——— (1935): "The Summation of Impulses Transmitted to the Moto-neurones through Different Synapses," *ibid.*, pp. 524–28.

NOTHHAAS, R. (1935): "Über Störungen der peripheren Zirkulation bei Ulcus pepticum und Asthma bronchiale," *Zeit. f. d. ges. exper. Med.*, XCVII, 296.

PAGE, J. D. (1935): "An Experimental Study of the Day and Night Motility of Normal and Psychotic Individuals," *Arch. of Psychol.*, CXCII, 1–39.

PROSSER, C. L., and W. S. HUNTER (1936): "The Extinction of Startle Responses and Spinal Reflexes in the White Rat," *Amer. Jour. Physiol.*, CXVII, 609–18.

RANSON, S. W. (1937): "Functions of Hypothalamus," *Bull. N.Y. Acad. Med.*, XIII, 241–71.

RANSON, S. W.; H. KABAT; B. S.; and H. W. MAGOUN (1935): "Auto-nomic Responses to Electrical Stimulation of Hypothalmus, Preoptic Region and Septum," *Arch. of Neurol. and Psych.*, XXXIII, 467–77.

BIBLIOGRAPHY

RATHBONE, J. (1936): *Residual Neuromuscular Hypertension: Implications for Education* (New York).

RATKOĆZY, N. (1936): "Neue Grundsteine zur funktionellen Röntgendiagnostik des Magens," *Fortschr. d. Geb. d. Röntgenstrahlen*, LIII, 343–53.

RIOCH, D. McK., and C. ROSENBLUETH (1935): "Inhibition from the Cerebral Cortex," *Amer. Jour. Physiol.*, CXIII, 663–76.

ROBINSON, G. C. (1937): "Relation of Emotional Strain to Illness," *Ann. Int. Med.*, XI, 345–53.

ROBINSON, S. C. (1937): "Rôle of Emotions in Gastroduodenal Ulcers," *Ill. Med. Jour.*, LXXI, 338–47.

ROSENBLUETH, A. (1937): "The Transmission of Sympathetic Nerve Impulses," *Physiol. Rev.*, XVII, 514–37.

ROSENBLUETH, A., and W. B. CANNON (1935): "The Chemical Mediation of Sympathetic Vasodilator Nerve Impulses," *Amer. Jour. Physiol.*, CXII, 33–40.

ROSENBLUETH, A., and R. S. MORISON (1937): "Curarization, Fatigue and Wedensky Inhibition," *Amer. Jour. Physiol.*, CXIX, 236–56.

ROSENBLUETH, A., and D. McK. RIOCH (1933): "Electrical Excitation of Multifibered Nerves," *Amer. Jour. Physiol.*, CIV, 519–29.

SACKS, J., and W. C. SACKS (1934): "The Rôle of Phosphocreatine in the Fundamental Chemical Changes in Contracting Mammalian Muscle," *Amer. Jour. Physiol.*, CVIII, 521–27.

———— (1935): "Carbohydrate Changes during Recovery from Muscular Contraction," *Amer. Jour. Physiol.*, CXII, 565–72.

———— (1935): "The Resynthesis of Phosphocreatine after Muscular Contraction," *Amer. Jour. Physiol.*, CXII, 116–23.

———— (1937): "Blood and Muscle Lactic Acid in the Steady State," *Amer. Jour. Physiol.*, CXVIII, 697–702.

SACKS, J.; W. C. SACKS; and J. R. SHAW (1937): "Carbohydrate and Phosphorus Changes in Prolonged Muscular Contractions," *Amer. Jour. Physiol.*, CXVIII, 232–40.

SCHLUTZ, F. W.; A. B. HASTINGS; and M. MORSE (1933): "Changes in Certain Blood Constituents Produced by Partial Inanition and Muscular Fatigue," *Amer. Jour. Physiol.*, CIV, 669–76.

SCHLUTZ, F. W.; M. MORSE; and A. B. HASTINGS (1935): "Acidosis as a Factor of Fatigue in Dogs," *Amer. Jour. Physiol.*, CXIII, 595–601.

SCHULTZ, J. H. (1932): *Das autogene Training* (Leipzig).

[475]

Seham, M., and D. V. Boardman (1934): "A Study of Motor Automatisms," *Arch. of Neurol. and Psych.*, XXXII, 154–73.

Siefert, A. C. (1937): "The Rôle of the Vegetative Nervous System in Production of Motor Phenomena Observed in the Upper Digestive Tract," *Radiology*, XXVIII, 283–300.

Smith, O. C. (1934): "Action-Potentials from Single Motor Units in Voluntary Contraction," *Amer. Jour. Physiol.*, CVIII, 629–38.

Smith, W. K. (1936): "Alterations of Respiratory Movements Induced by Electrical Stimulation of the Cerebral Cortex of the Dog," *Amer. Jour. Physiol.*, CXV, 261–67.

Spiegel, E. A. (1937): "Comparative Study of the Thalamic, Cerebral, and Cerebellar Potentials," *Amer. Jour. Physiol.*, CXVIII, 569–79.

Steblov, E. M. (1935): "Probleme der zentralen Regulation und Pathologie des Muskeltonus," *Monatschr. für Psychiat.*, XC, 345–64.

Steiman, S. E. (1937): "Factors Determining the Type of Response in the Fiber of Striated Muscle," *Amer. Jour. Physiol.*, CXVIII, 492–505.

Strehle, H. (1935): "Das Ausdrucksgebaren und seine Beziehung zu Affekten und Vorstellungen," *Med. Klin.*, XXXI, 873–76.

Travis, L. E. (1937): "Brain Potentials and the Temporal Course of Consciousness," *Jour. Exper. Psych.*, XXI, 302–9.

Travis, L. E.; J. R. Knott; and P. E. Griffith (1937): "Effect of Response on the Latency and Frequency of the Berger Rhythm," *Jour. of Gen. Psych.*, XVI, 391–401.

Travis, L. E., and M. Patterson (1933): "Rate and Direction of the Contraction Wave in Muscle during Voluntary and Reflex Movement," *Jour. Exper. Psych.*, XVI, 208–20.

Urban, F., and H. B. Peugnet (1937): "Oscillographic Study of the Cytochromes during Muscular Contraction," read before the American Physiological Society, April 23, 1937.

van Gehuchten, P. (1933): "Tubercules de la protubérance et du noyau rouge," *Rev. neurol.*, I, 74–87.

Weiss, P. (1936): "A Study of Motor Co-ordination and Tonus in Deafferented Limbs of Amphibia," *Amer. Jour. Physiol.*, CXV, 461–75.

White, B. V., Jr., and E. F. Gildea (1937): "Cold Pressor Test in Tension and Anxiety: A Cardiochronographic Study," *Arch. of Neurol. and Psych.*, XXXVII, 964–84.

Wittkower, E. (1935): "Studies on Influence of Emotions on Functions of the Organs Including Observations in Normals and Neurotics," *Jour. Men. Sci.*, LXXXI, 533–682.

AUTHOR INDEX
1690–1929

AUTHOR INDEX

[479]

AUTHOR INDEX

Pemberton, R., 1, 24
Penfield, W. G., 248
Perky, C. W., 185
Perroncito, A., 233, 234
Peters, W., 184
Peterson, F., 10
Pfahl, 252
Piéron, H., 127, 204
Pike, F. H., 255, 279
Pillsbury, W. B., 35
Piper, H., 221, 238, 242, 256, 259
Pottenger, F. M., 18
Pratt, F. H., 239
Prideaux, E., 203, 204

Rademaker, G. G. J., 248
Rahe, J. M., 205
Ranke, J., 223, 271
Ranson, S. W., 228, 229, 234
Rappleye, W. C., 259
Ray, L. H., 243
Read, J. M., 422
Rechède, J. V., 201
Redfield, A. C., 222
Reynolds, E. S., 10
Ribot, T., 165
Riddoch, G., 263
Riggs, R. A. F., 24
Roaf, H. E., 221, 222
Rogers, J., 205
Rogoff, J. M., 205
Rollet, A., 274
Romanes, G. J., 271
Rosenthal, I., 255
Rossbach, M. J., 271
Rossi, G., 247
Royle, N., 236
Ruffini, A., 232
Ruger, H. A., 140
Russell, J. S. R., 272

Samajloff, A., 273
Schroetter, H., 202

Schultze, O., 19
Setschenow, J., 243, 260, 270, 274
Sherrington, C. S., 78, 133, 160, 195, 212, 213, 223, 225–26, 227, 230, 243, 244, 247, 250, 252, 253, 260, 261, 266, 271, 272, 274, 280–82, 289, 294, 296
Shimberg, M., 186
Simici, D., 374
Simond, A. E., 205
Simonelli, G., 247
Singer, H., 10, 26
Skaggs, E. B., 107
Slaughter, J. W., 185
Smith, W. W., 204, 427
Snyder, C. D., 276
Sommer, R., 301
Sowton, S. C. M., 243
Spiegel, E. A., 224, 250
Starling, E. H., 221, 357
Steindl, H., 358, 359
Sternberg, M., 214
Stewart, G. N., 205
Stirling, W., 271
Strasmann, R., 19
Stricker, S., 184
Strümpell, A., 296
Symes, W. L., 239

Tarchanoff, J., 203
Tashiro, S., 240
Thiele, F. H., 250
Thörner, W., 240
Thomas, J. E., 359
Thompson, H. C., 261
Thorndyke, E., 140
Thorson, A. M., 186
Tiegel, 271
Tiegs, O. W., 224
Tilney, F., 279
Titchener, E. B., 196
Totten, E., 202
Tower, S. S., 236
Tschiriew, S., 233, 280

AUTHOR INDEX
1929–37

AUTHOR INDEX

SUBJECT INDEX

Accidents, 19

Acetylcholine, 240

Albuminuria, 22

Alcoholism, 20

"All-or-none law," 239, 267

Anemia, 23

Anger, 29, 207, 293

Anxiety states, 19, 20, 25, 26, 28, 217, 407, 417, 421

Apoplexy, 4, 23

Appendicitis, 358, 382

Arterial hypertension, 23, 200, 422, 423

Arteriosclerosis, 24, 358

Arthritis, chronic, 24

Association, 190–91, 300

Asthma:
    bronchial, 21, 22, 418
    cardiac, 22, 398–99, 417, 418
    renal, 22

Attention, 35, 110, 111, 165, 184, 192
    See also Mental activities

Aufgabe, 192

Augmentation, 270–91
    definition, 270
    in intact mammals, 273
    of knee-jerk, 271–72
    in reflex-arc conduction, 272
    rôle of fatigue in, 274–76
    Treppe, 271
    See also Summation

Autocondensation, 4

Autonomic nervous system:
    hypertension:
        in arousing cerebrospinal reflexes 32, 235–38
        evidence of, in, 14, 15
    relaxation:
        effect of, on, 164, 269, 296
        rôle of, in alimentary spasm, 359
    See also Colon, spasm of; Esophagus, spasm of; Muscle, smooth

Autosensory examination, 192–93, 196, 327–45

Bahnung, 270–71, 275
    See also Augmentation

Behaviorism, 110, 167, 190, 197

Belching, 18

Berger rhythm, 238, 245

Bladder contractions, 107

Blood-pressure, 201, 269, 422–23
    See also Apoplexy, Arterial hypertension, Arteriosclerosis

Brain waves, 238, 245

Bromides, 3, 42, 414, 423–24

Bronchitis, chronic, 22

Cardiac maladies, 1, 200, 396, 397, 419

Cardiac neurosis, 417

Cardiospasm, 358–59, 362
    See also Esophagus, spasm of

Cerebellum, 245–47; 267–68

Cerebral hemorrhage. *See* Apoplexy

Cerebrospinal-fluid pressure, 107, 201

Cerebrum, 111
    circulation of, 201
    emotion and, 213–14, 218
    facilitation in, 273
    inhibition and, 260–65, 269, 288–89
    knee-jerk as test of inactivity of, 127–28
    tonus and, 250–52; 254–59; 268–69, 296–98
    See also Mental activities

Cholecystitis, 358, 382

Colitis, 417, 418, 419
    See also Mucous colitis

Colon, spasm of, 21, 203, 231, 267, 381–94, 422
    See also Mucous colitis

Coma, 20

"Common path," 253–54, 268

[486]

SUBJECT INDEX

Compulsion neurosis, 353–54, 417
Conditioned reflexes, 299
Constipation, spastic, 361–62, 391
Convalescence, 18
Convulsive tic, 417
Cough, 16
Cyclothymic states, 414, 417, 418, 419

Debility, 419
Depression, 25
  See also Cyclothymic states
Diabetes mellitus, 20, 200
Diarrhea, nervous, 18, 386–91
Differential hypertension. *See* Hypertension, differential
Differential relaxation:
  aim of, to replace the "rest-cure," 81
  curve of, 139–41
  definition of, 34, 83
  economy of, neuromuscular energy in, 98
  examples of, in daily life, 83, 84
  insignificance of caloric saving, 98–100
  knee-jerk:
    methods of testing, during, 135–38
    as test for, 81, 82, 134–63
  in the physical arts, 81–83
  of primary and secondary activities, 97, 98
  rationale, 97–100
  technic of, 84–97
  in treating a particular neurosis, 96
  in treating speech disturbances, 95
Diminishing tensions, 53
Distraction, 97, 109, 125, 279, 307
Distress, 17, 18, 30, 94, 125
Dizziness, 18
Dreams, 24
Drugs, sedative. *See* Sedatives (drugs)
Dyspnea, 22

Effort, 49–50, 109, 110, 300
Emaciated states, 26
Emotion:
  adrenalin and, 204–5, 215
  in arterial hypertension and various

other medical conditions, 200–201, 430
  basal metabolic tests and, 202, 206
  blood-pressure and, 201, 206
  central control, 215, 216, 219
  cerebral circulation and, 201
  colonic spasm and, 203, 361
  effect of relaxation on, 28, 30, 31, 165, 217–19, 293–94
  energy requirements and, 202
  esophageal contraction and, 203, 361, 380
  exophthalmic goiter and, 21, 200, 205, 206
  facial expression and, 206
  in "functional nervous disorders," 24, 430
  gastric and intestinal movements and, 203
  gastric and intestinal secretions and, 203
  glycosuria in, 6, 204
  hyperglycemia and, 204
  hypothalamus, 216, 219
  investigations, 201–7
  James-Lange theory, 167, 207–17
  knee-jerk and, 202
  measures of, 218
  mechanism of relaxation of, 295–300
  oxygen consumption and, 202
  pituitary gland disease and, 205–6
  psychogalvanic reaction and, 203–4
  pulse-rate and, 201
  respiration and, 201–2
  salivary flow and, 203
  sex organs and, 205
  spinal-fluid pressure, 201
  theories of, 207–17
  tremor and, 202
  in untrained individuals, 32
Emotional shock as cause of nervous hypertension, 8
Endocarditis, 23, 396–97
Esophagitis, 358
Esophagus, spasm of, 21, 30, 203, 231, 267, 357–81, 417, 419, 422
Exercise, physical, 2, 223
Exhaustion, 414, 419, 421
  See also Fatigue

[487]

in arthritis, 1
basal metabolic tests of, 202–3
in cardiac maladies, 1
in gastrointestinal disorders, 1
general effects of, 2
as general remedy, 1–3
in hyperthyroidism, 1
in infectious diseases, 1
mental, 29
in nephritis, 1
in "ordinary relaxation," 28, 31
in peptic ulcer, 1
in surgery, 1
in treatment of ketosis, 20
See also Relaxation, ordinary
Rest-cure:
cause of failures, 2
frequent absence of relaxation in, 3
replacement of, by general and differential relaxation, 80, 407–11
Restlessness, 6, 16, 20, 22, 23, 28, 30, 65–69, 79, 84, 95, 113
See also Neuromuscular hypertension

Secondary activities, 97
Sedatives (drugs), 3, 5, 42, 396, 414, 423–24
Self-control, 105–8, 109, 163
See also Inhibition, voluntary
Sensations, 191–93, 197–98, 243
joint, *see* Joint sensations
See also Muscle-sense, Muscle-sensations, Tenseness
Sensitiveness, 105, 108, 126, 293–94, 364
Sex organs, 205
Singing, tenseness in, 17
Sleep, 29, 31, 111, 123–25, 127–28, 130–31, 179, 183, 297, 298, 375, 397, 399–403, 417, 418, 419, 422
See also Insomnia, Sleep-start
Sleep-start, 23, 24
Smell sensations, 109
Spastic esophagus. *See* Esophagus, spasm of
Spastic paralysis, 236–38, 251
Spastic paresis, 417
See Spastic paralysis
Stammering, 95, 417, 419

Start:
nervous, 17, 30, 101–11, 121–22
predormescent, 17
Static tension, 56, 57, 87, 89, 96
Stimulus error, 196
Strain, 37, 54, 55
Strychnine poisoning, 20
Stuttering, 95, 417, 419
Suggestion, 4, 31, 42, 237, 293, 300, 301–8, 414–15
Sunstroke, 19
Summation, 270–91
See also Augmentation
Suppression, 219, 300
Surgery, 26, 422, 431
Sympatheticotonia, 14, 357
Syphilis, 18

Tabes, 4, 358
Tenseness, experience of:
definition, 35
difference from experience of "strain," 54, 55
in esophagospasm, 360
Fechner's views on, 166
"meaning" and, 78
in mental activities, 74–80, 327–45
in nervous irritability, 112
in the nervous start, 101
recognition of, 182
See also Muscle-sense
Tenseness, objective. *See* Insomnia; Neuromuscular hypertension; Residual tension; Restlessness; Start, nervous; chap. xvi
Tension headaches, 88, 89, 419
Tetanus, 18, 298
Thinking. *See* Thought-processes
Thought-processes, 35, 77, 105, 165, 184, 190, 194–96, 327–45, 353
See also Imagery, Meaning, Mental activities
Tics, 79, 417, 419
Tonus:
central nervous control, 245–69
cerebellum and, 245–49
cerebrum and, 250–52, 254–59